D1621354

HOW *to* MAKE *a* SPACESHIP

www.penguin.co.uk

HOW *to* MAKE *a* SPACESHIP

A Band of Renegades, an Epic Race, and the Birth of Private Space Flight

Julian Guthrie

BANTAM PRESS

LONDON • TORONTO • SYDNEY • AUCKLAND • JOHANNESBURG

TRANSWORLD PUBLISHERS
61–63 Uxbridge Road, London W5 5SA
www.penguin.co.uk

Transworld is part of the Penguin Random House group of companies whose addresses can be found at global.penguinrandomhouse.com

First published in Great Britain in 2016 by Bantam Press
an imprint of Transworld Publishers

A CIP catalogue record for this book
is available from the British Library.

ISBNs 9780593078280 (hb)
9780593078297 (tpb)

Typeset in Chronicle Display
Printed and bound by Clays Ltd, Bungay, Suffolk.

Penguin Random House is committed to a sustainable future for our business, our readers and our planet. This book is made from Forest Stewardship Council® certified paper.

1 3 5 7 9 10 8 6 4 2

To the memory of my late father, Wayne Guthrie,

and to my mother, Connie Guthrie.

Thank you for your love and strength.

Contents

PART TWO

THE ART OF THE IMPOSSIBLE

PART THREE

A RACE TO REMEMBER

Foreword

by Richard Branson

Prizes have spurred great milestones and launched industries. The British government's Longitude Prize, offered in 1714, ended up saving both sailors' lives and ships. I was already a believer that prizes can make an incredible difference when Peter Diamandis came to see me about funding his $10 million XPRIZE. As Peter shared his idea about a prize to encourage small teams to jump-start space exploration, my instinct was to say yes. My nickname is, after all, Dr. Yes, and in those days I was running ahead of myself, spending money before I had it. But for some unknown reason, "no" came out of my mouth!

By the time we met again in the late 1990s, I had made quite a few trips to various places to see people who claimed they could go to space. Most were father-son types of operations and many had elaborate plans and no hardware to show. There was a rocket in the Mojave Desert in California called the Roton, which promised to "put NASA out of business." But the rocket appeared impossible to control, and looked quite perilous to me. So I kept looking.

Space was something that I had dreamed of for decades. I can still clearly remember sitting with my mum and dad and my two sisters watching Apollo 11 land on the Moon. I was nineteen years old and spellbound by these men who had traveled to another world. It went without saying that in my lifetime ordinary people would get to travel beyond the Earth's atmosphere. Then decades passed and governments were not sending the general public to space. In 1999, I registered the name Virgin Galactic, believing the right opportunity would come along.

Burt Rutan, who was already well known in aviation circles, and I worked on a ballooning project called Earthwinds. We were a small team trying to make the first nonstop circumnavigation of the globe in a balloon. Burt, whose shop was in the Mojave Desert, was helping to build the capsule. A few years later, while collaborating again with Burt and adventurer Steve Fossett on a plane to fly solo nonstop around the world, the Virgin Atlantic *GlobalFlyer,* Burt said he was building something "even cooler." He was secretly building a spaceship. And he was competing for Peter's $10 million prize. At that point, I thought, "This may be my dream come true." If anyone can pull it off, it is Burt.

The story of Peter Diamandis, Burt Rutan, Paul Allen, and a group of big thinkers and crazy dreamers—I use the word "crazy" here with admiration—is as entertaining as it is inspiring. It tells of a turning point in history, when entrepreneurs were offered the chance to do something only governments had done before. Whether you are nine years old or ninety-nine, this is a tale that will capture your imagination. The drama in these pages played out over many years and is filled with unforgettable people. There were high-adrenaline, high-emotion moments that I witnessed firsthand and will never forget. These moments, and the bravery, brought tears to my eyes. I feel honored to have been a part of this great history that set out to rewrite the rules.

Rules are meant to be broken. I left school at sixteen to start a magazine run by students to make a difference in the world. The Vietnam War was going on and I wanted to be a voice to stop it, to play some little role. It wasn't about making money or becoming an entrepreneur. Virgin began as a mail-order record retailer in 1970, then it was a record shop and a recording studio. Soon the biggest music acts flocked to our label. We signed the Sex Pistols and the Rolling Stones and became the biggest independent label in the world. No one thought any of this was possible. In an effort to beat the record for the fastest boat to cross the Atlantic, we ended up sinking the first time but succeeding the second. When we tried to fly a balloon across the Atlantic, we failed the first time but were successful the second. You learn by doing, by falling forward. There isn't much of a difference between being an adventurer and an entrepreneur. As an entrepreneur, you push the limits and try to protect the downside. As an adventurer, you push the limits, and protect the downside—which can be your life.

As you read Julian Guthrie's book, you will meet people who set huge and seemingly unachievable challenges and then rose above them. Without Peter, who is a pretty unique individual, commercial spaceship travel would simply not have happened. Thanks in part to the XPRIZE, billions of dollars have been invested in commercializing space. My dollars might not have gone to his initial prize but they have built Virgin Galactic, the fulfillment of a dream long held by me and countless others and an endeavor that, as you will read in this book, will forever be linked with Peter and the XPRIZE. If I'd said yes to Peter in those first meetings when he was pitching me on funding the prize, I don't know if I would have actually gotten into the spaceship business. Instead of spending $10 million to fund the XPRIZE, I will now end up spending half a billion dollars to commercialize it!

Our goal with Virgin Galactic is to open space to change the world for good. That includes realizing the dreams of thousands of people around the world of seeing the majestic beauty of our planet from above and the stars in all their glory. We believe there are untold benefits to this human experience and we want every country in the world, not just a privileged few, to have its own astronauts.

The story of the XPRIZE is the dramatic prelude of many more chapters to come, chapters that are being built now with some of the same people—like me and Paul Allen—who were inspired by the XPRIZE. Building our commercial spaceline has taken longer than we thought, and been more painful than we thought. We accept the risks and time line of commercializing flights to space that would otherwise be possible for only a few brave pilots. One of the messages of this book—and my own personal philosophy—can help provoke positive change in the world: Life is best lived looking forward—and up.

Sir Richard Branson

Founder, Virgin Group,
bestselling author, entrepreneur, and philanthropist

Prologue: Mojave Desert

JUNE 21, 2004

Alone in a spartan black cockpit made from carbon fiber and epoxy glue, sixty-three-year-old test pilot Mike Melvill rocketed toward space. He had eighty seconds to exceed the speed of sound and begin the vertical climb to 100 kilometers, a target no civilian pilot had ever reached. The rocket motor burned liquid nitrous oxide and a form of solid rubber fuel, generating a violent seventeen thousand pounds of thrust that knocked him back into his seat and screeched like metal scraping metal. Wind shear rocked the plane 90 degrees to the left and Melvill, right hand on the stick and feet at the rudders, tried to correct the problem but trimmed the plane 90 degrees to the right, banking a full 180 degrees, a move bordering on aerobatics. He was off course by 30 miles, shooting nearly straight up and closing in on Mach 1,* the chaotic, once-mythical region around 700 miles per hour known to pummel planes and kill their pilots. There was a chance he would not make it back

*Mach 1 is when the speed of a vehicle exceeds the local speed of sound. Below Mach 1 is subsonic. Above Mach 1 is supersonic. The region in between is transonic. The term *Mach* came from Austrian physicist Ernst Mach, who studied the shock waves of bullets at supersonic speeds. Mach speed decreases with higher altitude and lower air density. At sea level Mach 1 is about 760 miles per hour; at 60,000 feet Mach 1 is about 660 miles per hour.

alive. If he did, he would make history as the world's first commercial astronaut.

"Please, Lord, don't let me screw this up," Melvill said under his breath, paraphrasing the test pilot's prayer.

Melvill would lose stick and rudder control as he went faster than the speed of sound, as shock waves dampened control surfaces and the air refused to move out of the way. The self-described daredevil, known to kayak over waterfalls and do headstands on boulders at the edge of cliffs, was hurtling through the atmosphere in an air-launched, podlike rocket the size of a small bus, built by a team of about forty engineers in California's high desert. The idea was to do what only the world's biggest governments—the Soviet Union, the United States, and China—had done before: get people to space. More than twenty thousand people—Buzz Aldrin among them—had made their way by car, bike, plane, and motor home caravan to the Mojave Desert, 100 miles north of Los Angeles, to see the early morning flight of the winged ovoid called *SpaceShipOne*. Peter Diamandis, an entrepreneur who had dreamed up an improbable private race to space, with a $10 million prize for the team that made it there first, was watching from the desert floor. His life's work had brought him to this day, when a manned spaceship, built and flown without the government's help, would attempt to rocket out of Earth's atmosphere and return safely to a runway just a dozen feet away. So much was at stake, not only for would-be space travelers, but for Diamandis himself. Melvill's six-thousand-pound, hand-flown spaceship streaked through the sky nearly straight up, slashing the blue expanse with a jagged white line.

Very rough ride initially, a lot of pitching," Melvill said, his breathing labored as he talked to flight director Doug Shane in Mission Control overlooking the Mojave flight line. Directly behind Melvill's seat was the hybrid rocket engine with three thousand pounds of nitrous oxide and eight hundred pounds of rubber fuel. Melvill added, "Slowing down on me. The engine shut down. I did not shut it down. It shut down on its own.... It didn't run very well." The engine had cut off at around 170,000 feet after a seventy-seven-second burn, but inertia propelled his craft toward apogee, toward his target of 62 miles above Earth, or 328,000 feet. This was the Karman line,* named for Hungarian physicist Theodore von Kármán and widely accepted as the altitude above the Earth's sea level representing the start of space.

"Start the feather up" came the call from Doug Shane. The "feather" was the rocket plane's secret weapon, wings that bent in half to add drag—aeronautical concept designer Burt Rutan's promising but still unproven invention for delivering man and machine back to Earth. Rutan was a master of the improbable, creating flying machines out of unconventional composite materials and surfboard technology, moving wings forward and engines back, and delighting in defying symmetry and being a creative battering ram to establishment aerospace. But he had zero experience sending people to space. There were times in the program, especially on days like today, when Rutan thought to himself: *This is really out there. We are absolutely crazy to be taking this kind of risk.*

"Feather unlocked. Feather coming," Melvill said as the white rocket rotated in the thin air. "Trying to get it upright." Melvill had flown 9,500 hours in more than 150 different types of planes—even piloting one whimsical Rutan design by riding on top of it like a jockey rides a horse. But he had never encountered the violent power of a rocket. He peered out the small, round, double-paned plastic portholes—there were sixteen nine-inch-diameter windows around the nose. The inside window was made of Plexiglas, and the outside was the even stronger polycarbonate.

---

*While the Karman line offers a nice intrinsic definition as the putative boundary between the atmosphere and space, there is no real "boundary" of space, just a thinning of the atmosphere until, above 100 km, the air is so thin that trying to fly an airplane is pointless, requiring flying at rocket speeds to stay aloft.

During the building and testing phase, Rutan handed his pilots welding axes and challenged them to break the windows.

It was around eight A.M. in California, and from near the top of his parabolic arc,* Melvill could see frothy clouds along the Los Angeles coastline, browns and beiges of the desert, the shimmery coast of Baja California, and the forests and mountains of the Sierra Nevada—enormous peaks that from this height looked as flat as the desert to the south. The clouds were varied, in shades of white, platinum, and gray. Wisps turned thicker like silvery cloth, and waves of ethereal gray rolled in the sky like waves on an open ocean. Lakes and sinewy rivers glistened liquid gold. The Earth's thin blue line looked a million miles away. He now understood why astronauts were forever changed by "Earth gazing," by taking in how fragile and beautiful this little blue marble looked from above.

Melvill was not far from the skies above Edwards Air Force Base, where they had been given permission to fly in the tightly restricted area known as 2515. Edwards was the dry, hot, isolated Valhalla of test pilots and the Mecca of experimental planes, the place where the sonic boom was born, where pilots were tested for skill and mettle and some of the world's fastest, most powerful planes were let out to gallop. Melvill watched the energy height predictor, an instrument that gave a digital readout of the final altitude the plane would reach once the engine was off. His friend and mentor Albert "Scotty" Crossfield, the first pilot to fly twice the speed of sound, and the pilot with the most experience flying the military's X-15—a matte-black brute of a rocket plane that in 1963 first reached an altitude of 100 kilometers—told him he would feel

*Parabolic arc is a flight path that ascends until apogee—its highest point from Earth—and descends in a mirrored path, like a rock or ball thrown into the air.

disoriented after lighting the rocket motor and pulling back on the stick. "You will think the nose is coming up and you're going to go over on your back," Crossfield told him. "Everyone in the X-15 felt that."

"Doing a good job with RCS," Doug Shane said of the cold-gas reaction control system, small thrusters used to maneuver the vehicle's orientation.

"Everything is good here, Doug," Melvill reported.

From Mission Control came the announcement, "Three-twenty-eight," and the sound of clapping, which quickly dimmed. After that moment of euphoria, it was uncertain whether Rutan's *SpaceShipOne*, registration number N328KF, had made it to the start of space. They would have to wait for the data to come in to be sure. Rutan and his team settled back into their chairs. The toughest part of the mission lay ahead. Space shuttle *Columbia* disintegrated during reentry the year before, in 2003, killing all seven astronauts on board. The X-15—the only other winged vehicle to get to space—had ferocious loads when reentering Earth's atmosphere, traveling at Mach 5 and coming in at a forty-degree, nose-down attitude. X-15 pilot Mike Adams, a friend of Rutan's, was killed in 1967; after reaching a peak altitude of 266,000 feet, the thirty-seven-year-old Adams, a scholar and top test pilot, was at around 230,000 feet when he went into a violent Mach 5 spin and couldn't recover. The rocket plane broke apart, pieces scattered for nearly 60 miles on the desert floor.

Melvill looked at the instrument panel. Pilots were told to trust their instruments more than their bodies, but Melvill needed to *feel* the plane. He flew through the seat of his pants, literally feeling the plane through his rear end, the same way he once rode motorcycles in races. Planes, like people, had their quirks. Melvill flipped the switch on top of the stick to move the horizontal stabilizers, the movable flaps used for pitch and roll control. He reset the trim for reentry to thirty degrees on each side. He waited and watched. The feather had deployed perfectly; with the engine off, he could hear the feather make a thud against the forward tail booms. He looked again at the instruments.

Something was wrong.

"I'd like to see the stab trim here," Shane said quickly. Control of the plane's horizontal, fixed-wing stabilizers and elevons—the hinged flaps on the trailing edge of the stabilizers—was operated by sophisticated electric motors and gear boxes mounted in the tail booms and used at high altitudes and speed when stick and rudder were ineffective. The stabilizers had to be precisely set at plus-ten degrees for reentry.

Rutan studied the telemetry. For a moment, no one moved. No one said a word. The only sound in Mission Control came from more than 60 miles up, from Melvill repeatedly, quickly, flipping switches.

"Whoa! Pull the breakers!" said Rutan's chief aerodynamicist, Jim Tighe. The breakers initiated the backup motor. Melvill had tried that. Nothing. The stabilizers were unevenly positioned, with the left one at thirty degrees and the right at ten degrees. A twenty-degree difference would result in a high-speed, potentially fatal spin. Melvill knew enough about physics to know that his rocket motor took him out of the Earth's atmosphere at Mach 3—three times the speed of sound—and that gravity would pull him back at the same speed. There was little if any chance of surviving reentry with asymmetrical stabilizers. The only way out of this rocket was through the nose; unlike in the X-15, there were no ejection seats. In an emergency, Melvill would have to first depressurize the cabin, unlatch the front end of the plane by pulling a lever from the floor up, pop the nose of the plane right off, and somehow jump out the front—all while traveling faster than a speeding bullet. Scotty Crossfield had said that trying to punch out of a rocket plane was "committing suicide to keep from getting killed."

Melvill had the sensation of falling back. There was no panic, only sadness. *Man, all of this effort and this is how it ends,* he thought. A small team in the desert had a shared dream of a new golden era of spaceflight, of doing what most deemed impossible. The engineers and builders could not have worked harder. His wife of more than five decades, the cute blonde he'd run away from home with, was on the flight line below, probably clasping their son's hand. Wiry, watchful, and still very much besotted with him, Sally had pinned their lucky horseshoe on his flight suit—a piece of jewelry he designed for her in 1961, engraved "Mike and Sally." Sally was his first and only love. He tried the switches again.

The left stabilizer would not move.

Jim Tighe said darkly: "This is not good."

Rutan, sitting to Shane's right, grimaced slightly and hunched forward. Mike was his best test pilot and best friend. He was his first employee at Rutan Aircraft Factory. Sally had wanted her husband taken off the flight test program of *SpaceShipOne*. She had a bad feeling about the rocket and tried to make the case that Mike had done enough for the program already. Rutan had seen how Mike was uncharacteristically nervous before the morning's takeoff. Mike wanted to make history—for himself, for the team, for those who were never supposed to amount to much. There was also Peter Diamandis's $10 million cash prize dangled out there, offered to a team like theirs that could fly to the start of space twice within two weeks. Today was a day to make history, but it also got them one step closer to the prize.

Before the 6:47 A.M. takeoff, when the gusting wind and enveloping dust of the night before had calmed and the orange sun rose over the pale landscape, Rutan had reached into the cockpit and clasped his friend's hand.

"Mike, it's just a plane," he said. "Fly it like an airplane."

Part One

THE INFINITE CORRIDOR

1

Unruly

At around ten P.M. on July 20, 1969, eight-year-old Peter Diamandis positioned himself in front of the large television set in the wood-paneled basement of his family's home in Mount Vernon, New York. His mom, dad, younger sister, and grandparents were seated nearby. Peter, in pajamas and cape, aimed his mom's Super 8 camera at the screen, panned the room, paused on his white German shepherd, Prince, and returned to the television.

On the carpet next to Peter were his note cards and newspaper clippings, organized by NASA mission—Mercury, Gemini, and Apollo—and by rockets—Redstone, Atlas, Titan, and Saturn. The third-grader, unable to sit still under normal circumstances—his mother called him *ataktos*, Greek for *unruly*—fidgeted, bounced, and rocked in place. This was the moment Peter had dreamed about, a moment that promised to be better than all the electronics he could buy at Radio Shack, cooler than every Estes rocket ever made, more exciting even than the M80s lit on his birthday, sending his mom and friends diving for cover.

The Sears Silvertone TV was turned to *CBS Evening News* with Walter Cronkite, the seasoned newsman who was at Cape Kennedy, Florida. Peter, with the camera on, read the words "MAN ON THE MOON: THE EPIC JOURNEY OF APOLLO 11." He listened to a clip from a speech given by President Kennedy in May 1961: "I believe that this nation should commit itself to achieving the goal, before this decade is out, of landing a man on the Moon and returning him safely to the Earth. No single space project in this period will be more impressive to mankind, or more important for the long-range exploration of space; and none will be so difficult or expensive to accomplish." The onscreen countdown began for Apollo 11 astronauts Neil Armstrong and Edwin "Buzz" Aldrin to park their lunar lander on the surface of the Moon, a quest for the ages, a Cold War imperative, and a high-stakes contest between nations that had begun when the Soviet Union launched Sputnik, the world's first artificial satellite, on October 4, 1957. Now, almost twelve years later, America was trying to make history of its own. Astronaut Michael Collins, piloting Apollo 11's command module *Columbia*, had already separated from the lander and was alone in lunar orbit, waiting for his fellow astronauts to walk on the Moon.

If all went according to plan, Collins, Aldrin, and Armstrong would reunite in orbit in less than a day. About seventeen thousand engineers, mechanics, and managers were at the Florida space center for the launch. In all, an estimated four hundred thousand people had worked on some part of the Apollo program, from the women in Dover, Delaware, who did the sewing and gluing of the life-protecting rubberized fabric of the spacesuits, to the engineers at NASA, Northrop, and North American Aviation who worked for years on the clustering, three-chute parachute system for *Columbia*. The cost of the program was put at more than $25 billion.

Peter daydreamed constantly about exploring the glittering and dark expanse in his own spaceship, like the Robinson family in the television series *Lost in Space*, with the precocious nine-year-old son Will Robinson and the humanized and weaponized Robot. But on this night, the TV screen had his undivided attention.

Cronkite, in his deep voice and languid manner, said, "Ten minutes to the touchdown. Oh boy... Ten minutes to landing on the Moon." The program flashed between streamed images of the Moon and simulations of the landing done by CBS with NASA's help. The signal from the lunar camera had to be transmitted a quarter of a million miles to the Parkes Radio Astronomy Observatory west of Sydney, Australia, and then across the Pacific Ocean by satellite to the control center in Houston. From there, the images would go to television networks and finally to television sets in the United States and abroad.

In the first few minutes of flight, the Saturn V first stage—which had its design origins as a ballistic missile used by the Germans in World War II—had used four and a half million pounds of propellant, and the craft's velocity relative to Earth had gone from zero to 9,000 feet per second in ascent.*

Cronkite announced: "Go for landing, three thousand feet."

"*Eagle* looking great," said Mission Control in Houston, as grainy black-and-white images of a barren, rock-strewn landscape appeared on television sets.

"Altitude sixteen hundred feet," Cronkite narrated. "They're going to hover and make a decision.... Apparently it's a go. Seven hundred feet, coming down."

"Nineteen seconds, seventeen, counting down," Cronkite said. It was just before dawn on the Moon, and the sun was low over the eastern horizon behind the lunar lander.

Peter focused his camera on the screen. He had used his mom's camera to film NASA television broadcasts before. He had clipped countless

*This was for the first of three stages. At first-stage cutoff, the Earth-fixed velocity was approximately 7,900 feet per second, and the space-fixed velocity was approximately 9,100 feet per second. Orbital velocity was achieved after the third-stage cutoff, at a space-fixed velocity of approximately 25,500 feet per second. Total oxidizer plus propellant for all three stages was approximately 5.7 million pounds. Earth-fixed velocity is lower because the rockets are always launched in the direction that Earth is rotating, getting a free approximately 1,000 miles per hour of space-fixed velocity.

newspaper and magazine stories of space missions and written letters to the National Aeronautics and Space Administration. He had a "Short Glossary of Space Terms," issued by NASA, and he memorized terms like "monopropellant" and "artificial gravity." He won first place in a county dental poster contest with his drawing of the launch of Apollo to the Moon and the caption "Going away? Brush three times a day." He and his elementary school friend Wayne Root made their own stop-motion movies, using *Star Trek* models on fishing line as props. Peter learned that he could scratch the film in postproduction to make spaceships fire laser beams. On weekends, Peter loved to sit his family down in the living room upstairs and give lectures on stars, the Moon, and the solar system, explaining terms like "LEO," for low-Earth orbit.

The launch of the Saturn V rocket on July 16, four days before the scheduled Moon landing, had been to Peter every Fourth of July rolled into one. *Three men riding on top of a fiery rocket aimed at space! Five F-1 engines burning liquid oxygen and kerosene and producing 7.5 million pounds of thrust!* It was like sending the Washington Monument rocketing skyward.* Peter littered his schoolbooks with sketches and doodles of planets, aliens, and spaceships. He had drawn the Saturn V over and over, with its first stage, second stage, and third stage, its lunar module, service module, and command module.

At 363 feet, it was taller than a football field set on end, both beauty and monster, weighing more than 6.4 million pounds when prepared for launch. Peter had watched Neil Armstrong and Buzz Aldrin climb through the docking tunnel from *Columbia* to *Eagle* to check on the lunar module. The lunar module—the LM, pronounced "LEM" and originally called the Lunar Excursion Module—had never been tested in the microgravity of the Moon. Peter was not alone in wondering whether this spaceship would make it back to Earth. *Columbia* would return at more

---

*The Washington Monument is actually 50 percent taller than the Saturn V and weighs fourteen times as much (not including the monument's foundation).

than 17,000 miles per hour. If its descent was too steep, it would burn up; if too gradual, it wouldn't make it through the atmosphere back to Earth. Even when coming into the atmosphere perfectly—threading the needle at supersonic speeds—*Columbia* would be a fireball, with temperatures on the outside exceeding three thousand degrees Fahrenheit. Peter's father, Harry Diamandis, appreciated this moment in history and welcomed any news that wasn't about the Vietnam War or the emotional civil rights struggles of the day. But he couldn't understand his son's fascination with space, given the challenges of life on Earth. He and his wife, Tula, had come from the small Greek island of Lesbos, where he grew up tending goats and bartering for food—olives for almonds, kale for milk—and working at his father's café. Harry's mother, Athena, was a housekeeper who would bring home surplus bits of dough in her apron pockets to bake for the family. One of Harry's favorite Christmas presents was a red balloon. He was a village boy, the first in his family to graduate from high school and go to college. Harry had wanted to be a doctor, and passed his medical boards in Athens before setting his sights on America. He arrived in the Bronx speaking no English. Their journey from Lesbos to America, and Harry's path to becoming a successful obstetrician, at times felt like its own trip to the Moon, with improbable odds, an element of fear, and a feeling of being a stranger in a foreign land.

On the television screen in the Diamandises' living room, images showed a simulation of the lunar landing. Then Apollo 11 commander Armstrong radioed, "Houston, Tranquility Base here. The *Eagle* has landed." The *Eagle* sat silently on the Sea of Tranquility in the Moon's northern hemisphere. Mission Control radioed back, "Roger, Tranquility. We copy you on the ground. You got a bunch of guys about to turn blue. We're breathing again."

"The lunar module has landed on the Moon," Cronkite marveled. "We're home. Man on the Moon."

More than five hundred million people, from crowds gathered before screens in Disneyland to American soldiers in Vietnam, watched as the white-suited, tank-headed Armstrong, a ghostly, blocky figure, backed out of the module and made his way down the steps. Tula watched Peter, hoping her son remembered to breathe. Armstrong said, "I'm at the foot of the ladder. The surface appears to be very, very fine-grained as you get close to it. It's almost like a powder. I'm going to step off the LM now."

It was just minutes before eleven P.M. in the Diamandis household. From Earth, the Moon was in a waxing crescent phase. Slowly, Armstrong moved his cleated foot onto the talcum surface, becoming the first human to ever touch another celestial body. "That's one small step for man," Armstrong said, "one giant leap for mankind." The view was desolate but mesmerizing, a desert scrubbed clean. The sky looked thick and dark like black velvet.

Peter stopped filming. This was the difference between believing in God and witnessing God. It was both answer and question, new frontier, old Earth. It was NASA doing what it said it would do. The astronauts were modern-day Magellans.

Cronkite rubbed his hands together and dropped his paternal demeanor. "There's a foot on the Moon," he said, removing his black-rimmed glasses and wiping his eyes. "Armstrong is on the Moon. Neil Armstrong—thirty-eight-year-old American—standing on the surface of the Moon! Boy, look at those pictures—240,000 miles to the Moon. I'm speechless. That is really something. How can anybody turn off from a world like this?"

It was close to midnight when Tula finally got the kids to bed. Marcelle, who was six, was asleep before her head hit the pillow. Peter, still wired with excitement, told his mom once again that he was going to be an astronaut when he grew up. Tula's reply never varied: "That's nice, dear. You're going to be a doctor." Medicine was known; space was experimental. Besides, the first-born son in a Greek family always followed his father's path. Family friends were already calling young Peter the future Dr. Diamandis. Tula had given Peter a child's doctor's kit, and he would sometimes have her recline on the sofa so he could check her pulse and listen to her heartbeat. Being a doctor would be an honorable profession for Peter.

After Tula left the room, Peter turned on his flashlight and ducked under his tented bedspread. He made entries in his secret diary: The Moon was freezing in the shadows but baking in the sun. He would need a suit and the right boots—maybe his ski boots. There was no air to breathe on the Moon, so he'd need oxygen. He'd need food, water, and of course, a rocket. He drew more pictures of Saturn V, and of the astronauts. Late into the night, drawings and notes scattered around him, Peter fell asleep wondering how he could possibly be a doctor when he needed to get to the Moon.

In the years following the lunar landing, Peter began making his own rovers, among other machines. He was predatory in his pursuit of motors to hack. In one case, the lawn mower motor disappeared, turning up later on his go-kart. Then the bedsheets went missing, revealing themselves eventually as parachutes for the go-kart. The Diamandis family lived in the middle of the block on a middle-class street on the north side of Mount Vernon, New York, about thirty minutes from New York City and bordering the Bronx. Their house was a two-story white Dutch colonial with blue shutters, a big front yard, and a narrow gravel driveway where Peter liked to set up jumps for his bike. The house also had a side yard and backyard, with cherry trees and a swing set put together with great effort by his dad and uncle.

Peter drove his lawn mower–powered go-kart down the street from his house, turned onto Primrose Avenue, and pushed the cart to the top of an enormous hill. Wearing no helmet, he blasted down Primrose Avenue like a junior John Stapp,* the Air Force colonel who studied g-forces by famously riding rocket-powered sleds to a top speed of 639 miles per hour. Peter deployed his go-kart's "parachute" only when precariously close to the busy intersection.

Peter took particular delight in his sister's toys, eyeing them as a raven stares at a meaty carcass. When Marcelle received a new Barbie Dream House, Peter discovered that its motor was perfect for one of his projects, and the Barbie window shades provided the ideal chain to automate the arm of one of his robots. Marcelle and her parents went from amused to exasperated. Peter also hatched various weapon-related plans, including one that used a pipe cleaner fashioned as a projectile for his BB gun. When it didn't work, Peter mistakenly tried to suck it out of the barrel, only to have the discharged pipe cleaner shoot straight down his throat. He was rushed to the hospital and back to his experiments by nightfall. Peter got good grades, but his teachers wrote on his report cards, "Peter talks too much," and he could "work a little harder on settling down."

Every Sunday, Peter and his family drove to the Archangel Michael Greek Orthodox Church near Roslyn, where Peter was an altar boy, tasked with carrying the incense, candles, or the large gold cross and helping with communion. Confession wasn't required, but he talked openly with the kind Reverend Father Alex Karloutsos, telling him that he regularly took his sister's toys and too often made his parents worry. And he told him about his love of space; it was his "guiding star."

Peter shared with Father Alex his belief that they were all living in a biosphere, a kind of terrarium seeded with life by aliens. The aliens returned, Peter confided, to collect people as specimens or seedlings, but only in rural places like Nebraska where they wouldn't be noticed. Father

*Stapp famously coined an appendix to Murphy's Law called Stapp's Law: "The universal aptitude for ineptitude makes any human accomplishment an incredible miracle."

Alex liked listening to Peter and knew that he was not a boy who could be placated by statements like "God is love." Father Alex told Peter that the greatness of the universe was a reflection of God's presence in our lives.

In early spring, Peter was out riding his gold Schwinn Stingray banana-seat bike when he came across a neighborhood boy selling fireworks. Not long after, when it came time for Peter's birthday, Tula and Peter went over the party plan. Peter wanted to light off his new "fireworks." Tula, concerned about the noise, decided she could mute the sounds by burying the M80—Peter kept insisting these were everyday fireworks—under a pile of gravel in their narrow driveway. She said she would light the fuse herself. Peter's buddy Wayne Root was there with camera in hand. Tula told the kids to step back, nervously lit the red fuse, and scurried off. There was a long pause. The suburban neighborhood was quiet. Then—the sounds of gunfire. *Pop! Pop!* Tula yelled out, "Duck! Everybody duck!" Gravel flew, glass shattered, and she and the kids dove for cover.

When Tula finally looked up, there were clouds of lingering smoke and wide-eyed kids. Wayne was still holding his camera. Miraculously, no one was injured, and—at first glance—only a small side window of their house was cracked. Tula, heart racing, feeling as if they'd all just been shot at, gave Peter a you're-in-*big*-trouble look. Peter did his best to appear solemn, all the while thinking excitedly about the power and possibilities of projectiles powered by a fraction of a stick of dynamite.

The Diamandis family moved from Mount Vernon to Kings Point, Long Island, in the summer of 1974, when Peter was entering eighth grade. Harry Diamandis's medical practice was thriving in the Bronx.

They moved to Long Island for the schools, and because Tula fell in love with a century-old house she saw advertised in *The New York Times*, which had been on the market for three years. It was eight thousand square feet, at the bottom of a hill, with access to a community tennis court, swimming pool, and marina. Where others saw a white elephant and a lot of hard work, Tula saw possibility, and quickly set about restoring the house room by room.

Great Neck, a thirty-minute commute to Manhattan, was the fictional setting for F. Scott Fitzgerald's *The Great Gatsby*; it had sprawling verdant lawns, long driveways leading to estate homes, and nine miles of waterfront along Long Island Sound and Manhasset Bay. The Diamandis home was in Kings Point, the village at the northern tip of the Great Neck peninsula in Nassau County.

Peter claimed the third floor of the house for himself, posting a green and white "ADULTS KEEP OUT" sign, printed on his new dot matrix printer, at the top of the stairs. Peter's domain consisted of three rooms, one for sleeping and studying, one for projects—robots, rockets, chemistry, general experimentation—and the third for playing Ping-Pong, rerouting his electric train set, watching TV, and listening to music and studying.

Peter still decorated his bedroom with NASA posters, but now the posters were of the Apollo 17 astronauts Eugene Cernan, Ronald Evans, and Harrison Schmitt, NASA's first scientist-astronaut. Their mission in 1972, two years earlier, had spanned twelve days and included three days of exploration on the surface of the Moon. Cernan, who drove the Lunar Rover more than twenty miles collecting geologic samples, made a wishful statement before departing the Moon: "As we leave, we leave as we came and, God willing, as we shall return, with peace and hope for all mankind." The Apollo missions were over, but the new space shuttle program had begun, announced by President Nixon in 1972 as a rocket that would land like an airplane and would be "a reusable orbital vehicle that will revolutionize transportation into near space, by routinizing it." In Peter's mind, NASA could do no wrong, though he thought the name "space shuttle" was uninspired when compared with Apollo.

It didn't take long for Peter and a new friend in Great Neck, Billy Greenberg, to realize they were going to need more money for their projects and experiments. Cannibalizing household appliances and siblings' toys would get them only so far. They rounded up like-minded friends Gary Gumowitz, Danny Pelz, and Clifford Stober, pooled their cash, and set off on their bikes to the bank.

The boys explained to the teller that they wanted to open an account to pay for cool stuff for their club.

"Does your club have a name?" the teller asked.

The boys looked at one another quizzically.

"Well, what do you do?"

"I don't know," Peter said, "we build stuff."

"Like what?"

Rockets, trains, robots, remote-control planes, remote-control cars, boats.

"It sounds like you do everything," the teller said finally. "Why don't you call it 'The Everything Club'?"

The loosely formed Everything Club was officially launched. The boys met in Peter's tree house, intentionally built with a stepladder too rickety to support adults. And they met in Peter's project room. They ordered Estes rocket kits organized by skill level, beginning with the classic Der Red Max, which had red wood fins, a black nose, and a skull and crossbones. Standing sixteen inches tall, the rocket flew to around 500 feet and had parachute recovery. The boys had a schedule to work their way up the skill levels, from one to five, and then start building their own rockets and making their own propellant.

Peter and Billy and the rest of the boys joined the Great Neck North High School computer club, math club, and future physicians club. They started programming on Hewlett-Packard and Texas Instruments calculators, and then programmed on computers that were offered as vocational training for high school students. They learned electronics by building Heathkits, making small transistor radios with resistors, capacitors, diodes, transistors, a rheostat, and a small loudspeaker. Their classmate Jon Lynn was the first in their group to build a working computer, the Sol-20 by Processor Technology, similar to the early Altair. Their first "computers" relied on punch cards for programming, based on the same mechanical principle as the Jacquard loom, with the punch-card reader converting the perforations into on/off electrical signals, which the computer interpreted as numbers and instruction codes for the calculation. Carrying the punch cards around school was like being a part of a secret fraternity.

After school, the boys hung out at the Gold Coast video arcade, playing *Pong, Tank,* and *Speed Race.* One of their favorite games was *Lunar Lander,* where they used arrow keys to rotate the lander and change the thrust, with the goal of landing safely on an *X* on the Moon. Peter was on the high school diving team, and though he was never very interested in sports, he was muscular like a wrestler and could do a backflip from the standing position. He had thick and dark feathered hair, wore a gold chain with a cross, and got teased for his height—he had topped out at five feet five inches tall.

Peter and his friend Billy's outlook for building and flying powerful rockets improved greatly when they found themselves in the popular Mr. Tuori's chemistry class. Mr. Tuori, who had taught chemistry at Great Neck North for decades, favored experiments that woke kids up and left an impression. Peter and Billy were lab partners and watched attentively. This was learning they would use.

In class, lab coats and goggles on, Peter and Billy followed as Mr. Tuori took metallic-looking gray iodine crystals from a small jar and put them into a beaker. Mr. Tuori then relocated to the fume hood to pour a small amount of concentrated solution of ammonia over the crystals. He shook the mixture gently, explaining that the new compound, nitrogen triiodide, with three iodine atoms stuck around a single nitrogen atom, was pretty safe while wet. Once dry, though, anything could set it off, from a snowflake to a feather. After giving the chemicals time to react, Mr. Tuori filtered the mud-colored mixture to get rid of the excess ammonia. It was critical, Mr. Tuori again cautioned, to set it down before it had time to dry. When it came time for testing, Peter and Billy were front and center. Using a long pole, Mr. Tuori reached toward the charred-looking material. Peter noticed a fly buzzing just above the nitrogen triiodide. He gently elbowed Billy to look in the direction of the six-legged interloper. As Mr. Tuori's stick approached the compound, the fly landed on the powder—setting off a loud and sharp *snap*! A poof of purple smoke followed. The unfortunate fly was blown to smithereens.

Soon, shipments of explosives began arriving at Peter's door in boxes marked with a skull and crossbones and the warning "DANGER: EXPLOSIVES" stamped on top. The boys discovered they could find whatever they wanted through chemical supply companies advertised in the back of *Popular Science* magazine. They could have chemicals sent in bulk by UPS directly to their door. Peter secretly turned one of his third-floor closets into a chemical supply room, apprehending the boxes before his mom and dad made it home. Peter and Billy split the supplies in half, so if one of them was found out, they'd lose only half their supply.

The boys ordered equipment for their chemistry labs: beakers, Bunsen burners, flasks, stoppers, droppers, funnels, and thermometers. Peter was drawn to the alkaline earth metals, especially magnesium, which burned a bright white light. He ordered boxes of magnesium ribbons and powder, and he'd add barium to make it burn green and strontium to make it burn red. He did tests with calcium and—of course—loved potassium nitrate, sulfur, and charcoal, the mainstays of gunpowder.

The only thing that Peter didn't like was that potassium nitrate and sulfur needed oxygen to burn. He wanted to find something that wasn't saddled with that requirement. To Peter, chemistry pushed into the unknown, into what felt like the opposite of ordinary schoolwork. It held mystery, order, and logic. Chemistry reminded him of being a little boy again and jumping into rain puddles. Only now, he got to make the puddles *and* cause the ripples.

Peter began studying rocketry, reading books by the Russian teacher and physicist Konstantin Tsiolkovsky, who was born in 1857, was nearly deaf, largely self-educated, and introduced ideas about space travel and rocket science still in use more than a century later. In the late 1800s, Tsiolkovsky wrote about the effects of zero gravity on the body, predicted the need one day for pressure suits for space travel, developed Russia's first wind tunnel, envisioned rockets fueled by a mixture of liquid hydrogen and liquid oxygen, and developed the mathematical formula for changes in a rocket's momentum and velocity.* Peter also read about Robert Goddard, the American physicist who built and launched the world's first liquid-fueled rocket in 1926, an event likened in significance to the Wright brothers' flight at Kitty Hawk. Goddard was ridiculed when he stated his belief that a big enough rocket could one day reach the Moon, but he drew support from aviator Charles Lindbergh. Peter appreciated how Goddard's rocket experiments as an undergraduate at the Worcester Polytechnic Institute yielded explosions and smoke that sent professors running for fire extinguishers.

Peter learned about German physicist Hermann Oberth, also a believer in liquid-fueled rockets over solid-fuel rockets, and another German, Wernher von Braun, father of the Saturn V, which came out of his work for Nazi Germany on the V-2 ballistic missile during World War II.† Peter knew that

*Tsiolkovsky invented the rocket equation, the formula describing how much speed you gain from a rocket engine. It depends on the exhaust speed of the gas and the net change in mass of the rocket as fuel is expended. The rule of thumb is that the change in velocity ΔV = (exhaust gas speed) × (logarithm of the initial mass/ final mass). For example, if the rocket burns enough fuel to decrease the total mass by a factor of 3, then the velocity increase approximately equals the exhaust gas speed. If the mass decreases by a factor of 9, then the velocity increase approximately equals twice the exhaust gas speed. The takeaway is that you want to burn as much as quickly as possible, and expel it as fast as possible. The faster you burn, the lighter the rocket gets, and the easier it is to change the velocity.

†The entire Saturn rocket management team was composed of ex-V-2 engineers from Peenemünde.

if it weren't for von Braun and his group of German engineers, the United States would not have reached the Moon by the end of the 1960s.

On weekends, Peter and his rocket-making pals packed their creations, along with various remote-controlled planes, into their backpacks and hopped on their bikes. They rode to the nearby Merchant Marine Academy in King's Point. Sometimes they'd pick a football field just outside the gates to launch their Estes rockets. It never took long before the academy guards chased them away.

Other times, the boys persuaded one of their parents to drive them to Roosevelt Field, where Lindbergh took off in the *Spirit of St. Louis* to try to fly to Le Bourget Field in Paris. The field had a parking lot and vast open space. The boys filled their rockets with homemade gunpowder, sometimes getting a poof, other times a firework, and sometimes an unwanted ballistic missile that tore at them like a flying snake, coming close on at least one occasion to taking out an unsuspecting Harry Diamandis.

One of Peter and Billy's best creations was a rocket series they called Mongo, with Mongo 1, 2, and 3 getting progressively taller and driven by the most powerful motors they could get their hands on. They built an autonomous launch system that could fire up to three rockets in sequence using a circuit they designed around the 555 Timer IC. This way, when it was just two of them, one rocketeer could be downfield tracking the launch and the other taking photos. This won them first place in an Estes rocket design competition, awarding them certificates to buy more rockets. While working on their arsenal and making their way down the periodic table, Peter and Billy made an important discovery: potassium chlorate made better explosions than potassium nitrate.

Peter also discovered important properties about potassium perchlorate. Not only was it highly explosive, but it also produced its own oxygen when it decomposed. He bought the colorless crystalline substance—common in fireworks, ammunition, sparklers, and massive rocket motors—in five-pound boxes. He experimented by drilling holes in film canisters and covering the holes with automotive body filler. To make an explosion with potassium perchlorate or chlorate, he would need to combine it with something else that would combust, like sulfur or aluminum powder. The right concoctions shot out of the filled hole; some fizzled or failed altogether.

One winter afternoon, the boys met up at Jon Lynn's house. They filled film canisters with various potions, wrapped them in duct tape, and lit them off in the icy driveway. One flew directly at a boy's head, several worked as planned, and a few fizzled. After more experimentation, a new plan was hatched: They should take one of their potassium perchlorate film canister bombs and put it *underwater* to see what would happen. Potassium perchlorate didn't need oxygen to burn.

The boys ran to the back of the house, where the Lynns had a swimming pool, which was partially frozen over. They put one of the canisters in the water below the ice. The boys stood slightly off to the side, watching and waiting. Nothing happened. Seconds passed. Then they heard a muted *booosh!* The ice rose up an inch—the boys stepped back—and then it appeared to settle. Peter was relieved. But then came an enormous, clear *crack*. Jon Lynn's mom, Suzanne, inside preparing dinner, suddenly felt the house move.

It was becoming more and more evident that this suburban neighborhood was not a big enough canvas for Peter's rocket dreams.

2

Early Regrets

Peter sat alone in his dorm room at Hamilton, a New England–style liberal arts college of 1,800 students in Clinton, New York, almost five hours north of his home in Great Neck. The school had a beautiful campus, but after only a few weeks there, Peter was feeling like he had made a big mistake.

His love of space, chemistry, and rocketry was stronger than ever, yet he was on a premed track at Hamilton, not on a path to becoming an astronaut. Making matters worse, it didn't look like the school would permit him to double major in biology and physics, a compromise he hoped would allow him to continue his dual desires of space and medicine. This was not a place for a science-based double major, let alone a serious student of space.

Peter was at Hamilton because he hadn't felt smart enough to apply to Ivy League schools, though most of his friends had ended up there. The premed track was for his parents. The boy who couldn't be contained was now contained.

Not long after his arrival at college, the eighteen-year-old wrote in his journal: "I must admit I've been having second thoughts about Hamilton.

I wanted a larger selection in courses. I'm trying to get a major in biochemistry set up. Don't know if I can do it."

A page later: "I'm extremely bothered by this paradox I've encountered about Hamilton (though I strongly hope it's false). First I hear it to be extremely hard scholastically, harder than many of the larger 'better' well-known schools. Yet I don't know if it will be considered when it counts, on entrance to graduate work."

When Peter returned home to Great Neck for Thanksgiving break, he ran into Michael November, a former high school classmate who was now a college freshman at MIT. They set up a tennis match at the Shelter Bay tennis court adjacent to his home to get some exercise and catch up with each other. As they rallied on the cold but dry late-fall day, Peter admitted to Michael, who had been in his advanced-placement chemistry class, that he was already starved for science and technology. Whenever Hamilton hosted a science-related talk from a visiting professor, Peter was front and center taking notes. "I eat it up," Peter told his friend. "Philosophy is great and literature is fine, but I want more."

Michael, who had played football at Great Neck High and loved math the way Peter loved space, was having the opposite experience. Between tennis shots, Michael told Peter about a program at MIT called UROP, the Undergraduate Research Opportunities Program, giving undergraduate students the chance to work on research in fields as varied as nuclear science, urban planning, and solar-photovoltaic systems for houses.

Michael told Peter that he was working on fusion experiments involving the building of a scaled-down version of a tokamak, a vacuum inside a circular steel tube that used magnetic fields to confine fusion. His project leader was Professor Louis Smullin, who had helped create the school's Department of Electrical Engineering and Computer Science, and was head of the radiation laboratory in the early 1940s when the lab developed airborne radar used during World War II.

Peter stopped playing. He couldn't believe a *freshman* was getting to work on *fusion*. "Ohmygod, that's incredible," he said.

Michael was also taking a class on relativity from Professor Jerome Friedman, director of the school's Laboratory for Nuclear Science. His freshman physics class was taught by Professor Henry Kendall, who was working with Friedman on breakthrough research into subnuclear particles called quarks.* By the time the tennis match had ended, Peter couldn't get MIT out of his head. He spent the rest of Thanksgiving break working on his car, a Pontiac Trans Am with a V8 engine and a golden firebird on the hood. He had modified the carburetor intake manifold to take in more oxygen, and he toyed with the idea of putting in a nitrous injection system.

When Peter returned to Hamilton after Thanksgiving, he called MIT to learn about its transfer policies. Buoyed by what he learned, he scheduled a tour and interview in early January; the admissions application for transfer students would be mailed to him. He wondered whether he even had a chance; MIT was one of the most competitive universities in the world, and it was even harder to be accepted when transferring. In the meantime, though, Peter would try to make the most of Hamilton, while also reaching outside the campus for science and space. He started a biology study group to meet two to three times a week. He sought out professors and local authors with connections to space. He wrote letters to NASA, including:

> *Dear sirs:*
>
> *I am writing to you in regards to my education. I'm presently a college student, and eventually I hope to enter the space program.*

*The name "quark" was invented by Murray Gell-Mann, a colleague of Richard Feynman's at Caltech, reputedly based on a quote from James Joyce's book *Finnegans Wake*: "Three quarks for Muster Mark!" Gell-Mann wanted it pronounced as "quork." The quark model unified in a single stroke the whole disorganized pile of miscellaneous particles that had been observed up until that time.

First, however, I wish to enter graduate school to obtain an MD, hopefully in conjunction with a Ph.D. (probably in biochemical engineering). My question is, does NASA offer any educational programs which would be of interest to me?... Also if possible, please send me any information you can about entering the space program, astronaut training, etc., and an application if possible?

Sincerely,
Peter H. Diamandis

When January finally arrived, Peter and his mom drove to Boston and headed to the Massachusetts Institute of Technology in Cambridge, across the Charles River. Peter and Tula walked along Massachusetts Avenue, past the Harvard Bridge,* continued up a well-worn staircase, past a row of grand columns, and into the entrance of 77 Mass Ave., the marbled and domed rotunda at the heart of campus. Peter took in every detail, from the inscription in Greek under the dome to the long hallway stretched out ahead.

The winter sun streamed through the tall windows behind them, filling the travertine lobby with a muted buttery light. Students were bundled up in puffy jackets and carried books and backpacks, their chatter echoing in the cavernous space. The lobby led to a hallway running 825 feet in length, connecting to other parts of campus. Tula, who loved architecture, thought the building was as breathtaking as the Pantheon in Rome. Peter had a feeling he couldn't describe. Maybe it was how his mom felt when she spotted their Great Neck house, or how his dad felt

*The official length of the Harvard Bridge is 364.4 Smoots plus one ear. Distances on the bridge are indicated with a colored paint mark every Smoot and a number every ten Smoots. Oliver Smoot was a pledge at the Lambda Chi Alpha fraternity, and he got picked as the standard unit for the bridge in the 1958 pledge season. Some of his "brothers" laid him out three hundred times along the bridge until the cops came and chased them off. Smoot's cousin George Smoot became very famous for his work on the COBE satellite, which first measured anisotropy in the cosmic microwave background radiation.

when he saw his mom for the first time, wearing her Friday night dress, and *just knew.*

Peter and Tula walked slowly down the long hall, and Peter studied the posters and flyers on bulletin boards and in display cases. There was a sign for UROP, the undergraduate research opportunities program that Michael November had touted. A student told Peter and Tula about "MIT Henge," where for a few days every year in late January, the setting sun lined up with the string of buildings on the north edge of Killian Court and shined all the way through building 7 and reached building 8, with the best viewing from the third floor.

The hall, known on campus as the Infinite Corridor, ran through parts of buildings 3, 4, 7, 8, and 10. MIT was number-centric—students, classes, and buildings were all assigned numbers. Classroom doors had opaque creamy glass with black, hand-painted department and professor names, reminding Peter of doors he'd seen in old detective shows. Peter wanted to open every door and explore every subject. He was uncharacteristically silent, soaking in all of the details. He and his mom passed buildings 10 and 11 and stopped at building 8, the physics department. Created in the nineteenth century by MIT founder William Barton Rogers, the department had among its faculty and graduates a dazzling array of Nobel Prize winners and some of the field's greatest minds, from Richard Feynman (quantum electrodynamics), Murray Gell-Mann (elementary particles), Samuel Ting and Burton Richter (subatomic particles), to Robert Noyce (Fairchild Semiconductor, Intel), Bill Shockley (field-effect transistors), George Smoot (cosmic microwave background radiation), and Philip Morrison (Manhattan Project, science educator). Physics classes at MIT had been flooded with students in the years following the launch of Sputnik and the success of Apollo.

Peter and his mom made their way to the biology department. He had an understanding with his parents that if he was accepted at MIT, he would stay on his premed track. The biology department here had much more to offer. Peter and Tula doubled back through the corridor, studying more of the photos, posters, and signs for events and clubs, from salsa dancing to stargazing. Peter had two departments he had to see before leaving. The first was in building 37, devoted to astrophysics. It was a field both ethereal and real, where words and equations tried to interpret vibrant colors, patterns, and formations of the cosmos.

The very last stop on their tour was building 33, AeroAstro, which had produced more astronauts than any other place outside U.S. military academies. Military officers had received aviation training for World Wars I and II here. Breakthroughs in hypersonic flight testing were achieved. Buzz Aldrin got his PhD here in 1963. Other astronauts—Jim Lovell, Apollo 13; Ed Mitchell, Apollo 14—took MIT's astronautical guidance class.

There was a picture of a half dozen NASA astronauts visiting the Instrumentation Lab: three men were the Apollo 1 astronauts—Virgil Grissom, Roger Chaffee, and Ed White—who died during a prelaunch test. Next to them were MIT students and astronauts Dave Scott, Rusty Schweickart, and Jim McDivitt.

Peter studied the department time line: Charles Stark "Doc" Draper started at MIT in the 1920s and founded the Instrumentation Laboratory in the 1930s. Peter kept reading, stunned by what he was learning. The inertial guidance system for Apollo—the computer that got man to the Moon—was developed here in the Instrumentation Lab. *Right here!* The guidance system came to life at a time when computers took up entire rooms, when typewriters with carbon paper were the norm, and when television was black and white. A small team from MIT had figured out how to use a new technology, the integrated circuit, to send man and machine to the Moon and back. Baseball fans had Wrigley Field. Golfers, St. Andrews. Surfers, Mavericks. Climbers, K2. This was Peter's hallowed ground.

Peter looked at space relics, including parts from the Instrumentation Lab's Mars probe. Built in the early 1960s, the probe never launched, but its technology evolved into the guidance computer for Apollo. MIT was awarded the first NASA Apollo contract for the guidance computer in the months following Kennedy's famous speech. Jim Webb, the administrator of the newly formed NASA, knew Doc Draper, engineer, inventor of inertial systems, and a pilot who tested parts he made by flying the planes himself. According to Draper, Webb had called him and said, "Doc, can you develop the guidance and navigation system for Apollo?"

"Yes, of course," Draper said.

"When will it be ready?" Webb asked.

"When you need it," Draper replied.

"And how do I know it will work?"

"I will go along and operate it for you," Draper said, formally volunteering to be an astronaut at age sixty.

Draper could not have known that he and his team were capable of building a computer to get men to the Moon. It hadn't been done before. But Draper didn't hesitate to put everything on the line by saying yes to one of the most difficult technical challenges in history. He believed in himself, and in his team. Walking through the labs, Peter jotted down another note, this one involving one of his idols, Wernher von Braun, who was asked early in the Apollo program: "Wouldn't we do a lot better if we collaborated with the Russians?" Von Braun replied, "If there were collaboration with the Russians, there wouldn't be a program in either country." Peter wrote, *Competition got America to the Moon.*

As Peter and Tula headed out into the late afternoon cold, Peter's mind went over the array of classes, subjects, and breakthroughs in this one location. This was a place of limitless possibilities.

Back at Hamilton after his tour, Peter was antsy. MIT had been another reminder of the audacity of NASA, of what had been accomplished in less than a decade. He wanted those glory days restored. The 1970s were in many ways the opposite of the 1960s. Money had been redirected to the Vietnam War and to myriad social problems.

During the 1960s, NASA's budget was around 1 percent of the total federal budget, and the agency had more than 400,000 employees and contractors at its peak in 1965.* By 1979, NASA's percentage of the federal budget had been cut in half, and the agency counted around 20,000 employees. NASA canceled its mission to send a spacecraft to do a flyby of the famed Halley's Comet, set to pass close to Earth in 1986 and not return for another seventy-five years. Apollo 18, 19, and 20 were canceled even though most of the hardware had been bought and built. The Moon had been reached, and critics said the government was "shooting money into space." Design and development of the space shuttle were delayed, and plans for a U.S. space station in low-Earth orbit ended. Lovers of space wondered what was next. It was the withering away of a dream.

Desperate to do something space related at Hamilton, Peter created and circulated a pro-space petition to send to every elected leader he could find, from local representatives to President Jimmy Carter's adviser on space affairs. Peter expressed his concerns about the "slow but sure degeneration of the U.S. space program's goals and budget." He collected about two hundred signatures from Hamilton students and faculty—significant for a campus of 1,800. Peter then penned a letter that he hoped would be published in the science magazine *Omni*:

> *I am directing this letter to college students. Having learned of the death of both the Galileo and Halley's Comet/T II missions and the embarrassing delay of the shuttle program—and having seen the space program pushed aside by our government—it is about time that those of us who support our space program, our future, address the issue.*
>
> *The method for letting the government know how we feel is simple: begin a petition at your college, collect signatures, and submit them to*

*NASA's biggest total employment year was 1965, when the space agency employed 34,300 in-house employees and 376,700 out-of-house contractor employees.

the proper offices of the president and Congress. There are nearly 1,000 colleges and universities in the United States, with an average of 2,000 students per institution. We represent a powerful force and we can change our future.

Peter H. Diamandis
Great Neck, NY

In early February 1980, Hamilton hosted a lecture by visiting professor Jim Arnold, founder of UC San Diego's chemistry department, consultant to NASA, and one of the first to study rocks and soil returned from the Moon. Arnold talked about the rich and useful resources to be mined from the Moon and near-Earth asteroids. Peter had never heard of mining asteroids for metals like nickel, iron, and platinum. When the talk ended, he met two visiting students representing what they called the "international school of the future," who were intent on establishing a "space micropolis." Walking back to his dorm, Peter looked at the cards they had given him. This had been a good night, but the lectures were few and far between at Hamilton, and he wouldn't hear from MIT for at least another six weeks.

A key course required for premed students at Hamilton was intro to biology by Professor Frank Price. The course was tough and had a reputation for winnowing the number of students who continued on the premed track. The study and dissection of a fetal pig made up 50 percent of the student's grade.

Peter's class had around eighty students, and consisted of three hours of lectures and three hours of lab work per week. On the first day of class, Professor Price, a biology teacher at Hamilton for five years, gave a stern talk about the importance of treating the fetal pigs with respect and care. He cautioned that "under no circumstances are the pigs to leave the lab." There was one pig for every two students; the course instruction focused on physiology, organ functions, and the patterns of blood circulation through the heart, lungs, stomach, and liver.

Two weeks before the all-important dissection exam, Peter came down with the chicken pox and had to spend a week in the school infirmary. He missed crucial lab time and test practices. He knew that if he didn't do well on this test, he wouldn't have a chance at a top medical school, and he wouldn't be considered for MIT. Peter could not do poorly on this exam. The thought kept him awake all night. Finally, he hatched a plan: he would "borrow" a fetal pig to study over the weekend. He enlisted help from his lab partner, Philip, who was also a dorm mate. It was decided after class that Philip would casually *brush* the pig into a book bag, held open by Peter.

The following Monday, Professor Price asked for everyone's attention. He didn't look happy. "It has come to my attention," he said, "that a fetal pig has been stolen. I would like to have whoever has done this to please turn themselves in. And if you know who's done this, it's your duty on the school's honor code to turn him in."

Peter looked over his shoulder at Philip, *terrified*. Hamilton's strict honor code was signed by students at the start of school. Academic dishonesty resulted in expulsion or automatic course failure. The pig, about a foot long, was in a plastic bag with formaldehyde-soaked paper towels hidden at the back of their dorm's refrigerator. After class and back at the dorm, the worst scenario unfolded: Peter was told that a suite mate was going to turn him in.

My life is over, Peter thought. He asked the suite mate for a day to deal with it on his own. In a state of panic, Peter and Philip met in his room and decided to get rid of the evidence. They walked the campus, eyeing Dumpsters and secluded spots. They were searching for somewhere the pig would never be found. That night, Peter and Philip headed to the woods; Hamilton was in a rural area with hundreds of acres of wooded land. The pig was buried, and the burial spot marked. All Peter could think about was that he would be expelled; his family would be disgraced, and he would never get to MIT. He was physically ill.

Peter called his father. Harry and Tula were at a friend's house playing cards. Harry excused himself to talk in private, returning more than an hour later. When Tula asked what was wrong, Harry said it was taken care of. Peter told his dad about the mess he had gotten himself into, and said he intended to turn himself in the next day.

Harry Diamandis had listened closely to his son. After a long pause, he presented another idea. Peter was to call the doctor in the infirmary and explain the situation. Harry had visited Peter while he was sick, and found the campus doctor to be kind and smart. The physician should be enlisted as an ally, Harry said. The doctor would talk to Professor Price, and then Peter would return the pig to the lab and have a candid talk with his teacher.

Peter returned to the woods and exhumed the pig. Late in the afternoon, he entered Professor Price's lab. His hands shook as he took the pig out of his bag. Professor Price could see that Peter was ashen and appeared on the verge of tears. The professor had experienced the theft of his pigs before. He had seen them hanging from nooses in students' dorms, and pigs had been used for pranks in students' beds and bathrooms. Professor Price, who had talked with the infirmary doctor, asked Peter, "Did you learn your lesson?" Peter nodded, fighting back tears and stammering, "I'm so sorry." Peter looked at his professor. *Is my life over, or will I be shown mercy?* After a pause that seemed to go on forever, Professor Price said, "Good luck on your exam, then."

A few weeks later, Peter got his letter from MIT. His eyes filled with tears. *"Peter, On behalf of the Admissions Committee, it is my pleasure to offer you admission to MIT. . . ."*

Professor Price had given him a second chance, a lesson of grace and generosity Peter would never forget. There would be no more shortcuts, no more bending of rules. Getting to MIT was a gift. Peter would return to the Infinite Corridor, open some doors, join clubs—and maybe even launch something of his own.

3

Pete in Space

It wasn't long after his arrival at MIT in the fall of 1980 that Peter got a new nickname: "Pete in Space." His fraternity brothers at Theta Delta Chi called him "PIS" for short, and had fun with "space cadet" and "spaced out" jokes. Peter took the friendly ribbing in stride and started signing his name as PIS; he was happy to be at MIT.

With each passing day he grew more impressed with the university's diversity of programs in molecular biology, physics, computer science, electrical engineering, astrophysics, aeronautics, and astronautics. His new school was dazzling. But as he walked the halls of the Infinite Corridor, scanning announcements and posters, he discovered one thing missing: MIT did not have a student space group.

How could there not be a student space group at MIT, of all places? Peter went to the campus activities administrators to ask about space-related clubs for students. There were computer clubs and astronomy clubs, but no space club. He was told that if he wanted to start a club, he needed four signatures and a name.

Peter got the signatures from his fraternity brothers and a friend and drew up a list of potential names: Student Space Society; Children of

Icarus; Students for the Preservation of the Future; Students for the Exploration and Development of Space; Space Cadets of America; and Space Cadets at MIT. He nixed the space cadet options after hearing how the potheads in his house loved the name. He eventually settled on the name Students for the Exploration and Development of Space (SEDS), because it best explained the group's mission. He made a couple hundred flyers, blanketed the campus, and took special care in finding a prime spot in the Infinite Corridor. He used a transfer letter kit for the block lettering of SEDS, and wrote in thick pen, "If you care about your future in space, join me in the student center."

The nineteen-year-old Peter was already juggling a heavy class load and two undergraduate research projects, one space related, and the other for his premed track. Many nights, his lab work kept him so busy that he wouldn't get back to his fraternity house until three A.M. On the premed side, in the genetics lab of Graham Walker, he worked on plasmid instability of PKM-101 in *E. coli.* To feed his space-related passion, Peter had scored a UROP project in the Man Vehicle Laboratory (MVL) in AeroAstro, building 37. Peter worked in the windowless ground floor, where the design, cabinets, floors, and even some of the equipment felt like it hadn't changed in a half century. It all transported Peter to the days of Apollo.

Peter's work in the MVL was far from glamorous, but he loved it. MVL research focused on the physiological and cognitive limitations of humans in aircraft and spacecraft. Founded in 1962, the lab worked closely with NASA to study space sickness for astronauts in the early Apollo program. It now had another NASA contract to work with a new breed of astronaut, the payload specialist, scientists trained to conduct experiments in space aboard the shuttle. Peter helped design and build an electrogastrograph to record the electrical activity of the stomach during motion sickness. Later in the school year, he would begin to research and construct an experiment tracking involuntary eye movements called nystagmus, a part of motion sickness that astronauts faced. Peter was told he would get to work one-on-one with astronauts. He was also aware of the buzz that NASA needed more astronaut physicians for future shuttle flights and had plans for an eventual space station. Peter was officially premed, but passionately pre-astronaut.

On the scheduled Wednesday night of the first meeting of SEDS, Peter waited nervously in his reserved second-floor room in the Stratton Student Center. Only five people had signed up, and he feared that no one would come. He watched anxiously as students passed by. A few paused, as if they were about to come in, and then continued walking. He bit his nails, a habit he had been trying to give up. Within a few minutes, several people ventured in, then a few more. To his great relief, about thirty people gathered in the room—not a bad turnout.

Peter welcomed the group and talked about his background, saying he "drank all the Tang"—*Star Trek*, *Star Wars*, Apollo. He talked about why the time was right to start a student space organization: "This is our future we're talking about. We can't have myopic politicians say where it goes. We need to stand up for the future of space."

Peter talked about the momentum of the Apollo program in the sixties carrying through to a lesser degree in the seventies, with Apollo 17; the *Voyager* probes reaching the space between stars; and Skylab, the U.S. space station launched by a modified Saturn V, providing the first glimpse of space station technology. But progress had slowed, the space shuttle was delayed and over budget—it was being called the "$9 billion spaceship that refuses to fly"—and NASA had no new plans to send people to the Moon or beyond. Public interest in space had dropped. Peter enthused about all of the spinoff technologies that came from the space program, including cordless appliances, compact integrated circuits for navigation, implantable pacemakers, and freeze-dried food.

"Our goal," Peter said, surprised by his own fervor, "is to enlighten our government, private industry, and the general populace regarding the benefits of a strong space program."

Peter was asked whether he would consider making SEDS/MIT part of a national space group called L5, formed around the ideas of Princeton University physicist Gerry O'Neill. As author of *The High Frontier* and founder of the Space Studies Institute, O'Neill called for establishing a colony of about ten thousand people at L5, a gravitational sweet spot between the Earth and the Moon where a spacecraft could remain stationary, always more than 350,000 kilometers from Earth.

Peter shook his head. "I want an organization created for students run by students," he said.

A man at the back of the room raised his hand and introduced himself as Eric Drexler. "I think Peter made a case for a student-led organization," Drexler said. "I don't think this group should be part of L5." Drexler had worked for O'Neill at Princeton for two summers making a mass driver, an electromagnetic cannonlike device to shoot payloads of lunar material from the Moon. His master's degree was in aeronautical engineering from MIT, and his thesis was on a high-performance solar sail system for space. He was a PhD candidate studying the groundbreaking field of molecular nanotechnology.

The meeting closed as Peter gathered names and addresses. He stayed late answering questions and brainstorming the future of SEDS. When he finally walked outside, the air was still warm. The sky was clear with stars. The moment felt perfect. He'd had this feeling before, when he walked through the hallway of the Infinite Corridor and was sure he was entering something big, something real. Walking through campus, posters under his arm, Peter felt as though he could reach out and touch the future.

Chapters of SEDS were soon announced at Princeton and Yale. Scott Scharfman, a friend of Peter's from Great Neck High, started the Princeton chapter while another former Great Neck classmate, Richard Sorkin, initiated the Yale chapter. Peter, Scott, and Richard drafted a four-page constitution; started a national petition drive directed at President-elect Ronald Reagan and the U.S. Congress, urging funding for solar power satellite research; created a club logo that included the space shuttle; and sent off a carefully worded letter to *Omni*, known for its mix of hard science and pseudoscience, stating, "The steady deterioration of the U.S. space program's goals and budget endangers our future and demands an organized response from our nation's campuses. . . . We invite you and the other students at your college to begin a chapter and join us in our cause." The international headquarters of SEDS was 372 Memorial Drive—Peter's fraternity.

Omni published the letter in April 1981. That month, space shuttle *Columbia*, STS-1, was finally launched, generating international attention and a resurgence in national pride. The mission was the first of four planned orbital tests of the shuttle, which took off like a rocket, cruised in orbit like a spacecraft, and returned to Earth as a glider. It was the world's first reusable manned spacecraft and marked the first launching of American astronauts in nearly six years.

Thousands of spectators had poured onto the beaches across the Indian River from the Kennedy Space Center. As the countdown reached its final seconds, crowds chanted "Go, go, go!" As *Columbia* was pushed straight up, crowds roared, yelled, and prayed.

Returning to his fraternity after class one day, Peter slowed to a stop inside the front door. There in the foyer of his fraternity house, in the midst of the dozens of wooden slots, was his mailbox. It was overflowing with letters, packed in like a thick deck of cards, with a few jutting out at odd angles. *Was this a prank?* Peter carefully pulled the letters out and examined the envelopes, with all different handwriting, stamps, and postmarks. He sat down in the foyer and began to read. A student in Bombay wanted to form a SEDS chapter. A woman at Arizona State University was interested in meeting like-minded students to form a chapter in

Phoenix. A man in Lubbock, Texas, said he was studying "colony ecosystems, mass drivers, etc." to reach space if "Uncle Sam defaults." An engineering student from Toronto wrote, "It is hard to express my reaction to your idea; ecstatic is probably close. An organized student voice in support of the space program is long overdue, and your initiative has sparked a welcome feeling of optimism inside me." He suggested that the goal of SEDS become not just national, but international, and he offered his services as Canadian coordinator.

The letters poured in day after day. Peter's fraternity brothers took notice. Now when they called him "PIS," it was with an air of respect.

Over the next two years, SEDS grew from a three-campus group to a student association with close to one hundred chapters in the United States and abroad. Peter, who was now serving as chairman of SEDS, traveled to visit nearby chapters and juggled putting out newsletters and passing his finals. He worked on his public speaking, trying to improve his cadence and confidence, and got his first lessons in fund-raising when he set out to secure $5,000 to cover the cost of printing and mailing the SEDS newsletter to all of the chapters. Fund-raising meetings were set up through friends and faculty, but Peter had a hard time actually asking for the money. He feared rejection.

When he landed a meeting with administrators at Draper Lab, started by Doc Draper of Apollo guidance system fame, Peter knew he needed to do his best. He made his pitch with all the passion he could summon. Afterward, the Draper Lab folks said they loved what Peter was doing with SEDS, but the lab was nonprofit and unable to give him the money. Peter nodded with understanding, but as he began to walk out of the lab, he had another idea. He turned back and asked, "Those newsletters I'm trying to get printed, any chance you have the ability to print them here at Draper?" The response was yes. Peter continued, "Any chance you can mail them out to our chapters for us as well?" Again, the answer was yes. Peter learned a lesson that would stay with him: there is always a way.

Peter organized conferences at universities in the area—Tufts, Harvard, Boston University—attracting notables from academia, NASA, and other established space groups. The first annual international SEDS conference was held over four days in July 1982, with NASA deputy administrator Hans Mark speaking primarily on the government's military motivations to get to space. In another celestial coup, Peter was invited to attend a United Nations conference on space in Vienna. The meeting would focus on the peaceful, nongovernmental use of space.

Peter searched for the cheapest plane ticket he could find. He would fly to Austria with Bob Richards, chairman of SEDS–Canada, and the engineering student from Toronto who had written to him after reading his letter in *Omni*. Bob had graduated with degrees in industrial engineering and aerospace and had started a SEDS group in Toronto. He studied at Cornell, and was an assistant to space scientist and author Carl Sagan. Peter and Bob had become allies and close friends with another student, Todd Hawley, who started a SEDS chapter at George Washington University, where the 1982 international SEDS conference was held. Todd, who spoke Spanish, French, and Russian, was a double major in economics and Slavic language and literature at GWU. Hawley had introduced them to David Webb, who was chairman of the nongovernmental space conference at the United Nations.

Peter, Bob, and Todd had become so unified in their vision and presence that people often called them by one name, *Peterbobtodd*. They were even the same height: Peter had feathered brown hair, parted down the middle; Todd had dirty blond hair and round, wire-rimmed glasses; and Bob had strawberry blond curls and a cherubic face. Todd believed that space was where differences could be erased. Bob looked at the cosmos as the next step in humanity's evolution. Peter was interested in the hardware of space, and the adventure.

Peter and Bob found their cheap tickets on an Austrian airline, Arista

Air, while Todd made his own way to Vienna with his girlfriend, MaryAnn. About nine hours into their flight, Peter and Bob were awakened by a terse announcement that they would be landing in Budapest, Hungary, not Vienna. Upon landing in Hungary—still a part of the Communist Eastern bloc—their plane was boarded by military police with guns and dogs. Bob was convinced they were being hijacked. Peter got out his camera and began taking pictures, until an officer told him to put the camera away. The plane baked on the tarmac, and Peter and Bob waited, worried, and sweltered. This was not how Peter envisioned the first international mission of SEDS would unfold. Eventually they were told that all Viennese citizens on board had to get off the plane and take a bus. There was reportedly an issue with someone's not paying the landing fees in Austria. Peter and Bob remained in their seats, and the plane eventually took off for Vienna. The two wondered whether the rest of their trip would be as strange.

The next day, Peter and Bob rendezvoused with Todd and MaryAnn, and the four made their way to the United Nations conference. At the front of the majestic building—adorned with flags from dozens of countries—horse-drawn carriages were parked next to satellite trucks. Peter took a picture of a satellite truck with the lettering "MOSKVA, USSR." He attended sessions called "Tomorrow's Peacemakers," "Remote Sensing Centers," and "Land Use Information Through Space." He shared a table with a man who told him that he had left his job as a scientist working on Reagan's Star Wars initiative because he believed the program was dangerous. He told Peter that he'd had his passport stolen and had a KGB woman try to befriend him.

On their second morning in Vienna, Peter, Bob, and Todd stood in the lobby and studied the roster of speakers. The three of them were scheduled to speak the following day. Suddenly, Bob whispered excitedly, "That's Arthur C. Clarke!" Todd couldn't believe it. Peter, unaware of Clarke's godlike status, said, "So?" Clarke was the author of *2001: A Space*

Odyssey, an originator of the idea of geostationary satellites, and a futurist known as the "prophet of the space age."

As Bob and Todd stared at Clarke, Peter said, "Let's go talk to him." Before the starstruck Bob and Todd could respond, Peter headed straight for Clarke, who was flanked by an entourage. Peter got close enough to Clarke to extend his hand and gesture back toward Bob and Todd.

"We are from SEDS and . . ."

Clarke walked away. Peter shook his head. He couldn't leave it at that. Bob was embarrassed by Peter's brashness. As the crowd slowly made its way into the auditorium to hear Clarke speak, Peter moved quickly and snagged front-row seats. Clarke talked about the future of the telecommunications industry. Decades earlier, Clarke had written a famous paper in *Wireless World* that defined the geostationary orbit and proposed using space satellites for global communications.

Peter was fascinated by Clarke's concept of geostationary satellites and was determined to connect with him. He whispered to Bob, "We're going to have dinner with him." Bob rolled his eyes.

After the talk, Peter intercepted Clarke—again.

"Mr. Clarke, we're from Students for the Exploration and Development of Space, and we'd love to take you to dinner tonight," Peter declared, just as Clarke was being shuffled off to an interview. "Here's the telephone number of our hotel and our room number. We'd love to tell you what we're up to."

Clarke looked at the three young men and said in his deep British accent, "I'll call you."

Back in their shared hotel room, Peter and Bob sat on their beds and stared at the phone. Peter and Bob were now making bets on whether Clarke would ever call. Peter was confident. Bob was skeptical. They watched the clock: 5:30. 5:35. 5:50. *Rriiiinnngg.* Peter grabbed at the phone. Bob held his breath. Peter said, *Uh huh, yes, okay, the InterContinental, sure.* He put the phone back on the receiver.

"Well?" Bob implored.

"That was Arthur," Peter said impassively. "He said he can't allow us to take him to dinner."

Bob sighed.

"He wants to take *us* to dinner!"

Bob didn't know whether to hug Peter or punch him.

That night, Peter, Bob, and Todd met Clarke in the lobby of the Inter-Continental Hotel. They went to dinner and listened, rapt, to Clarke's stories of growing up in the 1940s, reading pulp fiction, and his early days with the Planetary Society. He talked about how he came up with his ideas around geostationary communication satellites, and spoke passionately about what he saw as the unifying bond that came with a shared interest in space. He had met all of the major rocket scientists from space programs in the Soviet Union, China, and Japan. "They all have a common sense of space," Clarke said, urging Peter, Bob, and Todd to think of students from all languages, nationalities, governments, and ideologies coming together over a shared love of space. "Focus on young people," he said.

He used a line the group loved: *Any sufficiently advanced technology is indistinguishable from magic.*

Toward the end of dinner, Peter said, "If you don't mind, can we call you 'Uncle Arthur'?"

The answer was yes, and soon Uncle Arthur, the prophet of the space age, had signed on to become an adviser to SEDS.

Back in the Man Vehicle Lab at MIT, working with Professor Chuck Oman, Peter affixed adhesive electrode patches to the face of a man he considered royalty—Byron Lichtenberg, test pilot, Vietnam fighter pilot, MIT-trained mechanical engineer and biomedical engineer, and new breed of space traveler.

Selected as a payload specialist in 1978, Lichtenberg was excited by the space shuttle's promise of about forty-eight flights per year, doing science, satellite deployment, and building of the space station along the way. But it had already taken three times as long as planned to launch the shuttle, and the astronauts were expecting to do only a third as many flights. It was the winter of 1983, and Lichtenberg was scheduled to fly the shuttle's first Spacelab mission by the end of the year.

One of the key areas of research for the payload specialists was space sickness, something the tough-guy astronauts of Mercury, Gemini, and Apollo—all former military test pilots—were loath to admit. Russia's Yuri Gagarin, the world's first astronaut, reported no sickness in space. Russia's second cosmonaut, Gherman Titov, who orbited Earth seventeen times, earned the distinction of being the first man to vomit in space. Apollo 9 astronaut Rusty Schweickart suffered space sickness on his first day in orbit. Buzz Aldrin told his friends in the MVL that he had felt extremely queasy returning to Earth.

With electrodes attached to his head and stomach, Lichtenberg was seated in Peter's rotating chair. He was spun in one direction, moving his head in another until the symptoms of nausea began. Data were recorded and responses compared. Peter and his professors also spent a great deal of time in the chair.

There were interesting findings—once you feel the onset of nausea, full recovery takes thirty-five minutes—and new questions. Lichtenberg planned to wear a head-mounted accelerometer made by the folks in the MVL on his shuttle flight and keep detailed notes of how he felt. The work in the MVL was about "bringing space sickness out of the closet." The director of the MVL, MIT professor Larry Young, was selected as principal investigator to perform a series of space experiments on the crew of Spacelab 1—the first pressurized laboratory to be carried into space on the shuttle.

When Peter and Lichtenberg weren't doing experiments, they talked about the life of an astronaut. Peter wanted to know what was asked during the interview process. He wanted to be ready. Lichtenberg told him that the standard questions focused on the hardware and software of spaceflight, the training, and effects on his family life. But there were some obsure and silly questions, including: "What is the lifetime of an average red blood cell in your bloodstream?," "Don't you think we faked the Moon landings?," and "We hear aliens visit the space shuttle and hop on board with you guys. What do you think?"

Peter peppered Lichtenberg with questions about the odds of becoming an astronaut. Lichtenberg said that NASA received about six thousand applicants for every round of hiring, and usually no more than ten of those applicants became astronauts—1 out of 600, less than .17 percent. Selections could be "random or political," and the vetting process was brutal, he said.

It suddenly occurred to Peter to mention that he had a small tear in his retina, the result of getting kneed in the eye while playing football.

"Would that get me thrown off?" he asked.

"Yep," Lichtenberg said. "That would get you tossed out of the selection process."

Peter was stunned. He didn't know what to say.

"NASA is risk averse." Lichtenberg shrugged. "Heck, even if you are selected, it doesn't mean you'll actually fly. Most astronauts are called penguins: they have wings but don't fly."

Peter had dreamed of being an astronaut for as long as he could remember. What would he do if he couldn't reach space through NASA? What would he try? What would he risk? Was it even possible to get to space without the government's help?

Nearly three thousand miles away, in California's high desert, an airplane designer disenchanted with government-run programs was asking similar questions. For now, he was working on a low-flying plane that he hoped would do laps around the best that the U.S. military had to offer. But like Peter, this dreamer in the desert had high hopes of one day reaching the stars.

4

Mojave Magic

Burt Rutan stared down one of the longest runways in the world, the seven-and-a-half-mile partially paved stretch at Edwards Air Force Base in the California desert. He was sitting in a chase plane, and just to his left was one of his most audacious creations yet: the *Voyager*.

Weighing about 2,500 pounds when empty, the aircraft was flat and white like stretched taffy, with the twin booms of a catamaran, and 7,000 pounds of fuel packed into long, spindly wings.* Within minutes, the *Voyager* was scheduled to take off for its death-defying mission, a flight around the world without stopping or refueling. Aviation experts gave the *Voyager* little chance of success, because of the compromises Burt had made in its design and construction, and because his aim was nothing less than to *double* an aviation record that had stood for almost a quarter century. There was every reason to believe the *Voyager* would

*The *Voyager* design bears some resemblance to that of the famous World War II Lockheed P-38 Lightning. Burt moved the horizontal stabilizers forward, relocated the two engines into the center fuselage, and made one a puller and the other a pusher, thereby allowing him to lighten up and extend the wingspan.

land for no other reason than pilot exhaustion. This was a test of flying skill, physical endurance, and breakthrough design.

Just as Charles Lindbergh had done with the *Spirit of St. Louis* in 1927, Burt had pared down the *Voyager* to its lightest possible weight. The in-flight tools, whether wrenches or screwdrivers, were hollowed out. The top of the plane got only a light coat of white paint, to keep the structure cool in the hot desert sun. The plane's skin consisted of only two layers of graphite fiber composite, with a paper honeycomb in between. There were virtually no redundancies: if a part failed, there was no backup.*

Burt was off to the *Voyager*'s right on the runway. He wondered if the plane would be controllable with such a heavy fuel load.

But Burt had a habit of finding breakthroughs where others saw nonsense. He'd been dreaming up and engineering planes since the time before computers, relying on pencils, slide rules for calculations, and French curves for airfoils. The maverick with the Elvis sideburns and the glint in his blue eyes, who could look at a plane and guess its weight within a pound or two, moved in the contrarian footsteps of the self-taught R. T. Jones, inventor of the theory of the swept-back wing, basic to commercial airliners today.† Like Jones, Burt delighted in silencing his skeptics in the aviation world.

Socially indifferent, mischievous, and antiestablishment, Burt lived in a three-level hexagonal pyramid house among the Joshua trees of the Mojave Desert, not far from Edwards Air Force Base. His pool table, dining table, and conference table in the energy-efficient bermed house—most of it was underground—were custom built to have similar hexagonal angles. His favorite mural in the house was of the Egyptian pyramids, and included Egyptian girls, palm trees, and a pharaoh. He challenged

---

*Fortunately, the *Voyager* had a backup attitude indicator. The primary one failed before they were halfway around the world.

†There had been numerous experiments and even production aircraft with swept-wing designs, but the theory was missing. Jones experimented with thin delta wing aircraft, which is an extreme case of a swept-wing design.

visitors to try to find two tiny and unexpected things in the mural: a wind generator and a flying saucer. He had spent years studying the building of the pyramids, concluding that the pyramid builders were likely able to cast and machine limestone and even granite; he had also spent eight years studying the Kennedy assassination, and came to his own contrarian conclusions. His mailbox, across the driveway from his sagebrush-dotted helipad, was the tail section of a scaled-down version of an experimental military plane called SMUT (Special Mission Utility Transport).

Born in 1943, Burt grew up in the small town of Dinuba in California's Central Valley. His father George was a dentist, and his mom, Irene, took care of the kids on their small farm. The Rutans were strict Seventh-day Adventists, observing the Sabbath from sundown Friday to sundown Saturday. Burt couldn't play team baseball, basketball, football, or any other sport requiring weekend practices or games. He couldn't see movies, chase girls, or race hot rods. Instead, he flew, chased, and crashed airplanes.

He and his mother would drive to San Francisco, about four hours north of Dinuba, to shop at a small hobby store. Burt wanted parts, but he didn't want planes built from a kit. He had no interest in building something that he already knew would fly. At age eight he designed and built a model of an airplane that had engines under swept wings, looking a lot like the Boeing 707 years before it took flight. He began entering designs in model aircraft shows, and by the time he was sixteen, Burt was off to the nationals with nine different entries. Mom Rutan, as she was known, hitched a trailer to the family wagon, loaded the large models into the back, and drove Burt to competitions as far away as Dallas.

The tall and lanky teenager with the flat-top haircut was also interested in rockets, and closely followed the Soviet Union's launch of Sputnik, cosmonaut Yuri Gagarin's first flight, and the promise of America's space program. At age twelve, his favorite television program was the futuristic "Tomorrow Land" segment on the *Walt Disney's Disneyland* TV series, hosted by Walt Disney himself, which asked questions about life on other planets, weightlessness in space, the origin of stars, and lunar exploration, and featured interviews with scientists including Wernher von Braun. Burt learned to fly at age sixteen, paying Dinuba's radio station DJ Johnny Banks, who moonlighted as a flight instructor, $2.50 an hour. The plane rental cost $4.50 an hour. After logging less than six hours of flight instruction, Burt soloed in an Aeronca 7AC Champ. But Burt's passion was design. He graduated from California Polytechnic State University at San Luis Obispo halfway between Gagarin's flight and the Apollo 11 Moon landing.

After getting his degree in aeronautical engineering from Cal Poly in 1965, Burt took a job as a civilian flight test engineer at Edwards Air Force Base, where he was responsible for figuring out why one of their jets, the McDonnell Douglas F-4 Phantom, a workhorse fighter bomber, was having so many accidents. More than sixty accidents were characterized as "departures from controlled flight," where the plane would enter a stall or spin by itself, without pilot inputs. So in 1968, at Edwards, a test program was run to investigate the problem, and Burt was the project flight test engineer. He flew in the backseat of an F-4 with pilot Jerry Gentry, and the two subjected themselves to life-threatening spins. From these test flights, Burt came up with major revisions to the F-4 pilots' manual to help them avoid departures and spins, and to help them recover from them. He also developed a training film titled *Unload for Control.* He and Gentry then personally briefed every USAF F-4 pilot in the world, going on a whirlwind tour of forty-eight locations as far away as Incirlik, Turkey, and Bangkok, Thailand. Edwards, of course, was where some of the military's fastest and most experimental jets were tested, where Chuck Yeager was the first to fly faster than the speed of sound, and where the X-15 reached suborbital spaceflight.

Burt left Edwards in 1972 for what turned out to be an ill-fitting corporate job with Bede Aircraft in Kansas. He quickly realized he wanted to follow his own calling and, in 1974, he returned to California to open the Rutan Aircraft Factory on the Mojave flight line, designing and developing prototypes for small planes to be put together by amateur builders. The business of kits for homebuilt planes took off, but by the early 1980s, aware of the risks in the business, where he could be sued for a builder's mistakes, Burt folded his company and founded Scaled Composites to build a range of plane types and create prototypes.

Now, in 1986, as Burt scanned the hallowed Edwards runway on a cold December morning, he was concerned whether his new creation could get off the ground, let alone make history. The light plane had never carried this much weight. Burt calculated the mechanical reliability of the craft: His crew had done sixty-eight test flights with a total of 375 flight hours over a period of two years. They had serious mechanical failures on seven of those flights, ranging from a cockpit fire to a failed propeller that ripped the engine from its mounts. Now their flight plan would have the plane up in the air for more than 200 hours nonstop. The first part of the flight, carrying the heaviest fuel load, was westward over water toward Hawaii. The plane didn't do well for long in turbulence. Burt concluded the flight was improbable.

But if there was any pilot who could pull it off, it was Burt's pilot. His brother.

Dick Rutan handed off his black cowboy hat before climbing into the *Voyager* cockpit, about the size of a phone booth laid on its side. His copilot, Jeana Yeager (no relation to Chuck), petite and known to be fearless, was behind him wearing no seatbelt—another weight-saving measure. The two pilots had once been lovers, but now they were barely on speaking terms as they set out to fly around the world.

Dick, forty-eight, was sure that he was going to die in this terrible flying machine. Two days earlier, he had made a "death tape," saying good-bye to his team. He handed the cassette to his crew chief, Bruce Evans, to be played when he died. The decorated fighter pilot, labeled "aggressive among the aggressive," and his copilot, thirty-four, a skilled mechanical draftsman who had worked for Navy inventor and rocket engineer Bob Truax, were trying to double a record set twenty-four years earlier by U.S. Air Force pilots in a Boeing B-52 Stratofortress. Dick and Jeana had lived this dream for five years. They had put everything they had into the *Voyager,* including so much of their own money that they had none left over to pay their rent. They had pitched dozens of possible backers, including American businessman Ross Perot, who was close to sponsoring their flight, but declined in the end because Dick and Jeana were not married. The owner of Caesars Palace was willing to fund them if the flight started and ended in the casino's parking lot, an impossibility given the need for a long runway. The fragile aircraft, constructed in a hangar donated by the Mojave Airport, was funded through hundred-dollar contributions, and built with donated parts and a mostly volunteer crew.

Dick had fed on the adventure of flight since childhood. His mom took him to an air show, where he got his first ride. The plane was an old two-seater yellow Piper Club, the tires had no tread, someone came out to hand-prop the plane, and takeoff was from a grassy field. As soon as they were aloft, Dick unfastened his safety belt and stood up in the back of the tiny plane. He needed to see what the pilot was seeing. He knew in that moment this was the view he wanted for life. He got his pilot's license the day he turned sixteen. When Dick wasn't flying, he was tearing through narrow rows of vineyards on his motorcycle, delighting in his ability to evade the local police, who were always warning him to slow down. Dick dreaded the Sabbath even more than Burt. He snuck into movies, sat way in the back where no one would see him, and left through the back door instead of the front. He was told throughout his childhood that the end was coming. The only end he lived for was the end of the Sabbath each week. Beginning Saturday afternoon, he'd sit and watch the horizon, waiting for that great moment when the sun had set and purgatory was over.

In 1958, at twenty years old, Dick enlisted in the Air Force Aviation Cadet Program even though Seventh-day Adventists were conscientious objectors. He flew the military's F-100 Super Sabre supersonic jet over North Vietnam, returning from aerial combat in planes riddled with bullet holes. He thrived on the adrenaline. The more he flew, the more he sought out dangerous missions. He saw colleagues and friends killed. Some were burned alive in cockpits. He saw a colleague on the ground being chopped up by a machete. In the middle of a mission, he felt no fear. When it was all over, back on the ground, he would occasionally walk over to the edge of the runway and throw up.

It seemed natural for him to sign on to the *Voyager* flight as soon as Burt had proposed it. By the early 1980s, Dick was working for Burt as a test pilot. He had already set distance records in his personal Long-EZ—one of Burt's kit designs—and was looking for a new adventure. Burt, Dick, and Jeana were having lunch at the Mojave Overpass Café when Dick said he was interested in a new type of plane for aerobatics. Burt said he had a better idea, something he'd been thinking about for years, given the advent and availability of carbon fiber and composite materials for planes. Between bites of teriyaki steak, Burt grabbed a napkin and sketched a plane with a single long wing. Burt, Dick, and Jeana immediately grasped what it would mean to do a nonstop, nonrefueled flight around the world. They saw it as a milestone, the last first in aviation.

Burt calculated that it would take seven pounds of fuel for every pound of airplane structural weight to go around the world. The challenge was how to squeeze in enough fuel to fly nearly 25,000 miles and yet keep the aircraft light enough to take off. Part of the answer came from carbon fiber, which would make the plane half the weight of conventional aluminum construction, but would give it the needed strength. The aerodynamic efficiency, as well as the propulsive efficiency, would have to be better than that of any other light aircraft. The propeller had to be more efficient, and the engines had to do a better job of turning gasoline into energy. The flight would be low altitude because the cockpit would not be pressurized.

Dick and Burt battled over weight versus stability. Dick said he couldn't turn the plane. Burt said he was flying around the world and didn't need to turn at all. Dick said the plane would come apart in a rainstorm. Burt told him not to fly in the rain. Burt questioned the need for radar. Dick said he wasn't about to fly blind. Most people knew to walk away when the Rutan brothers went nose to nose.

But the two siblings eventually found compromises. From childhood, Burt designed planes and Dick flew planes. Their sister Nell was a stewardess. The *Voyager* would have long, slender wings, so frail they would flap like a bird's. The airplane comprised nineteen separate fuel tanks. Seventy-three percent of the total takeoff weight of the plane was fuel—Dick would basically be piloting a flying gas tank.

"Edwards Tower, this is *Voyager* One," Dick Rutan said from the runway of Edwards Air Force Base on the early morning of December 14, 1986. "We are ready for takeoff."

"Cleared for takeoff, and Godspeed" came the reply at eight A.M.

Nearby, in the chase plane, Burt took a deep breath as Mike Melvill, a world-class pilot in his own right, prepared to follow the *Voyager* on the initial stages of its flight.

Jeana called out speed: *forty-five, sixty-one, sixty-five—*

As the *Voyager* gained speed, the fuel-laden wingtips began to droop, like the long leaves of tulips. Then the ends touched down on the runway and began to scrape.

"The wings are *grinding on the ground,*" Mike said, following the plane. "There is fuel right at the wing's tip."

"Tell him to pull back on the stick!" Burt yelled. "Tell him he's got wings *on the ground*. Pull back on the stick!"

Jeana, in a monotone, continued to call out the speed: *eighty-four, eighty-seven, ninety.* Mike was afraid there would be a fire. The *Voyager* had to reach one hundred knots to have adequate climb capability after takeoff.

Slowly, as the *Voyager* gained speed, the wings lifted just an inch. Then two. Then three.

"The wings went up!" Mike said. "Wow. They bent up!"

Jeana called out: *ninety-four, ninety-seven, one hundred.* The plane took flight.

"One hundred knots! Hot damn!" shouted Burt. "We got one hundred knots!"

Mike, Burt, and Mike's wife, Sally, sitting in the back of the chase plane, were quiet. It had taken so much just to get to the starting line. Flying in Burt's twin-engine Duchess, they would follow the *Voyager* until they could no longer see California. The wingtips looked bad. Wire dangled from the end, and the winglet—the small vertical fin at the end of the wing—was barely holding on. As Burt had Dick perform stability tests, the right winglet tore off. The other one looked poised to do the same.*

Burt was telling Dick, "Don't abort. Don't abort." Dick was listening and shaking his head. He wasn't about to quit. He had "W" in his compass, signaling "west," and he was on his way. As the fuel gauge in the Duchess dipped lower, Burt, Mike, and Sally needed to turn around immediately if they wanted to make it back to Edwards. Mike banked the plane to wave good-bye. Burt was silent, his eyes searching the horizon. There was water for as far as he could see.

The *Voyager* team in Hangar 77 in Mojave, led by Bruce Evans—one of the few people Dick listened to unconditionally—stayed in contact with Dick and Jeana around the clock. Burt and Mike spent their days flight-testing their newest business plane, the Beechcraft Starship. Burt was on

*The right winglet landed in the backyard of a woman in Lancaster, who had been watching the *Voyager* takeoff on television. The left winglet tore off over the Santa Barbara area.

call to talk to his brother, or to give him space—whatever he needed. Mike spent nights tracking the *Voyager*'s journey, talking to Dick and Jeana throughout. He also ran interference, keeping Burt away when he knew he would only agitate his brother by giving Dick advice on things Dick didn't need to hear.

When they had first lifted off from the Edwards runway, when the fragile wings were finally taking flight, Dick said excitedly, "I can fly the airplane! I can fly the airplane!" Jeana, who was lying on her back looking up at him, said, "I knew you could." Dick didn't know whether she was being confident or naïve, but the encouragement was important. "The *Voyager* is flyable by the velvet-arm test pilot," he said happily. He had known that if the tip of the wing continued to tear, if the fuel tank ruptured, they were dead. He had put that information in what he called his I-don't-give-a-shit box.

The sense of danger never went away. He and Jeana spoke on an as-needed basis. Jeana managed the fuel log but never flew the plane. She had built sections of the plane by hand and had been key to the mission's success. She had passed certification requirements to fly the *Voyager*, including getting her multiengine rating and instrument rating, but she did not learn other key systems to fly the plane. She hadn't learned how to use the radar or set up autopilot or navigation or talk on the radio. Dick slept for occasional two-hour stretches, weather permitting, by using the autopilot. He flew in a semisupine position, looking at instruments he and his team had built by hand, going over the checklists that Jeana had painted in tiny, perfect, artistic script. The mission could not have happened without her, but he was now the only one capable of keeping them aloft and alive.

Even during short periods of rest, Dick feared the autopilot was going to do something terrible. Every time he tried to close his eyes, the plane pitched up in his mind. They encountered storms, low visibility, and no visibility. On day two, they flew through Typhoon Marge. On day three, the autopilot failed. Day five, crossing central Africa, they ran into fierce thunderstorms and monsoons, the worst storms Dick had ever encountered. The plane pitched and heaved as he flew in and out of clouds, and the airplane rolled into a ninety-degree bank. *Here it is,* he thought. But then came the reprieve, the kind pilot Ernest Gann wrote about in *Fate Is the Hunter*: "The peril was instantly there and then almost as instantly not there. We peeped behind the curtain, saw what some dead men have seen, and survived with it engraved forever on our memories." When the worst of the African storms had passed, they flew in a black haze that was so thick and inky that Dick imagined walking on it. On another day, they almost flew right into a mountain peak. Over Sri Lanka, they had an engine coolant leak. Over the Pacific, they lost oil pressure. Dick was taking a twenty-minute nap when Jeana woke him and pointed to the red oil pressure light indicating the plane was starting to overheat.

He felt like he was playing Russian roulette every day. Pull the trigger. Click. Survive another day. His body focused his mind on what he needed to do to stay alive. But every day, he looked out and thought, *Out there someplace is disaster. It may be tonight over water or tomorrow during the day, but at some point, this mission is going to end and we are going to die.* If faced with a structural error, they wouldn't be able to get out of the cockpit. The centrifugal forces would wind up the plane and prevent them from moving until impact. They wore small parachutes with nylon harnesses and had a vacuum-sealed raft about the size of a football. Dick put their chances of opening the hatch and jumping out at slim to none. His whole world had collapsed into the cocoon of their cockpit. After a couple of days of flight, he couldn't imagine anything outside.

Nine days after takeoff, news arrived that Dick and Jeana were not far out from Edwards, having flown up the California coast after a night crossing of Costa Rica. Burt and Mike jumped into the Duchess, hoping to join up with the *Voyager* over the ocean west of San Diego. About sixty miles off the coast, in the dark, early morning hours of December 23, Burt and Mike spotted a strobe light. They couldn't be sure it was the *Voyager*. The plane had no running lights; they had been lost with the winglets.

Mike radioed, hoping it was his wingman Dick, and instructed him to cycle the strobe on and off. The strobe went off. When the strobe flashed back on, Mike and Burt did something neither one expected: they began to sob. There they were, bawling like babies. Collecting themselves, they smiled and wiped their eyes. Soon, in the gray light of the rising sun, they could make out the silhouette of the *Voyager*. When they'd last seen the plane, it was bent and damaged. Now it was flying straight and slow, looking more like an ethereal boat floating on currents of air.

In the *Voyager* cockpit, Dick wondered whether he could land back at Edwards. About twenty minutes out, he still had his doubts. *They're flight-testing big bombers and fighter jets and I'm a little homebuilder from frickin' Mojave. Who is going to care about us?* Then he thought, *What if I can't land there?* If they couldn't land at Edwards, their record wouldn't count. The Fédération Aéronautique Internationale, the world sanctioning body for aviation records, required closed-course record seekers to take off and land from the same place.

Dick called the tower at Edwards. "We're twenty minutes out," he said. "I know you're real busy. Could you let me land in the restricted area?" He figured he could land in a remote area of the restricted dry lake—so as not to bother anyone.

The tower came back: "This is Edwards Tower. Sir, we have canceled flying today and we are all here waiting for your return."

Dick was dumbfounded. *They canceled flying for us?* None of it made sense, but then he had just gone nine days on no real sleep, stuffed into a capsule with the noise level of a freight train. Dick had no idea the *Voyager* was on the cover of that week's *Newsweek*, with the headline: "THE INCREDIBLE VOYAGER, HEADING AROUND THE WORLD WITHOUT A STOP."

There was a solid cloud layer as Dick flew over the San Gabriel Mountains. Closing in on the southern end of Edwards, he looked down, expecting to see the beige canvas of the base. Instead, he saw black and silver and beige. He saw thousands of people. There were trucks, satellite dishes, and motor homes framing the runway. The 747 that carried the space shuttle to the Cape was in front of the massive NASA hangar. The sight was unforgettable. Using Mike as their wingman to count wheel height, Dick and Jeana touched down. Nine days, three minutes, forty-four seconds: 26,358 statute miles. The event was televised live worldwide.

Dick had told the *Voyager* crew that he didn't want anyone running up to the plane before the FAI inspector arrived to certify the flight. He had nightmares of doing the flight and not getting the credit for it. Dick opened the cockpit and saw all of the commotion, with cameras and microphones and a crowd closing in.

Now he faced a new challenge: he wasn't sure he could walk. He pulled himself up and out of the cockpit and took a seat on the top of the plane. His legs felt like noodles, atrophied from nine days of no exercise. He pushed himself farther back on the fuselage, moving and stretching his legs. He would not let himself be carried off the plane; his fighter pilot credibility would be shot. He figured he'd sit and wave, and jostle and tense his legs. Postponing the move for as long as he could, Dick—black cowboy hat back on—carefully made his way onto the ground.

Burt was the first in line to give him a hug. It was December 23—the best Christmas present ever.

That year had started in a very different way. The space shuttle *Challenger* had disintegrated seventy-three seconds after liftoff on January 28, 1986, killing the seven astronauts, including a high school teacher, on board. The accident was found to be part mechanical failure, part management failure. Members of the Rogers Commission, appointed to determine the causes of the disaster, found among other things that NASA managers had not accurately calculated the flight risks. Commissioner Richard Feynman, the Caltech physicist and Nobel laureate, concluded, "The management of NASA exaggerates the reliability of its product, to the point of fantasy."

Burt had followed the commission's findings only from afar. One of the more memorable things to come out of it was something he heard from test pilot Chuck Yeager, who would swing by to visit on occasion. Yeager was appointed to the commission and attended the first meeting, where discussion centered around the shuttle's O-rings, which failed due to the cold temperatures on the morning of the launch. (In one meeting, Feynman took an O-ring and put it in ice water, showing how the material was compromised by the cold.) Yeager, big on courage but short on patience, listened as discussions dragged on for hours. He could predict how NASA was going to limit shuttle flights for years to come. He walked out of the first meeting with no plans to return. Asked in the hallway why he was leaving, he shot back, "Give me a warm day and I'll fly that sonofabitch."

There were plenty of things not to like about Yeager—he was all about Yeager—but Burt admired his spirit. Yeager got into a plane shaped like a .50 caliber bullet with a needle nose and flew into the unknown, pushing the X-1 faster than any pilot had flown. Burt knew that his brother was cut from the same cloth. All Dick ever wanted was to do something significant in a plane. He had made that happen. A scrappy team in the desert had defied reason, ignored skeptics, and made history. Burt was given the moniker "Magician of the Mojave." He had a lot more tricks up his denim sleeve.

5

Space Medicine

Peter drove his black Trans Am along Mass Ave., slowing as he passed the entrance leading to the Infinite Corridor. He turned onto Memorial Drive, home of the Stratton Student Center, where SEDS was launched on a perfect starry night. Across the way was building 37, the place of real-life astronauts and cutting-edge space technology. Boston's "Don't Look Back" played on the radio. Peter blinked back tears. MIT was in his rearview mirror.

Peter graduated from MIT in June 1983 with a bachelor's degree in molecular biology. He had worked on some dazzling things, ranging from space-bound experiments to genetic engineering. He was now headed to medical school. Peter had been accepted to Stanford on early admission—and moved out to Palo Alto, loving the weather—only to learn over the summer while backpacking in Greece that he was accepted to Harvard Medical School as well. Specifically, he was admitted to Harvard's MD-HST—Health Science and Technology program—something Peter embraced as a "geeky medical degree." The Harvard HST program was Peter's first choice, but the odds had been against him, given that only twenty-five students were admitted to the program each year.

Harvard took some getting used to. He quickly discovered that telling someone he attended Harvard Medical School was "dropping the H-bomb." It felt impossible to say the school name without sounding like a snob. On the positive side, it could prove effective in picking up girls, though a social life in medical school was as hard to come by as sleep. The medical school was about fifteen minutes across the Charles River from the Harvard University campus. Its epicenter was the Quad, a grassy area surrounded in a horseshoe shape by five marble-faced buildings. Peter lived in the student dorm, Vanderbilt Hall, on campus his first year. Surrounding the medical school buildings were hospitals. Peter loved the school's history and its miles of underground corridors. On the rare occasion he landed a date, he would suggest they go to the museum above the medical school administration building. The Warren Anatomical Museum detailed the history of health care and was filled with artifacts—many macabre—including nineteenth-century surgery kits, early microscopes and microscopic imagery, plaster casts of faces for phrenology and of the right hands of great surgeons, and sterilizing devices from the earliest days of surgery. The coolest exhibit was the skull of Phineas Gage, the man who survived a terrible accident that drove a steel bar straight through his skull.

The first real eye-opener of medical school, though, was anatomy class. Peter and his lab partner were given the choice of working on the cadaver of a small elderly lady or the cadaver of a large man. Peter chose the elderly female cadaver, hoping the smaller frame would make the dissection easier. Peter named her "Aunt Molly." The woman had died at age seventy-three of pancreatic cancer. On dissection day one, Peter and his lab partner covered Molly's face and hands with cloth to depersonalize the experience. They cut into her thorax, and then had what Peter considered the brutal experience of sawing her ribs with a handsaw. Not long after, they were tasked with dissecting Aunt Molly's groin area. This was followed by the study and dissection of the arm and hand. Before work began on the arm, the professor carried from table to table a perfectly dissected arm on a tray. Peter and his lab partner placed a nickel in the open outstretched hand, generating some laughs. Dissection under way, Peter was struck by the brilliant biomechanics of the arm, wrist, and hand, with the delicate and dexterous arrangement of muscles in the palm and forearm, the bands of connective tissue, tendons, joints, and nerves. He was reminded of the prosthetic hand Luke Skywalker was given in *The Empire Strikes Back*. And he was reminded of some of the robotic arms he had made in his upstairs room in Great Neck. He had always been enamored with the physics of the hand.

When the day came to uncover Aunt Molly's face, Peter paused. This felt like the most psychologically difficult part of the course. The cloth was set aside and students had to remove the skin of the face. In a later class, Peter had to open up Molly's skull, sawing it with a rotary saw in a line running from about an inch above the ear all the way around her head. Peter worked methodically to extract her brain to examine the cranial nerves, cerebral veins, and various components, from neurons and dendrites to synapses. Her tissue, he found, was like semisoft cheese. The smell was terrible, but seeing the inside of a skull was something he would not soon forget. What struck him was how this small brain, gray and slightly slippery like a rock lifted from a river and weighing only three pounds, held all of Aunt Molly's thoughts, memories, skills, loves, wants, and desires. Molly's entire life was lived in this matter until one day it just blinked off. Those memories, housed in the patterns of synaptic connections between neurons, were locked in this meaty human hard drive, never to be accessed again. Peter had recently bought one of IBM's first desktop personal computers, with two five-inch floppy drives. He could back up his files, but there was no way to back up the human mind. He was awed and saddened by what he held.

Early one morning on his first third-year medical rotation, while on a shift in the Baker Building at Massachusetts General Hospital, Peter had his first life-or-death encounter with a patient. It was around three A.M. and Peter was catching some sleep on a cot. He was jolted awake with "Code Blue Baker five." Someone on the fifth floor was having a cardiac arrest. Peter raced up a flight of stairs and was the first to reach the patient, who'd had open heart surgery the day before. He began chest compressions. The man's sternum cracked with each compression, and all Peter could think was, *Ohmygod this is real.* It was one thing to do this on a dummy, another on a human. This night ended well. There were other times when Peter felt overwhelmed and overloaded. He met an angelic sixteen-year-old girl who had lymphoma and was given very little chance of living. He saw premature babies fighting for life. He developed a friendship with a rheumy-eyed homeless man, an alcoholic with liver failure, and discovered that he was a Harvard graduate. The two had great conversations when the man was sober. Peter worked with a frail and elderly patient with impacted bowels and had to don gloves and disimpact him. He did his first spinal tap and first breast exam. He also began to see previously healthy men come in with debilitating infections, including pneumonia and Kaposi's sarcoma. The men were among the first diagnosed cases of HIV/AIDS. A mix of anxiety and hysteria surrounded the disease—the religious right called it "the wrath of God against homosexuals"—and Peter experienced it all from the front lines. With every patient, whatever the problem, Peter thought the same three things: *If only they knew how little I know, I'm doing my damnedest, I hope I don't fuck up.*

The problem was that Peter didn't love what he was doing. He sometimes fell asleep in class—from exhaustion, but also from finding many of the lectures tedious and uninspired. He thought the school's anatomy lab looked like it dated to the Civil War. He took a pathology class and found the subject matter similar to what he'd covered in seventh grade. One of the few subjects that intrigued him was something they spent too little time on—atrial fibrillation, where the class did an in-depth math analysis that left Peter thinking, *Now that was actually interesting.* He and his lab partner rushed out of anatomy class one day to attend a student-faculty council meeting only to find that no faculty showed up. His fellow HST students—an eclectic group that included a professional surfer, a former Ringling Brothers circus clown, and an eighteen-year-old whiz kid from Columbia—all knew of what Peter called his *space affliction*.

To Peter, one of the best things about the HST program was that Harvard Medical School collaborated academically with MIT, incorporating engineering principles into traditional medical school teachings. They would study the human heart and then work with MIT professors to build an electronic circuit diagram to represent the heart.

But Peter made a beeline whenever he could for MIT's Man Vehicle Lab, volunteering for research and gleaning details on new space projects. He was still running SEDS, which had grown by the mid-1980s to around 110 chapters, including a new chapter he'd started while at Harvard Med. He held big and impressive space fairs at MIT, inviting students, faculty, and executives from NASA, Boeing, and Lockheed. As he rode the bus back to Harvard early one evening, he wrote in his journal, "Oh how I wish I could cut loose of expectations. Have I been doing what I do to impress Dad?"

Though medical school was duty over passion, a family obligation, Peter believed on some level that there was a chance a medical degree might get him closer to space. He had done the research and learned that if he wasn't going to be a fighter pilot, then an MD might be the ticket to getting into the Astronaut Corps. Peter rationalized that at the very least, it might help him learn how to extend the human life span long enough so that technology could catch up with his dream of accessible space travel. The longer he lived, the better his chances were of getting off planet. He also liked the idea of being a polymath, the boy who played three roles in his Cub Scouts play. Peter's newest fictional hero was Buckaroo Banzai of *The Adventures of Buckaroo Banzai Across the 8th Dimension*. Buckaroo was a top neurosurgeon, a particle physicist, a race car driver, and a rock star. Buckaroo had perfected a wide range of skills; why couldn't he?

Sitting in Amphitheater A on the Harvard Medical School campus, Peter listened as the teacher went over the forms being handed out. The paperwork was to be filled out with information on where students wanted to be "matched" for internships and residencies after their fourth year. Peter had passed part one of his medical boards, and would take part two at the end of his fourth year. If he passed, he would begin his internship. At the end of the internship, he would take part three of the medical boards to get his license to practice. Then the residency would begin, ranging from a two-year to a seven-year commitment, depending on the chosen branch of medicine.

Peter looked at the forms. He was supposed to know his top choices of hospitals? He needed to declare a specialty? He thought, *I just want to go build rockets.* Through his first and second year, and even now in his third year, in the midst of clinical rotation, he had managed to get away with medical school as a part-time undertaking. He was running SEDS, doing his MVL research, and hosting national space conferences. And space was becoming more real. Shuttle missions were flying with increasing frequency, from two flights in 1981 to nine now in 1985. Looking at the papers on his lap, Peter felt a rising sense of panic. There was no way he could keep juggling medical school *and* space as an intern or resident. He would make a mistake that would kill someone. At the very least, he would turn into a frustrated doctor not living his dream. Sometime later, lost in his thoughts, he looked around and noticed that everyone had left the cavernous room. The lights were being turned off. Stuffing the papers into his backpack, anxious and worked up, he headed to the administration office. He had a long-shot idea.

He asked the receptionist if he could make a call to the Man Vehicle Lab at MIT. He needed to talk to *his* Obi-Wan Kenobi, lab director Dr. Larry Young. Peter was lucky to find Dr. Young in the lab. Peter quietly explained his situation at Harvard and asked, "Is there some way I can come back to MIT and do my master's or PhD in sixteen?"* He wanted to take a leave from Harvard, enter the aeronautical and astronautical engineering program, and defer this critical decision about medicine. Dr. Young told Peter that he would have to apply for admission, but that he would help fast-track the application. He had seen Peter's dedication in the lab, and wondered when Peter would move from the science of medicine to the science of space.

Larry Young's own love of space began the day the beach-ball-size Sputnik satellite orbited Earth on October 4, 1957. He was traveling by ship to France, where he had a Fulbright scholarship to study applied mathematics at the Sorbonne. On the night he arrived in France, everyone was staring up at the sky and listening to handheld radios. Young decided in that moment to shift his studies to space. When asked why space, he would reply, "It's a little like falling in love. You can't explain it rationally; but you know it." The license plate on his car read 2MARS. His student Byron Lichtenberg had become America's first payload specialist, and many of the experiments developed in the MVL were flying on shuttle missions. Young saw Peter as smart and enthusiastic and filled with one big idea after the other. Peter had more excitement in his eyes than just about anyone he'd ever met.

Where Harvard was interesting, being back at MIT was joyous. Peter was now doing exactly what he wanted to do: aerospace engineering. He reassured his parents that he was simply adding another degree, but would return to medical school. He was surrounded in labs and classrooms by real turbines, flight simulators and models, and photographs of planes and rockets. Even the make-or-break two-semester course, Unified Engineering, taken by sophomores, was a welcome challenge. Peter

*Course sixteen is an introduction to aerospace engineering and design.

was four years older than his classmates and sat in the front row, learning the underlying math equations and principles behind solid mechanics and materials, fluid mechanics, thermodynamics, and propulsion. Because of the sheer amount of interdisciplinary material covered in Unified Engineering, students were required to work collaboratively on problem sets. Peter returned to his old stomping grounds, the Theta Delta Chi fraternity, and found a new group of fraternity brothers to study with. There were also break nights, with movies—the latest James Bond film, *A View to a Kill*—and parties that lasted into the early morning hours with dancing to the Go-Go's. There were times when Peter longed for a girlfriend, and other times when he realized love would have to wait.

Shortly after returning to MIT, Peter had met with Dr. Young to talk over his idea for his master's thesis project. Peter was interested in somehow creating artificial gravity to alleviate the muscle deterioration, loss of bone calcium, and other known maladies that came with weightlessness. He told Dr. Young he was thinking about a small-radius rotating bed that could create gravity while the astronaut slept.

Any prolonged stay in space would require humans to figure out how to create artificial gravity or some other countermeasure, such as extensive exercise. The astronauts of Skylab, the NASA space station that orbited Earth between 1973 and 1979, had come home in worse shape than when they'd left. Owen Garriott and William Pogue—who set records at the time for the number of months spent in space—were examined upon their return, and the flight surgeon's official report read, "Capable musculoskeletal function will be threatened during prolonged space flight lasting one and a half years to three years, unless protective measures can be developed." In addition to muscle atrophy and bone density loss, the astronauts had balance disorders that persisted after the other conditions were fixed. Some stumbled in the dark, lacking visual clues as to which way was vertical, and continued to try to float things around them as they had done in the space station.

Both Konstantin Tsiolkovsky and Wernher von Braun had imagined a large, circular space station that would rotate to generate artificial gravity. A NASA Ames research study begun the year after the Skylab program ended also imagined a space station shaped like a wheel that would rotate at one revolution per minute, fast enough so the centrifugal force at its rim—where living accommodations were to be located—would be the equivalent of normal gravity on Earth. In the movie *2001: A Space Odyssey,* the *Discovery One* spaceship included a large spinning centrifuge inside the crew compartment.

Dr. Young liked Peter's topic but needed details and was worried about the effect on an astronaut's inner ear. Fortunately, inspiration came to Peter. Out on a walk with his mom, dad, and sister, who were visiting MIT, Peter found his answer on a playground. Tula, who had a habit of asking Peter the moment she saw him if he'd met any nice Greek girls, liked the idea of Peter's looking wistfully at kids. But Peter was stopped by something other than children. He was looking at the playground roundabout—the flat disk with handrails that spins like a merry-go-round. His mind started spinning: *centrifugal force, the force created by spinning,* is a function of the velocity squared and the radius from the center of rotation. At the center, the acceleration would be zero. And if an astronaut's head, more specifically his vestibular system, was placed at the center, there would be no axial acceleration to mess with the inner ear. Peter grabbed his sister's hand and ran toward the play structure. After persuading the playground kids to take a break, he instructed his sister to lie down with her head at the center. She protested to no avail. Inwardly, she smiled. This was Peter, the boy who often had a distracted, I-am-elsewhere look. First she was spun around, eyes closed. Then it was Peter's turn. He stayed on until a few moms began standing next to the play structure, throwing unhappy looks their way. Back at school later that night, he sketched out his idea. He needed something small enough to fit inside the space shuttle and eventually live at the International

Space Station, outlined the year before by President Reagan and under construction with segments reportedly close to launch. Peter wrote in his journal, "What if you can give a dose of gravity, like medicine, while someone sleeps?" The next Thursday, Peter was in the Man Vehicle Lab and talked with Dr. Young about the specifics of his plan. He had done three sketches: of a rotating plank on the floor, a plank on the side of the craft in a storage area, and a positionable beam. He dubbed it the Artificial Gravity Sleeper, or AGS.

"If you put people on a centrifuge, they can't do anything else," Peter said excitedly. "If they centrifuge while they're sleeping, they can work their cardiovascular system and stimulate their immune system."

Dr. Young looked at the drawings and said, "It's not a dumb idea." Still, he remained skeptical that someone could sleep while being rotated. But Peter told him he had spent time at the playground, and had even dozed off while spinning. Peter reasoned that if one could sleep in a gravity field, gravity could be thought of as a medicine and could be given in doses of four hours, six hours, and so on. In space, the bed's centrifugal force would act like gravity, making the body work harder than when it was weightless. The astronaut's head would be at the center, like the center of a record player, with no gravity.

Peter was able to get $50,000 in grants from the Space Foundation, NASA, the National Institutes of Health, the American Heart Association, and the AeroAstro department of MIT. He went to work building his motorized rotating space bed. He did drawings, calculations, and the hands-on building of the device, which was a two-meter-radius centrifuge. The bed, which could run at a rate as high as 40 rpm, providing 3 gs, would be made of a honeycombed aluminum and had counterweights and telemetry signals flowing through gold-plated slip rings to monitor the person being spun. The AGS sat atop concentric six-inch and eight-inch–diameter steel piping connected by sealed ball-bearing fittings. Peter offered MIT students thirty dollars if they would spend the night sleeping while spinning. His friend Todd B. Hawley, a leader at SEDS, was his first and best volunteer, spending nine nights in a row in the device. Peter noted in his observation log: "What amazes me is that it's working! TBH is actually getting a good night's sleep on the device. Todd has been really fantastic about cooperating, and has not really complained @ all." Peter evaluated sleep through observation, and by the data readouts. One day, Dr. Young stopped by and volunteered to get into the contraption. He stayed on for a few minutes before climbing off and saying he thought it was great. The next time Peter powered up the machine, the piping unwound off the flange and the whole thing toppled down. Peter was mortified, realizing he could have killed the Obi-Wan Kenobi of the MVL. Peter corrected the problem and continued with his experimentation. He also built the device for exercise, adding a "cycle ergometer," with pedals for use while the astronaut was supine. Peter loved every second of it: He was the little boy back building Estes rockets and experimenting with chemistry. Only this time, he was building something that might help people live in space.

In April 1987, Peter, Todd Hawley, and Bob Richards hosted a conference to realize a new and even bigger dream: the founding of a space university dedicated to space studies in every discipline. The gang of three, *Peterbobtodd,* had been exploring the formation of a graduate-level university for several years. The goal was to create the world's first nonprofit, nongovernmental university focused on space studies. The three men were counting on their body of work—SEDS and their popular national space fairs—to give them credibility and contacts. It worked: over three days in April, Peter, Bob, and Todd hosted hundreds of people at MIT's Stratton Student Center, the birthplace of SEDS. Peter, who was ten months shy of earning his master's degree and one year shy of an MD, again had the support of MIT president Paul Gray, and the conference drew space delegations and top leaders from the Soviet Union, Canada, Japan, China, the European Space Agency, and NASA.

The opening session drew more than five hundred people. Peter welcomed attendees to the Founding Conference of the International Space University Project and gave an impassioned talk about the state of the nation's space program. He noted that after the Apollo 11 landing, NASA had appointed a task force to chart three courses—fast, medium, and slow—for continuing beyond the Apollo program.

"The time line set forth in the fast-paced scenario reflects what we might read in today's science fiction stories," Peter said. "The slowest of the three scenarios goes as follows," with dates of completion:

Space Station and Shuttle: 1977
Space Tug: 1981
Lunar Orbiting Station: 1981
Lunar Surface Base: 1983
50 Person Space Base: 1984
Mars Expedition: 1986
100 Person Space Base: 1989

Peter asked the group, "What has happened in the last eighteen years? . . . It is time to revitalize the motor and the vision which will put humanity in space—and that motor is composed of yourselves, of the students around the world. This is where I see ISU playing its greatest role."

Closed sessions were led by notables including Byron Lichtenberg; John McLucas, the former U.S. secretary of the Air Force; and Joe Pelton, a director at Intelsat, the governmental consortium managing a constellation of communications satellites. Subjects of discussion and debate ranged from the university's mission statement to classes to be offered. From the start, it was Todd who insisted the university be truly international.

Peter, Todd, and Bob sat at the end of long tables, surrounded by distinguished-looking executives and professors and attentive groups of students. Their plan was to open their university the following summer, in 1988, with an inaugural eight-week session. They would borrow the bricks and mortar—a campus—and hire faculty from their own universities. Their goal was a permanent campus of their own, and their dream was a campus in space. They had already signed on a notable to be chancellor—their Uncle Arthur—Arthur C. Clarke.

For the second time, Peter walked out of the student center believing he was onto something big. Step by step, all of this effort had to make a difference. He had to excite and inspire a new generation of space dreamers. Since the space shuttle *Challenger* disaster the year before—Peter had watched the launch from a closed-circuit TV in the Man Vehicle Lab—risk seemed to go from something laudable to something lamentable. *Failure is not an option* solidified as NASA's mantra. At least for now, nothing was flying; the shuttle program was grounded. There was only eulogizing, finger pointing, debating. There were commissions formed, reports written, and town hall meetings held. Fear supplanted courage.

Late that night, the three-day conference over, Peter picked up the book he had just finished, *Atlas Shrugged.* Todd had given him the tome, and it became almost a playbook. Peter was moved by the story of what happens to the world when the most productive members of society—the thinkers, the engine of creation—go on strike. *What happens,* Peter pondered, *when a small group of thinkers and builders create a vision of a future they want when the politics of government let them down?*

Peter looked through his dog-eared, marked-up copy. It was interesting, he realized, how he had said in his opening speech earlier in the week: "It is time to revitalize the motor. . . . *You* are the motor." Flipping through the book, he came across a quote he'd underlined. It was by Hank Rearden, the indefatigable industrialist he most identified with, a man who spends a decade working to invent a new type of metal, and who undergoes a transformation over the course of the journey. Peter read the quote by Rearden aloud:

"All that lunacy is temporary. It can't last. It's demented, so it has to defeat itself. You and I will just have to work a little harder for a while, that's all."

Peter would also have to work harder to figure out his own path. His dream was always about space. His parents' dream for him was always about medicine. He wanted to honor his family legacy, yet also be true to himself.

A similar struggle for identity was playing out on an icy mountain across the country. There, a man a few years younger than Peter was also searching for his true self, but weighed down by a famous last name.

What Peter couldn't know was how this man—and his legacy—would change his life.

6

Being a Lindbergh

Erik Lindbergh woke at two A.M. and peered out of his tent on the icy slopes of Mount Rainier. The black sky was milky with stars—radiant, muted, cloudlike, an ethereal carpet unfolded above, a connect-the-dots of constellations all around. The air was still, the mountain slumbering. The twenty-one-year-old closed his eyes and inhaled the cold air, knowing the day ahead would be anything but calm.

Erik's goal on this early August morning in 1986 was to summit the 14,411-foot mountain, the highest peak in the Cascade Range. Climbing Rainier wouldn't be a history-making event like his grandfather's plane flight, the 1927 journey across the Atlantic to Paris that made Charles Lindbergh a hero and at the time arguably the most famous man on Earth. But—for now—scaling Rainier would be Erik's Paris, his milestone. He was intent on staying clear of anything too predictably "Lindberghian." A friend was trying to persuade him to get his pilot's license, but Erik found it way too obvious. Family members had learned to fly, but no one was a pilot. The only flying Erik planned to do was off the cornices at his favorite ski mountains.

Erik began packing up his gear. He was at Camp Schurman, 9,440 feet above sea level on the eastern side of Mount Rainier, an active, icy volcano about an hour southeast of Seattle in Washington state. He would start before three A.M. in a bid to get off the icy part of the mountain before the sun warmed the snow and increased the avalanche risk.

Mount Rainier was postcard beautiful, with glaciers toward the top, and wildflowers, lakes, and old-growth forests on the valley floor. But the peak was dangerous year round. It was riddled with crevasses. Rocks tumbled, the weather turned, and the glaciers were always shifting. Five years earlier, in the worst recorded climbing accident in American history, eleven hikers were killed on Mount Rainer, their bodies never recovered. A massive piece of the icy mountain had exploded like dynamite and rained down on the climbers.

After a quick breakfast of oatmeal, Erik, his brother, Leif—four years older and a more experienced mountaineer—and cousin Craig Vogel set out in the darkness carrying fifty-pound packs. Leif took the lead, and Erik carried his old green external-frame nylon pack nicknamed the "meat wagon." Stained and beat up, the backpack had accompanied Erik on family deer hunting trips when he was a teenager. He and his siblings and father would climb the steep trails of the North Cascades high country, returning with the meat wagon filled with deer meat wrapped in cheesecloth.

But on this cold morning, Erik had much higher aspirations: he was trying to reach the summit for the first time. Roped together and wearing battery-powered headlamps, the Lindbergh group made its way up the Emmons Glacier trail one booted step at a time in hopes of arriving at the destination in eight hours. From the top, they would still have to hike back down to Camp Schurman to gather the rest of their gear, and then continue to the trailhead at the White River Campground at 4,400 feet.

Erik had tried the same climb twice before and failed. He had made novice mistakes of packing too much gear—his dad's cast iron crampons, skis, extra boots—and taking the wrong route and overshooting Camp

Schurman by 1,000 vertical feet, arriving at camp at two A.M., just as other hikers were waking up to start their climb.

Tall, wiry, and athletic, Erik was unaccustomed to failing at any physical challenge. He had been the Washington state gymnastics champion at age twelve. His strongest discipline was the floor exercise, but he won for all-around performance. Shelves in his bedroom were filled with water-skiing trophies. He could climb a rope in gym class with just his hands, held his high school record for chin-ups, and as a student of tae kwon do, would do a jumping sidekick at head level.

While in high school in a small island town a short ferry ride from Seattle, Erik was a major ski bum—working as a dishwasher and shoveling snow at the Crystal Mountain Resort, not far from Mount Rainier. He also spent a ski season in Sun Valley, Idaho. He now lived in Olympia, Washington, and attended Evergreen State College, where he was studying political ecology. College, like high school, didn't engage him the way it should. He was thinking of opening his own backcountry luxury skiing company. He preferred working his body over his brain, favored action over intellectualizing.

Erik and his siblings had learned from early childhood to keep their heritage to themselves. As a boy, Erik had little understanding of the significance of his family name until a classmate told him he was reading his grandfather's Pulitzer Prize–winning book, *The Spirit of St. Louis*. Erik knew his grandfather not as the world-famous aviator Charles Lindbergh, but simply as "Grandfather." He was the tall and balding man who offered Erik fifty cents if he could learn to wiggle his ears (as he could), bought him a toy model of a Sikorsky helicopter,* and seemed more at ease with children than adults.

---

*Erik learned much later that his grandfather had worked with Igor Sikorsky, the Russian inventor of the first helicopter, who became a close family friend. The first helicopter was genius. It had a cyclic (essentially a joystick that controls the pitch of the helicopter by controlling the pitch of the main rotor blades), a tail rotor (which prevents the helicopter from spinning around with the main rotor), and landing skids (which effectively distribute the load of a vertical landing). A modern helicopter pilot would easily recognize the craft, and could very probably fly it.

Known to his children as a perfectionist, list maker, and lecturer, Charles could also be affectionate when away from the public eye. There was a tacit understanding among the Lindbergh clan that one should not ask Grandfather about his 1927 flight. Erik's uncle Land Lindbergh—the third child of Charles and his wife, Anne—did ask and was directed to "read the book."

In high school, Erik was struck by how people treated him differently when they discovered his lineage. They would clutch his arm, stand too close, and tell him stories about the man they idolized, the man who risked everything for a dream he believed in; who galvanized commercial air travel; who even inspired America's first astronauts through his dangerous journey into the unknown. There was also a controversial side to his grandfather, a man who was anti-interventionist during World War II, was impressed by the military might of Nazi Germany, and made anti-Semitic statements. Occasionally, the idolatry from strangers would focus on Erik's grandmother, Anne Morrow Lindbergh, herself a pioneering aviator—the first American woman to earn a glider pilot's license—and acclaimed writer. But Grandmother's legacy was gentler—she was gentler. Where Charles was tall, forceful, and peripatetic, Anne was petite, contemplative, and grounded.

Erik's father, Jon Lindbergh, was the second child of Charles and Anne. Their first son, Charles Jr., was the cherubic, curly-haired twenty-month-old toddler who was kidnapped from the Lindbergh home in 1932, held for ransom, and killed. The baby, taken from his second-floor nursery, was found dead in the woods close by, killed by an apparent blow to the head. The fame from the 1927 flight—ticker tape parades, autograph seekers, photographers following the family's every move—had made the Lindberghs a target. The media circus around the crime, and the con artists who emerged to make false claims, sent the Lindberghs into exile in England, where they lived for a period under assumed names. The message to descendants was clear: you put yourself out, you pay.

On the corridor of the Emmons Glacier, the first hints of dawn had lightened the sky from black to pale blue and rimmed the horizon with vibrant orange. Soon the sun rose above the clouds and dusted the white landscape in pink and mauve. Erik studied the massive mountain underfoot. The immense beauty of where they were was eclipsed by the personal agony of taking thousands of uneven steps. He and his brother were on a softly sloped area, but the sound of falling rocks was as constant as the crunching of snow. Leif, a skilled storyteller with an ironic sense of humor, was quiet today. Breathing grew more difficult as the air thinned. The brothers snaked their way up the glacier. Their goal was two breaths per step.

Pausing to rest, Erik looked back down and saw clusters of hikers at Camp Schurman looking up at them. They were now ten hours into their climb, which should have taken them no more than nine hours. They passed dramatic-looking seracs, giant columns of intersecting glacial ice, which could topple with no warning. They were on a painfully indirect trajectory, and this final pitch was grueling. Erik had been strangely sluggish from the start, and was feeling worse and worse the higher they climbed. His backpack was heavy, and his legs were uncharacteristically weak. He rubbed his forearms and wrists. They were now within sight of Columbia Crest, the mountain's official summit. Erik called up his strength for the final push.

Finally, short of breath, weighed down by their packs, and operating on little sleep, Erik, Leif, and Craig reached the summit of Mount Rainier. Leif wanted to stay and explore the summit, with its 360-degree view, caves, false summits, and calderas. The view was like a moonscape: different from anything Erik had ever seen before, filled with dangerous couloirs and turquoise crevasses hundreds of feet deep. There were places on the mountain that man would never be able to touch.

After about twenty minutes, the three men began their trek back down the mountain, aiming for a far more direct route. Within ten minutes of the descent, Erik, who had felt lousy for much of the climb, started to breathe better. But then he developed an epic nosebleed, and his nostrils felt sunburned. His feet throbbed, and his wrists were swollen and sore. The pain would not go away.

Erik finally arrived back at Camp Schurman, where he and Leif gathered up their gear, adding ten more pounds to their packs. His feet and shoulders ached. His head hurt. He had always pushed his body beyond its comfort zone, but this time felt different. It felt far worse.

At eight thousand feet, they left the ice and were on rocky terrain, no longer roped in. They saw powerful-looking Canada geese flying in arrow formation above, and agile black-tailed deer darting through the trees. The valley floor was laced with purple and yellow flowers, and snow-fed rivers glistened in the dwindling sun. As a student of ecology, Erik worried about losing pristine places like this. Nature was his religion; the place where he found peace, inspiration, and answers. He felt allergic to organized religion. As a boy, he spent hours watching eagles and herons swoop and hunt. He built sandcastles, marveled at the sea life of Puget Sound, and collected oddly shaped driftwood for its beauty. He and his five siblings often slept outside on the upstairs porch, watching for shooting stars.

Erik's grandfather had the same love of nature. Charles Lindbergh, by all accounts a restless spirit, traveled extensively after World War II and became as fixated on the environment as he had been on his history-making flight. He supported and spearheaded conservation efforts, became a director of the World Wildlife Fund, and fought for the protection of parkland and endangered species around the world, from the gray whale in Baja to a small buffalo called the tamaraw in the Philippines. He lived among indigenous tribes in the Philippines, Brazil, and Africa, eating stewed monkey and sleeping in huts with palm-thatched roofs. He was instrumental in working with locals to secure land for the establishment of Haleakala National Park in Hawaii. He had a particular love affair with the Philippines, where he worked to save the endangered monkey-eating eagle, the largest eagle in the world. Erik's grandmother was equally interested in the interplay between humans and the environment, writing a cover story for *Life* in February 1969 about nature in the midst of America's new spaceport. Anne wanted to understand whether the natural beauty and abundant wildlife of Cape Canaveral—where she and Charles had camped with their children decades earlier—could persist next to the brawny and fiery technology of space. The title of her story, published the week of the Apollo 9 mission, was "The Heron and the Astronaut," and concluded, "without the marsh there would be no heron; without the wilderness, forests, trees, fields, there would be no breath, no crops, no sustenance, no life, no brotherhood and no peace on earth. No heron and no astronaut. The heron and the astronaut are linked in an indissoluble chain of life on earth." The story, with photos of the indigenous wildlife—egrets, wood ibis, pelicans, alligators, armadillos, and rattlesnakes—ended with, "Through the eyes of the astronauts, we have seen more clearly than ever before this precious earth-essence that must be preserved. It might be given a new name borrowed from space language: 'Earth shine.'"*

Now close to the trailhead, where their painful adventure up Mount Rainier had begun, Erik took another break. He saw a marmot poke its furry head above the rocks. Sunlight steamed through the fir, pine, and cedar trees. Had he not been a Lindbergh, Erik might have worked in logging. It was outdoors and physical. But he had embraced a statement that his grandfather had made to *Life* in the sixties: "The human future depends on our ability to combine the knowledge of science with the wisdom of wildness." The Lindbergh Foundation, established three years after Charles's death in 1974, was focused on the intersection between technology and conservation.

---

*Anne Morrow Lindbergh wrote of a "cheerful" meal she and her husband shared with a group of astronauts, with Charles marveling over the amount of fuel consumed by an Apollo launch. "In the first second, he figures out, the fuel burned is more than ten times as much as he had used flying his *Spirit of St. Louis* from New York to Paris." Charles typed the *Life* story up for Anne, using two index fingers at a time.

It was around eight P.M. when Erik made it back to the parking lot near the Mount Rainer trailhead. He set his pack down on a rock. When he tried to pick it up again, he couldn't use his hands.

Had he sprained his wrists? He tried to make light of it, saying mountaineering wasn't his sport. But something wasn't right.

Months after the Mount Rainier climb, Erik finally went to see his family doctor. He would be fine for weeks, only to have the pain suddenly return. After competing in a waterskiing competition, both of his knees hurt and were swollen. The pain was intense, as it had been in his wrists. His doctor did some tests and said he wasn't sure whether Erik was just overexerting himself or had something more serious.

The doctor told Erik there was a chance that he had a degenerative condition. There was a *chance,* the doctor said slowly, that Erik had something called rheumatoid arthritis. One symptom of the chronic disease was a mirroring of pain in the body: both knees, both wrists, both feet, both ankles. It could be triggered, the doctor said, by something physically excruciating, whether a marathon or—for women—a particularly difficult childbirth. The doctor said the disease could damage and deteriorate joints, but he wouldn't give Erik a worst-case scenario. He wanted to watch Erik over time.

Erik walked out of the office shaken but telling himself there was no way he had rheumatoid arthritis, or any kind of arthritis. That was something afflicting the elderly, not an accomplished twenty-one-year-old athlete.

Erik's mother, Barbara Robbins, was quiet when she heard the news. She knew nothing about rheumatoid arthritis. Now divorced from Erik's father, Barbara sometimes felt the Lindbergh name had a curse to it: her in-laws lost their beautiful baby Charles Jr. Her sister-in-law, Reeve (Charles and Anne's youngest child), had lost her first son, Jonathan, when he too was just a toddler. Reeve had been staying at her parents' new house in Connecticut when the baby died of a seizure in the night.

For Barbara, living with Erik's father, Jon, had been full of ups and downs. When they married—having met as students at Stanford—everything had to be done in secrecy. The newlyweds were out of town when they began reading concocted stories of their wedding and the gifts they had received. Syndicated gossip columnist Walter Winchell wrote about Charles Lindbergh's gift to his son and new daughter-in-law, saying he had given them a new sports car, when in fact they drove an old blue Ford wagon that turned purple in the sun. *Redbook* ran an interview with them that had never happened.

Jon was a Navy "frogman," an underwater combat and demolition specialist (frogmen were the precursors to the Navy SEALs). After the Navy, he went into business diving on gas lines and off oil rigs, reaching unprecedented depths and experimenting with breathing mixed gases for decompression. He volunteered to fight fires and was one of the world's first "aquanauts," having lived underwater for more than twenty-four hours.

Now Barbara contemplated the prospect that her son Erik might have a degenerative disease. Barbara had met only one person who had rheumatoid arthritis—he was a young man who spent his days in a rocking chair.

Erik had been the easiest of her six babies; as an infant, he would just sit and smile and kick his legs happily. Erik looked just like Barbara's father, Jim Robbins, a Swede with blue eyes and angular features, but Barbara knew that Erik was a Lindbergh, and like his father and grandfather before him, he would define himself by what he could do physically. Her son had his whole life in front of him, but she worried whether Erik would get the adventure he needed, the adventure that was part of the Lindbergh DNA.

7

A Career in Orbit

It was after midnight when Peter and his friend John Chirban settled into a booth at the Deli Haus in Kenmore Square, across the Charles River from MIT. John loved the chicken livers, matzo ball soup, and cheese blintzes, and Peter liked the huge pastrami sandwiches, burgers, and cheesecake. The two drank coffee, listened to music, and watched the colorful scene that flowed in and out. They marveled that there were still punk rockers in the late 1980s, though the discussion for the night—as it often was for the two men—was not music, fashion, or food. It was the connection between the truth of science and the truth of faith.

John, ten years older than Peter, was friend, mentor, and consigliere. He admired Peter's passion for space but didn't share it. A Harvard-trained psychologist and professor—and fellow Greek American with a second doctorate in theology—John was committed to understanding and advancing spirituality. He had spent nearly two decades interviewing the famous and influential behaviorist B. F. Skinner, discussing similar topics of free will versus determinism, and whether spirituality could be rational. While John argued that feelings could be independent of behavioral conditioning, Skinner denied the existence of the soul, saying it was "silly" and "prescientific."

Peter had been introduced to John through his parish priest. The space-loving college student was questioning the Greek Orthodox traditions that had suffused his youth, and though Peter still attended church when he was home with his family, he rarely went on his own. He told John that he felt "weird" about not being actively religious and still found the smell of incense comforting. In the same way medical school was more about duty than passion, church was tradition. It was what his family did on Sundays.

Science, though, was Peter's rock. What he questioned was what existed *beyond physics*.

"Do I believe there's something spiritually out there?" Peter asked, sipping black coffee. "Do I believe there's energy beyond the physical matter? Where does energy that drives life come from?" He looked at John and said, "Did life on Earth originate as a result of aliens?"

John smiled, knowing Peter was serious. Peter's view of the energy of life came closer to the Force in *Star Wars* than what he was taught in church. John found in Peter an innocent competing and excelling in the academic big leagues, a young man with a great family and a foundation of love. Peter was his favorite type of person: a truth seeker. Peter had the zeal of a missionary, only his gods were Goddard and von Braun and his bible was a cross between *Atlas Shrugged* and *The High Frontier*. John believed in God's ability to transform lives, something he articulated to Skinner, who had a deep disdain for religion because of his upbringing. Skinner's work as an adult was rooted in the idea that there is neither choice nor freedom, and that all behavior is shaped by environmental conditioning—negating any role of God. John told Peter that Skinner had made *science* his religion. John chuckled remembering the day he invited Skinner to go with him to listen to Mother Teresa of Calcutta, who was visiting Harvard. Skinner declined, saying he thought Mother Teresa was "very narcissistic." Where John and Skinner came together—and where Peter and John came together—was over central questions of truth: *What is the nature of existence? How should one live life, given that so much remains unknown and perhaps unknowable?* With Peter, John made the case—quoting Albert Einstein—that "science without religion is lame, religion without science is blind."

When Peter was in high school and still an altar boy, his priest asked him to give a Sunday sermon. Father Alex often invited parishioners to give prepared talks, and on this day, Peter stood in front of a congregation of about two hundred people. Peter talked for quite a while without mentioning the word "God." Instead, he spoke of "meta intelligence," a generative force beyond words. Peter said that this meta intelligence was "beyond our comprehension, beyond human terms." Understanding, he said, would come through the exploration of space. He also said, "We may not be the only beings in this universe. We may not be the most intelligent species." Father Alex listened with interest, believing Peter was using nontraditional language to describe his relationship with God. He likened Peter's use of "meta intelligence" to Aristotle's cosmological "first cause" principle, where God is the uncaused cause. Father Alex smiled to himself, even as he saw perplexed looks on the faces of parishioners.

Peter was always drawn to the deeper philosophical concerns presented by science. He had once written a paper on why it was okay to clone people. In medical school, he struggled to understand the animating material of life, and what happens when the vital force goes dark. Now, in the early spring of 1988, Peter was at a crossroads. He had received his master's degree in aeronautical engineering from MIT. Dr. Larry Young had signed his thesis. And NASA had committed funding for three years of continued research involving his Artificial Gravity Sleeper.

Sitting in the Deli Haus diner, Peter poked at his cheesecake. It was after two A.M., and he and John were still talking about faith and truth seeking. John pointed out that many great scientists and engineers, including Einstein and Nikola Tesla, recognized "different kinds of truth and shifting levels of consciousness." Famous founders of science held Christian beliefs, including Blaise Pascal (who was a Jansenist), Galileo, and Johannes Kepler. It was Einstein, John noted, who said, "Whether you can observe a thing or not depends on the theory which you use. It is the theory which decides what can be observed." And Tesla, John continued, wrote about the relationship between matter and energy, saying: "If you want to find the secrets of the universe, think in terms of energy, frequency and vibration. What one man calls God, another calls the laws of physics."

Peter told John that he had recently spent six months reading the Bible every day. "I tried to think of the meaning," Peter said. "I was studying it for myself versus reading the gospel chosen for me on Sundays in church." Had he been asked in high school whether he believed in God, he would have said yes. "Now? I'm agnostic. I'm searching." But he acknowledged, "When the shit hits the fan, like when I had stolen this fetal pig at Hamilton and I thought I was going to be expelled, I prayed."

As the two stood to leave, Peter tossed out his own quote by Konstantin Tsiolkovsky: "You know what I believe?" Peter asked. "Tsiolkovsky said, 'Earth is the cradle of humanity, but one cannot remain in the cradle forever.'" Peter was two months shy of his twenty-seventh birthday. He had one big hurdle to go before he could focus full time on moving humanity out of its cradle. He needed to finish up at Harvard Medical School, either by graduating or by gracefully bowing out. And he had a certain ambitious side project—one that was drawing attention from all over the globe.

At the world headquarters of the International Space University—a tiny second-floor office above a bagel shop on Beacon Street—Peter and Todd Hawley marveled at the postmarks on the applications: the

Soviet Union, China, Japan, Kenya, Switzerland, Germany, France, Poland, India, Saudi Arabia. In all, there were more than 350 applicants from thirty-seven countries vying for one of 100 spots at ISU's inaugural summer session, just months away.

Hardly a day went by without Peter and Todd scrutinizing letters and résumés for clues to which students had the best chance of helping them get to space. They looked for leadership and engineering skills, creativity, and hints of a willingness to set aside political and nationalistic differences. Peter and Todd lined up big-name sponsors and visiting professors, and worked on endorsements, seeking Senators Edward Kennedy and John Glenn, and Elizabeth Dole, former U.S. secretary of transportation, among others.

Peter and Todd drew support from astronauts and scientists, including Byron Lichtenberg, Rusty Schweickart, rocket pioneer Hermann Oberth, and *The High Frontier* author himself, Gerry O'Neill. The inaugural summer session would be held at MIT, with eight fields of study: space architecture, business and management, space engineering, space life sciences, space sciences, policy and law, resources and manufacturing, and satellite applications. The first lecture would be on June 27 on space resources and manufacturing by Charlie Walker of McDonnell Douglas Astronautics Co. The final discussion would be on August 16 by Roger Bonnet, head of the European Space Agency. In all, there were speaking commitments from more than one hundred visiting lecturers. Peter and Todd winnowed the hundreds of applicants to 104 and worked on getting visas for all of the international students. More than $1 million in funding came in from a mix of foundations, individuals, companies, and governmental agencies. Tuition was set at $10,000.

Peter and Todd started taking flying lessons together in Piper Cherokees. They came up with pages and pages of sketches and ideas for an array of inventions. When they went skiing together, they concluded that the resort's setup was inefficient: too much time wasted riding the ski lift, not enough time spent skiing. They sketched a new lift and slope design, with a 180-degree slanted turnaround tunneled into the mountain, leading to an elevator to pop them back to the top.

Like Peter, Todd kept a daily journal. But instead of writing the month, day, and year above entries, he wrote the number of days he had lived—by now more than nine thousand. Peter soon adopted Todd's "lifetimer clock," and the two envisioned a clock they could sell that would "keep track of how many days/hours/minutes you've been alive!" They signed and dated their clock concept, with Peter on his 9,791st day and Todd on his 9,828th day. Another idea came one day while the two were having lunch at the Trident Café on Boston's Newbury Street and a waitress stumbled and dropped a tray of food. Peter proposed a food tray with an internal spinning magnetic disk. Peter jotted ideas: "This device is a gyroscope stabilizing serving tray. What is unique about this device is that it does not carry along with it a motor or battery. When the tray is placed on the counter to load the food, the magnetic disk is spun up to speed." The two devised their own written language called ALFON, derived from the UNIFON alphabet, which matched the most important sounds in the human language with a symbol.

But despite their shared creativity and passion for space, there were days when the two squabbled and disagreed, and Bob Richards had to step in to play peacemaker. Todd was the unrelenting idealist; Peter was the driven pragmatist. There were days when Peter found Todd inexplicably quiet. But there was no time to dwell on personality differences; the two had a university to launch.

A few months later, on the night of Friday, June 24, 1988, 104 graduate students from twenty-one countries gathered for the opening ceremony of the International Space University at MIT. Students from the Soviet Union stood next to American delegates, and students from the People's Republic of China shared dorms with attendees from Saudi Arabia.

After a weekend of socializing and settling into dorms, classes and lectures began early Monday morning. Students formed teams to work on the summer's chosen design project, which was to come up with blueprints for a working lunar base. They would tackle issues including transportation, surface operations, human behavior, environmental impact, and how to control machinery in real time on the Moon. Peter, Bob, and Todd bounced between classes and lectures and checked in on the lunar base teams. They hosted parties, directed faculty, and handled housing mix-ups. For the first week, there were constant questions and little sleep. Their crazy dream had come true. Peter also met people he was sure would become friends for life, including two men: Harry Kloor and Ray Cronise. Harry had earned simultaneous bachelor's degrees in physics and chemistry and intended to do the same with simultaneous PhDs in physics and chemistry. Ray was a scientist at Marshall Space Flight Center and had done countless hours aboard NASA's zero-gravity plane. Almost as impressive to Peter were the attractive women whom both Kloor and Cronise seemed to surround themselves with. The inaugural ISU summer session came to a close eight weeks later. A graduation ceremony for the class of 1988 was held on Saturday, August 20, from ten A.M. to noon at Kresge Auditorium, and featured opening remarks by David C. Webb, the eloquent trustee and early backer of both SEDS and ISU, who said, "This is about much more than a learning experience. It is a creative experience: You are creating your future. In your future lives the future of the human species."

Chancellor Arthur C. Clarke was beamed in via a double satellite bounce—courtesy of Intelsat—from his home in Sri Lanka. Uncle Arthur congratulated the students and gave credit to what he called "the gang of three: Peterbobtodd."

In his sonorous voice, Clarke said he hoped the students would continue to collaborate and create "lasting ties to foster international goodwill." He urged the graduates to go forward to "shape the destiny of many space efforts and lay the groundwork for new evolutionary and perhaps revolutionary fields." It was education and the world's first universities, Clarke noted, that moved humanity out of the Dark Ages and into the Renaissance. Clarke closed with a Chinese proverb: "To plan for a year, plant a seed; to plan for a decade, plant a tree; to plan for a century, educate the people." Peter listened and smiled. It was Uncle Arthur who had planted the seed for ISU. From their first meeting in Vienna, Clarke had talked about how political and cultural differences could be erased when there was a shared dream of space.

Finally, Peter, exhilarated and exhausted, took the stage. He said that the past eight weeks had been the most "intense and challenging" period of his life.

"It has, without a doubt, also been the most rewarding and exciting time," Peter said. "The experience has changed me in fundamental ways. Never again will I look at the many nations represented here as 'foreign.' From now on, these nations are the homes of my friends." He went on, "You attended over 240 hours of classroom lectures, typically equivalent to more than a full semester at MIT. In addition, you put about 280 hours each into the lunar design project. And you spent at least 300 hours in cross-cultural training, better known as partying."

At this, the graduates cheered.

Back at Harvard, Peter juggled the continuing demands of ISU—the next summer session was in the works, to be held in Strasbourg, France—with his return to medical school. He completed his neurology rotation, which had him on call at Mass General every third night, and required him to work seven A.M. to seven P.M. daily. He was miserable, writing in a notebook, "I have before me, in living color, all the reasons I'm not going into clinical medicine—an environment with continuous demands, little room for creativity, little monetary return, low success rate. In any event, it is not the fulltime career for me."

In late September of 1988, he finished up his OB/GYN rotation. He had participated in the delivery of twenty-two baby boys and fourteen baby girls. November and December would see him in surgery rotations, requiring him to set his alarm for 4:30 A.M. daily. He was assured by his sister, Marcelle—who had gotten married earlier in the year and was a second-year surgery resident—that he would get used to the hours. Peter didn't believe that he would get used to 4:30 A.M., and he was incredulous that his little sister was now married. He barely had time to go on a date. Marcelle often gave him a hard time about Harvard, saying he knew so little and was getting away with so much. Peter, for his part, was impressed by the academic rigor of Albany Medical School. He would say, "Thank God I'm at Harvard. It's hard to get into and harder to fail out of."

The difference was, his sister wasn't running a university on the side. And Peter piled on yet another passion-driven challenge: he was cofounding his first for-profit company, called International Microspace. The idea was to do what only the government or huge corporations had done before: launch satellites into orbit. A NASA engineer named Bob Noteboom, who had a unique ultra-low-cost multistage launch vehicle, was hired as the chief engineer, and funding came from a space lover and communications entrepreneur named Walt Anderson, who had risen from college dropout to multimillionaire, and sometimes preferred to pay people in gold bullion over cash. Anderson was a dreamer like Todd Hawley, an Ayn Rand fan, an extreme libertarian, and an early financial backer of ISU. The new company was based in Houston, home of the Johnson Space Center, and testing of their rocket engine was already under way. With each passing day, Peter's to-do list was getting longer and longer.

One of Peter's functions at ISU was chairing the curriculum committee, which met twice per year to update what was taught each summer. Peter reserved a room in the posh Harvard Club in Boston's Back

Bay neighborhood to host the upcoming curriculum meeting in preparation for the 1989 Strasbourg summer session. All of the department chairs that he'd hired were confirmed, including Harvard Medical School professor Susanne Churchill, director of ISU's life sciences department; Giovanni Fazio, head of the Harvard-Smithsonian Center for Astrophysics; and Dr. Larry Young from MIT. Peter was free of medical school commitments on that particular Saturday and looked forward to the planning meetings, which often turned into semifriendly battles over which departments could secure the most lecture slots.

Then, just days before the long-planned session, Peter got unwelcome news. He was told he had psychiatry rotations on the day of the curriculum meeting. The medical school controlled his schedule, and rotations were mandatory. He was required to go from hospital to hospital, visiting psychiatric wards and meeting with patients and doctors.

He could not miss the rotations, and he could not cancel the ISU curriculum planning meeting. He considered his dilemma. He sat down and drew up a map, charting distances from the Harvard Club on Commonwealth Avenue to the area hospitals. It was only 1.7 miles from the club to Mass General, but would take about twenty minutes to walk, seventeen minutes on public transit, and under ten minutes by taxi. He did calculations for time and distance between the hospitals on his rotation and from the hospitals back to the club.

On the morning of the meeting, Peter stuffed his doctor's coat, scrubs, medical notes, and gear into his backpack. Arriving at the Harvard Club, he stashed his backpack in the bathroom. He checked in with the professors and outlined the day's agenda. As soon as the planning was up and running, with discussions about how to organize lectures, Peter slipped out of the room. He walked as fast as he could without drawing attention to himself, went into the bathroom, took off his suit, stashed it in a corner, got into his white coat, and ran out of the building. He hailed a cab and headed to the first psych ward of the day. As soon as he made sure that his resident or staff physician noticed him, he skulked out to hit the next hospital. When he couldn't find a taxi, he took off on foot, often arriving at his next destination damp with sweat. This pattern continued until late morning, when he needed to be back at the Harvard Club. After lunch, the clothing change and taxi hailing began again, making Peter feel like Clark Kent changing into Superman, only this Superman was lacking a much-needed ability to fly.

Inside the Harvard Club, Professor Susanne Churchill was unfazed by the frenetic blur of Peter Diamandis. She knew him well enough to know he rarely sat still. She wouldn't forget the night Peter phoned her at home to tell her his idea for an international graduate-level space university. She knew him as an MD student at Harvard and saw he had a burning interest in space. The two had talked about her work developing an instrumented primate model to simulate the effects of microgravity on the cardiovascular system. That night on the phone, Susanne, standing in her kitchen, listened as Peter talked about his and his colleagues'—all in their twenties—vision for ISU. When Peter asked whether she would be the founding chair of the life sciences department, her reaction was, "Are you out of your mind?" By the end of the conversation, though, she said yes. She had found the first ISU summer session to be the most mind-expanding time of her life, watching as a gaggle of students from the Soviet Union attended with KGB minders in tow. The Berlin Wall had not yet come down, and America's communication with the other space-faring giant was closed off. She was awed, too, when Peter and crew managed to get the first Soviet physician cosmonaut, Oleg Atkov, to attend ISU, also with his KGB minder. At the end of the day—after the rush between space and medicine—after the hospital visits were done and the curriculum meeting was over, Peter sat alone in the Harvard Club. He was exhausted, and uncertain that he'd actually pulled this off. He was aware during his rotations at the psych wards that a few of the doctors were looking at him as if *he* were the one needing treatment.

A few weeks later, a letter from the dean's office showed up in Peter's student mailbox. Peter tore into it, his heart racing. He flashed to a recent surgery rotation where he had darted out of the room to take a call on his brick-sized Motorola cell phone. Bob Noteboom, his chief engineer at International Microspace, had been calling from Houston to brief him on their latest rocket-motor tests. Dan Tosteson, dean of the Harvard Medical School, wanted Peter to stop by his office. Peter held the dean in high regard. He always invited first-year students to his home for a welcoming barbecue, and was a skilled scientist in his own right. But Peter knew this gentle summons could not be good.

When the day came, Peter sat across a desk from Dean Tosteson. Pleasantries were exchanged, before the dean got right to it: "Listen, your interns and residents are telling me you're not paying attention the way you should be. I'm told you're showing up late or not showing up at all, and that you're on your phone all the time."

Peter was sure the color had drained from his face.

The dean told Peter he was concerned and wanted to know what was going on in his life. Were there financial problems? A family crisis? Was he in some sort of trouble?

Peter was suddenly back at Hamilton, feeling like his life was over because of a borrowed fetal pig. He had put in more than three out of four years of medical school. How could he *not finish* when the end was in sight? He would never be able to face his parents.

Peter exhaled. He told the dean everything: "I'm running an international space university," he began. "I've got this rocket company in Houston, where I'm CEO. I've taken in millions of dollars of investor money and

I have to make it work. And there's no way I can't finish medical school. It would devastate my family."

The dean shook his head, offered a bemused half smile, and said, "Only at the HST program." He looked at Peter for confirmation: "You're running a university, a rocket company, *and* trying to finish medical school?"

Peter nodded.

"What do you want to do?" the dean asked.

"I'm interested in medicine, but I'm really interested in putting humans in space," Peter said. "I thought getting my medical degree would make my parents happy, and it might be my avenue to space."

The dean asked, "Do you want to practice?"

"No," Peter answered honestly and sheepishly. "But I do want to finish the degree."

The dean paused. "Okay, I'll make you a deal," Tosteson said. "If you complete your rotations and can pass part two of the medical boards, I'll let you graduate."

As the two stood to leave, Tosteson looked Peter in the eyes and added, "Peter, you have to promise me one thing."

Peter nodded. "Anything."

"You have to promise me you will never practice medicine."

Peter had dedicated nine years to his college and graduate studies. The finish line was in sight, but elusive. He wrote in his journal, "Even as I spend my days at the hospital learning the healing arts, my mind and heart remain in the heavens. I know I will not practice medicine. It seems almost comical after I've come so far, but I truly hope that I have the strength of will to finish this degree." He spent three weeks studying around the clock for the board exam, resenting the time it took away from working on ISU and International Microspace.

In the weeks following the exam, Peter would run to his brass-plated mailbox to see whether the results were in. He was unable to sleep, worrying he hadn't passed. And he had other concerns: Todd Hawley had been acting erratically, often not showing up at the ISU office, other times wandering in late. Peter believed that ISU wouldn't work without Todd. But he couldn't figure out what was wrong, and Todd wasn't talking.

Finally, Peter opened the mailbox. There it was. The letter from the National Board of Medical Examiners, dated September 1989. He carefully opened it. His eyes went to the pass-fail box on the right side of the results page. There was just one word: Pass.

He stared at the piece of paper for a long time, his eyes welling with tears. He would graduate from Harvard Medical School. He was done. The passing score on the exam was 290, and he had gotten 360. His highest scores were in psychiatry—and OB/GYN. He smiled and thought, *That's for you, Dad.*

Peter was awarded his medical degree on November 21, 1989. The expression on his face that day as the dean handed him his diploma on stage was, *Really? Are you sure?*

After all of the celebrations were over, Peter—now well past his ten thousandth day on Earth—climbed into bed. He wrote in his journal: "Today is the beginning of the rest of my life. I am free, finally free, of my intellectual and emotional obligation to medicine and now my spirit and my mind focus with laser-like intensity upon the frontier of space."

8

Struggles in the Real World

On a beautiful fall morning in 1991, Peter went to Hopewell, New Jersey, to build robots with his pal and business partner Gregg Maryniak, a director at International Microspace, on the faculty of ISU, and Peter's adopted "big brother." Peter had met Gregg almost a decade earlier, when Gregg was invited by Todd Hawley to give a talk at the 1982 SEDS meeting at George Washington University. In more recent times, Peter and Gregg got together to design and make things for the fun of it. The creative sessions in Gregg's basement workshop helped get their minds off the troubles of their satellite launch company, International Microspace.

But on this day—Saturday, October 19—the skies were unusually clear, the weather unseasonably warm, and the rich and fiery fall palette too irresistible for Peter and Gregg to remain indoors. So they headed to the Princeton Airport, about fifteen minutes away, and rented an old Cessna 172, registration number N65827. Peter was copilot and sat in the right seat in the single-engine, four-seat, high-wing plane. It was a pilot's dream day. "CAVU all the way," Gregg said, using the aviation term for "Ceiling and Visibility Unlimited."

Gregg got the Cessna to an altitude of about 2,500 feet, flying northeast from Princeton to the mouth of Raritan Bay, and then east with Staten Island to the north and the Jersey Shore to their right. Gregg turned left to fly over the Verrazano-Narrows Bridge and continued up the Hudson River corridor. Small towns tucked in by blankets of foliage in hues of orange, red, burgundy, and green were behind them. The gray and shimmering Manhattan skyline, a mix of beauty and swagger, with buildings jammed shoulder to shoulder, was ahead. The green-cloaked lady, the Statue of Liberty, was to their left, Manhattan to their right. They flew lower than the top of the Twin Towers of the World Trade Center.

This was the most fun Peter had experienced in months, given his worries about International Microspace, which was $300,000 in debt and had no immediate source of cash. Some employees were "asked to seek alternative employment," and the board raised the possibility of filing for bankruptcy. Just when Peter was sure that his satellite company was at its bleakest hour, he would land an investor willing to put up $50,000 or $100,000, just enough money to keep the firm afloat. But how long could he keep this up?

No longer burdened by medical school, Peter had expected International Microspace to take off quickly, in ways similar to SEDS and ISU. In his mind, the goals of the satellite company were straightforward: reduce the cost of getting something to space, offer an alternative to the government monopoly, and use the low-Earth-orbit launchers as a stepping-stone to reaching the stars. But nothing in this enterprise had been simple; in fact, most everything had been tedious—and contentious. For starters, fundraising for the satellite company was harder than the launch of ISU and SEDS combined. And in stark contrast to the impassioned, idealistic space discussions that he had savored at MIT, Peter now found himself immersed in talk of legalities, contracts, licensing partners, strategic partners, vendors, customers, finance, government regulations, and valuations.

Peter tried to keep his spirits up, likening himself to Batman and Gregg to Robin, saying, "Batman is in an inescapable trap set by his archenemy.... Will Batman and Robin ever get out of this sticky mess? Or will it be curtains?" He told Gregg that he was confident financing was just around the corner. "I am on the line," Peter said. "This *has* to succeed."

Flying around La Guardia airspace in the Cessna, Peter and Gregg decided to take a short detour before heading back down the Hudson corridor. They wanted to fly over the Diamandises' home in Great Neck. Peter was nostalgic as he pointed out the house, with its sprawling lawn, long driveway, and nearby tennis courts. The house looked somehow different, the way an elementary school does when you go back inside as an adult.

Peter enjoyed being caught up in youthful memories, but it was tough to escape the realities of his current situation. He was no longer in the protective environment of academia, a cocoon in which he had stayed until he was almost thirty. Though he remained publicly optimistic about International Microspace, he asked himself, "What is it that continually drives me to set goals beyond my reach? Do I have what it takes to pull this off?" The tug of parental expectations remained, but it was different now. He was in the unusual position of having graduated from Harvard and then decided not to pursue what surely would have been a lucrative career in medicine. Until he succeeded in space, Harvard and what-could-have-been would hang over him. He could hear his mother's words: *Medicine is secure, space experimental.* After nearly two hours of flying, Peter and Gregg returned to the Princeton airport. Peter, animated by the flight, told Gregg that he would get his pilot's license at last. Gregg had gotten his pilot's license at seventeen—he was now thirty-seven—and had a dozen friends who'd become pilots after flying with him. Gregg wanted to do whatever he could to get his space-loving friend aloft.

Gregg had been running the Chicago Society for Space Settlement when Todd Hawley called in 1982 to introduce him to a new student-run space group, SEDS, started at MIT. Gregg had just walked into his house in Oak Park, Illinois, and set his briefcase down when he found himself on the phone with an impassioned stranger. Todd was so fascinating, his enthusiasm so palpable, that the two ended up talking for two hours. At the time, Gregg was working as a trial lawyer, though his passion had always been for space. In Todd, Gregg met a kindred spirit—a fellow space geek who would inspire him to give up pencil pushing and focus on something he really loved.

Gregg agreed to become a senior adviser to SEDS and quickly bonded with Peter, Todd, and Bob. Three years later, in 1985, Gregg, who had been teaching orbital mechanics and dabbling in space science in his free time, officially made the leap from law to space. He went to work with scientist and author Gerry O'Neill, running the Space Studies Institute at Princeton. Gregg's wife, Maureen, joked that O'Neill had promised: "You could earn as much as *some* poets!" Gregg, involved in the founding of ISU, was now a director with Peter, Todd, and Walt Anderson in International Microspace.

Gregg's thinking around space was influenced by three books: *The Limits to Growth, The Population Bomb,* and O'Neill's *The High Frontier.* The first two books made dire predictions about unsustainable growth and an exploding population chasing after finite resources, resulting in mass starvation and societal upheaval. O'Neill agreed in *The High Frontier* that Earth's resources were limited and that a population explosion was inevitable. But his book provided an exit strategy and a path to sustainability. Well known as a physicist before he became an author, O'Neill detailed in his 1977 book the coming age of space settlements and the limitless and valuable energy and materials of space. O'Neill wrote, "The concept of the humanization of space can stand on its own merits, survive detailed numerical checks, and survive logical debate. To support it requires no act of faith, only the willingness to study unfamiliar ideas with an open mind." The book was for many the first to articulate a path for a private-sector push into space, without the government's help. For Gregg, *The High Frontier* was an awakening. It was an antidote to doomsday scenarios—positive, pragmatic, and egalitarian. Working as a trial lawyer had been about who got what piece of a finite pie. As Gregg saw it, O'Neill was trying to expand thinking and expand resources. He wanted to make more pie.

Gregg believed that humanity was capable of addressing its challenges. The nonprofit Space Studies Institute at Princeton was established to realize the vision and concepts of *The High Frontier.* The institute held space manufacturing conferences and inspired students to build equipment for space, including mass drivers to catapult materials off the Moon for colonization of space. The prototype mass drivers, with electromagnetic drive coils and a payload container, were made at Princeton and MIT.*

Now, in late 1991, with a recession and high oil prices in the wake of Iraq's invasion of Kuwait the summer before, Gregg and Peter talked almost daily about the financial challenges of International Microspace, which had been relocated from Houston to Washington, D.C. Houston was where NASA's human launch systems were based; the Washington area was where satellite companies were finding financing.

Gregg knew better than anyone how Peter was struggling with International Microspace. Gregg felt that his role in the business was to be the steady force, the constant source of calm. If International Microspace was the starship *Enterprise,* Peter was the Captain Kirk, and Gregg was his Mr. Spock. More and more, Gregg was becoming concerned about his

*O'Neill invented the storage ring technique for particle colliders, which led to the building of the Stanford Linear Accelerator Center. He also invented a mass driver to move materials mined on the Moon into Earth orbit.

younger friend, who put a lot of pressure on himself—and who was about to face some very tough decisions.

Peter prepared for a private social meeting at his home in Rockville, Maryland—but in the back of his mind, he knew it was part of a mission to rescue International Microspace. He also realized that this strategy could come back to haunt him on a personal level, but there was no choice: the company's very survival was in question.

Already, Peter had met with potential investors across the globe. He had even approached a group he referred to lovingly as the "Greek Mafia," his parents and twelve of their friends. At this point, his satellite launch company had several failed rocket firing tests and was now building a small launcher of a new design to take one-hundred-pound payloads—experiments, signaling, imaging, whatever the customer wanted—to low-Earth orbit.

Peter likened the company's trajectory to a staging rocket: it had let go of its original unneeded weight and ignited new engines. The company had new designs, new management, and a new board of directors that included the last man to set foot on the Moon—Gene Cernan—and Andy Stofan, former director of launch vehicles at NASA and former associate administrator for the space station. Peter had pulled in $10 million in financing. Even the name of the rocket had changed from the Galt vehicle—after John Galt in *Atlas Shrugged*—to Orbital Express.

But International Microspace couldn't add to its momentum and was struggling to meet payroll. One major reason for this standstill was that investor Walt Anderson was vehemently libertarian and antigovernment, describing himself as a "pretty rampant pacifist." Walt had contributed about $80,000 to finance ISU and put $100,000 into founding International Microspace. As a condition of Walt's early backing, Peter was not to pitch anyone with government ties. This had turned out to be a major impediment: the few private companies that had done small launchers were now a part of governmental agencies or relied on government contracts.

The most successful of the companies, the darling of the private launch business, was Orbital Sciences, started by MIT and Harvard business school graduates. Their rocket, the Pegasus, was designed by former MIT professor Antonio Elias, who as a boy growing up in Spain loved nothing more than to search the sky for planes. The Pegasus, built to carry payloads of up to one thousand pounds to low-Earth orbit, was modeled after some of the X-series planes and the McDonnell Douglas F-15 Eagle fighter jet. It was air-launched horizontally from a NASA B-52 aircraft at 40,000 feet before engines ignited and shot it toward space. On April 5, 1990, Pegasus made history as the first privately developed, all-new space launch vehicle. Its first customer was DARPA, the Defense Advanced Research Projects Agency of the U.S. military. The rocket's delta wings, fins, and wing-body fairing were designed by Burt Rutan of *Voyager* fame, in Mojave. Rutan, who decades earlier had spent several months working on the F-15 Eagle, was impressed with the air-launched system. He saw how relatively inexpensive and flexible it could be in delivering payloads to the start of space.

Now, on this day in 1991, Peter was hoping that International Microspace could tap into the government's considerable payload of money. He had scheduled a social meeting with a man he considered a friend, yet someone whom Walt Anderson would consider the devil himself. The friend was Pete Worden, the newly appointed head of technology for SDI, the Strategic Defense Initiative, otherwise known as "Star Wars," the program begun in 1983 under President Reagan to build a new missile defense system.

Worden was an astrophysicist and a straight-talking Air Force colonel who had enough clout to be hired by NASA even after criticizing the agency as a "self-licking ice cream cone" and saying the name meant "Never A Straight Answer." Worden and Peter had met in the late 1980s, when Peter was running ISU and Worden was working as director of new initiatives in the White House's National Space Council. At the time, Worden had no money to offer—only contacts. These days, though, Worden had a multibillion-dollar budget at his discretion. And he was a believer in the need for small, low-cost satellites.

After Worden arrived at Peter's home, they had a long and winding discussion about humanity's expansion into space. Worden was no longer a fan of the space shuttle, saying it had been "an interesting experiment that didn't work." It didn't allow man to go back to the Moon or on to Mars, and it failed in its primary mission of providing routine and affordable access to space.

Soon, Worden and Peter agreed to set up a follow-up meeting about International Microspace. That face-to-face occurred in Worden's office in Washington. The discussion ranged from the space shuttle to Brilliant Pebbles, a program under design that would send a swarm of small and smart satellites into orbit to be used as a missile defense. Peter and Worden both concluded that there needed to be a vibrant private sector, but one that involved the government. "You build the equipment and work from the outside, and I'll work from the inside," Worden told Peter. By the time the meeting ended, Worden made it clear that if International Microspace had a viable vehicle that ideally was "somewhat cheaper than Pegasus," then the "government has a good reason to support a second supplier." Peter walked away confident that he had a deal. It was great news—salvation for the company—but was it selling out?

Predictably, when Walt, then chairman, learned of Peter's likely deal with Worden, he blew up and announced he wanted out. He told Peter and members of the board that he would sell back his shares at fifty cents to the dollar—getting $50,000 on his $100,000 investment—in order not to be involved. He said he never wanted his name associated with the company again, and never wanted anything to do with Peter Diamandis. This was not the starry-eyed Peter he had met and admired. Now Walt labeled him a "liar" and worse. Todd Hawley, another idealist, had an equally strong but more personal reaction.

Todd sent Peter a six-page handwritten letter that began "I am as DISTRESSED now with my own role in [International Microspace] as I am with your drive to succeed at all costs. I fear that you're now set to create another incremental institution that won't ultimately change much. The present company vision and approach is so much like the old guard that it essentially is all but another, small cousin of existing giants. Nowhere is a radical new approach in sight." While Todd acknowledged that Peter was leading the company to a "place of possible profit," the final product would be diluted at best. "The countless hours of unusually driven, visionary and idealistic people will never show up on the bottom line. I believe you've abandoned any and all of our non-economic objectives." The letter was signed, "Sincerely, Todd."

The words—and final signoff—stung. Peter's friends were saying that he had sold his soul. It wasn't lost on him that the founders' original vision, hatched over coffee-and-excitement-fueled all-nighters, had been severely diluted. Their dream was to unleash rapid experimentation in space, sending out hundreds of private Sputniks. But Peter was responsible to the shareholders. Without this deal with Worden, which was moving forward fast, they had nothing. Peter knew this decision was pragmatism over passion. It was Harvard Medical School over MIT. It was the real world over academia.

Peter sat down several times to write Todd a letter, taking more than a month to get his thoughts straight. He finally wrote: "Since January 6th of 1982, some 9 years, 2 months (3,344 days), since we first met, we have shared emotions, triumphs, experiences, adventures and challenges like no one else I know. I feel safe in saying that we have spent the highlight of our lifetimes together, side by side. I love you as a brother and respect you as a colleague—together we have accomplished many great things.

From the moment we met, I knew that our energy and vision would enable the development of space. I have been greatly saddened that our friendship has drifted apart. I speak here of friendship—not business or work relationships. Todd, it *is* okay to disagree, to hold differences in opinion, to learn from each other. I want to put the time and energy into something very valuable to me—our friendship. With love, your brother, Peter."

Gregg saw the situation from both sides, but in the end stood by Peter. It was Peter who had thrown everything he had into trying to make the company successful, asking family, friends, former professors, colleagues, fraternity brothers, and astronauts to invest. He traveled the world pitching any investors who would listen. He had nearly moved the company to Alaska after being wooed by the governor to settle the company in Poker Flat, north of Fairbanks. Gregg saw Peter and Todd as the "Damon and Pythias of space," referring to the ancient Greek story of friendship, where two men are each willing to risk their lives to save the other. On other days, when Peter and Todd were feuding or refusing to speak, Gregg would tell them: "Quit this bullshit. You have one of the most amazing friendships I've ever seen."

But for months, even after Peter's letter, Todd was silent.

Finally, happily for Peter, the silence ended. Todd wanted to meet Peter at the Deli Haus near MIT. On the appointed afternoon, the two found a booth and settled in. Peter was relieved to see his friend, but Todd was somber and said he had some news to share. Peter assumed it was about International Microspace. He had thought about what he wanted to say. But as he sat sipping his coffee, Peter could feel something else was going on. Todd looked away, avoiding eye contact. He moved his plate, picked up his silverware and put it back down again. He looked at Peter, frowned slightly, and said it: "I've been diagnosed with AIDS."

Peter was physically jolted back in his seat. He looked at Todd and for a moment said nothing. He closed his eyes. Peter had known that Todd was gay for some time now, though for years Todd was closeted, bringing his "girlfriend" MaryAnn to events and on trips. When Todd first told him he was gay, Peter had reacted negatively. He didn't know how to handle the news, was homophobic—he and Todd were always sleeping at each other's homes—and he shut Todd out. Peter realized months later that he was being an idiot, called Todd, and told him he loved him unconditionally. Now, in the Deli Haus, with loud music playing and dishes and silverware rattling, Peter fought back tears. He had treated AIDS patients at Mass General. AIDS was a death sentence. Since the early 1980s, about a hundred thousand people—mostly young men ages twenty-five to forty-five—had succumbed to AIDS. Magic Johnson had just told the world that he was HIV positive. There were one million confirmed cases of HIV infection in the United States. These men were emaciated, pockmarked, and unfairly ostracized. Todd was brilliant, handsome, and full of life.

Peter collected himself and vowed to find Todd the best treatment possible. His childhood friend and rocket-making buddy Billy Greenberg, now a doctor, was involved in an experimental AIDS treatment. Peter would contact him. The anti-HIV drug AZT was on the market. Todd listened to Peter and smiled slightly. As the moments passed, the two found a way back to laughter, recalling the meetings where they were in way over their heads but acted so cool and confident. They marveled at how they had talked brilliant professors into working for them, and how they had brought together like-minded people from across the globe at a time when personal computers didn't exist and electronic mail was not yet conceived. Through Todd's insistence, they had gotten America's putative enemy—the Soviets—to attend their university before the Cold War's end. It was Todd who had said, "We're not excluding fifty percent of the space-faring part of Earth because Americans have a phobia about Russians." Todd asked Peter whether he remembered the day they got the call from their bank, and were sure they'd be told they were overdrawn. Todd reluctantly took the call. The banker said, "Mr. Hawley, I want to let you know the international wire transfer you expected has come in." Todd perked up and asked whether it was from Spain or Sweden. The man responded, "It's from the USSR Ministry of Education—for $120,000." Todd nearly fell out of his chair. The Soviets were sending twelve students to the first ISU summer session. Cash from the Cold War villains saved ISU!

After a few hours at the Deli Haus, Peter and Todd hugged good-bye. Peter promised to support him in whatever ways he could. He told Todd that he loved him: "You are strong. You will persevere."

That night, Peter wrote in his journal, "Like a superconducting magnet, this news has brought us back together."

It was a beautiful fall day in 1993 when Gregg was in New Hope, Pennsylvania, on the Delaware River just west of Princeton. He and his family, spending the day there relaxing, wandered into a bookstore. Gregg was soothed by the filtered light and sense of quiet, like sitting in a church pew on a weekday afternoon. He meandered through the shop, and stopped when he accidentally kicked a book left on the floor. He picked it up and dusted it off. It was like an old friend to Gregg: *The Spirit of St. Louis* by Charles Lindbergh. Gregg was fourteen when he read Lindbergh's story of his dangerous, history-changing flight from New York to Paris.

Gregg opened the book to an earmarked page. He smiled as he read the words, "Suppose I really could stay up here flying; suppose gasoline didn't weigh so much and I could put enough in the tanks to last for days. Suppose, like the man on the magic carpet, I could fly anywhere I wanted to—anywhere in the world." A few pages later, Lindbergh wrote of how he considered making the transatlantic feat, something no one had ever done. "Why shouldn't I fly from New York to Paris? I'm almost twenty-five. I have more than four years of aviation behind me, and close to two thousand hours in the air." A few paragraphs down, the aviator wrote, "The important thing is to start; to lay a plan, and then to follow it step by step no matter how small or large each one by itself may seem." Lindbergh knew he didn't have enough money to buy the right plane for the trip. He considered raising money. He wrote of a tantalizing prize that had been offered to the first person who could make the treacherous journey: "Then there's the Orteig prize of $25,000 for the first man to fly from New York to Paris nonstop—that's more than enough to pay for a plane and all the expenses of the flight. New York to Paris nonstop! If airplanes can do that, there's no limit to aviation's future." Lindbergh was right, Gregg thought. He changed the world's mind about air travel, seeing a future that others didn't see. Before Lindbergh flew, Americans were afraid to fly. After he landed in Paris, the world wanted to fly. In 1929, nearly 170,000 paying passengers boarded U.S. airliners, nearly three times the 60,000 that had flown the previous year.

At the bookstore cash register, Gregg's daughter asked why he was buying a book he already had. "It's for your uncle Peter," he said. Gregg hoped the book would inspire Peter to finally get his pilot's license. More than that, he hoped it would remind Peter of the importance of impossible dreams.

9

Meeting the Magician

Looking out the window of the plane, Peter saw carless roads and grids of streets for development that had never been developed, all on a flat stretched canvas of beige and sand. His eyes followed an empty railroad line running through the desert, with a sprinkling of arthritic-looking trees on either side. A gaping crater turned out to be the Mountain Pass rare-earth mine. As Peter's plane descended toward runway 30 of the Mojave Airport, he took in the bleached hangars to the left and what looked like a plane boneyard to the right. Mojave was a place where planes were born; Peter hadn't known it was also a place where some old birds came to die.

After the plane taxied to a smooth stop, Peter climbed out to get the lay of the land. The peaks and slopes of the mountains to the west and south gave the lapis-blue sky a jagged border. Heat radiated up from the runway. It was not exactly John F. Kennedy International. The only signs of life were a handful of crows stationed above the Voyager restaurant, and a tortoise that tucked its head into its shell, camouflaged by rocks and burrow-weed. Peter half expected Gary Cooper to appear as a sheriff facing off against four killers in the Mojave version of *High Noon. We are in the middle of nowhere,* Peter thought.

Desperate to escape the corporate drudgery of International Microspace, Peter had cofounded another new venture in 1993 called Angel Technologies, which sought to jump on the nascent and restless commercialized Internet, opened only a few years earlier to the public domain from the academic and scientific community's ARPANET.* Peter and his Angel business partner, Marc Arnold, wanted to provide low-cost, high-speed Internet access to the developed and developing world. Instead of laying cables underwater, under roads, or on telephone poles, their plan called for a faster, cheaper alternative. They wanted to deliver broadband from above the stratosphere.

Peter had been introduced to Marc through their mutual friend David Wine, the ISU supporter who had invested in International Microspace. Marc had made his money by selling a hospital services company to Smith Barney in 1991. After the sale, he pursued soaring (also known as gliding) as a sport, buying a Stemme sailplane and becoming the North American dealer for the plane. His love of soaring led to his interest in high-altitude projects.

Peter and Marc's Angel Technologies plan had competition. Cable operators, software companies, and start-ups were looking at a range of broadband delivery methods, from launching hundreds of satellites to using low-voltage electricity grids. A company called Sky Station International, a project of former secretary of state Alexander Haig, envisioned

*The ARPANET was the first packet-switched network. Packet-switched networks were the work of many hands: Leonard Kleinrock (UCLA) and Paul Baran (RAND), as well as Bob Kahn (DARPA), who is related to futurist and nuclear strategist Herman Kahn, and Vint Cerf, who connected with Kleinrock at UCLA, worked with Kahn at DARPA, and works at Google. ARPANET was all about breaking down messages into little self-contained packets like postcards that have a "from" and "to" address and can shuttle through a heterogeneous network of cooperating computers. As long as all the computers share enough information about what the "from" and "to" addresses mean, they can forward the little packets to their eventual destination. It's a robust and organic process. There is no center, no controller, and no simple point of failure. A truly brilliant technology that completely changed the world.

beaming Internet service to cities using football field–size balloons hovering in the sky. Marc and Peter's plan was to send solar-powered, high-altitude airplanes to circle above populated areas at 61,000 feet to provide news, entertainment, and information. The vehicle's name, HALO, stood for "high altitude, long operation," but it was also intended to evoke positive feelings as customers looked up and saw their planes flying in circles, leaving halo-shaped contrails of water vapor.

Before they could go further with their venture, Marc and Peter needed a plane that would be able to do routine operations at high altitude. They needed a world-class engineer, an original thinker and a dreamer. There was only one person for the job—Burt Rutan.

Rutan emerged from his office, looking as tall as a cactus, dressed in denim, with sideburns shaped like Idaho, and grinning ear to ear. Burt stood six feet four, making Peter—at five feet five inches tall—feel dwarfed. After exchanging brief hellos on the runway, Peter, Marc, and Burt made small talk around Marc's plane, a twin-engine turboprop Cessna 421 Conquest. Nearby were a few Pipers, Beechcraft, and another Cessna. The plane graveyard across the way was apparently where aircraft were retired, dismantled, and recycled. The planes resembled giant beached whales that had been pushed up and onto a sandy beach from all directions. Locals sometimes heard explosions and artillery fire coming from the boneyard in the early morning hours—the military occasionally used the planes for hostage rescue simulations.

The three men walked into Scaled Composites, which was a collection of buildings and hangars in various states of upgrade, looking out onto the flight line. Inside the main entrance were pictures of Burt's planes and his many awards. Peter marveled at the range, from the homebuilt planes and an AD-1 with its pivoting wing à la Robert Jones, to the Grizzly, built for camping and landing in meadows, and the

Solitaire, a self-launching sailplane. Hanging upside down from the rafters was the Catbird, first flown in 1988. Burt's brother Dick had recently broken a world speed record in the light aircraft, flying a 1,243-mile-long course at 246 miles per hour. Some parts of the shop were curtained off so visitors couldn't see what was under development. Not far away were several delta wings for Orbital Sciences' Pegasus rocket. Peter could have spent hours just talking about the development of those wings.

Burt's reputation as an aviation concept designer was well established, but he was also becoming an innovator in the rarefied air high above the clouds. Peter had talked on the phone with Burt once before, about potential launch vehicles for International Microspace. He knew of Burt's remarkable design of the *Voyager*. Peter was also following the progress of the DC-X, the Delta Clipper Experimental, a vertical-takeoff, vertical-landing reusable "rocket for the people," being funded by the Department of Defense and supervised by Pete Worden. Burt and the crew at Scaled Composites had built the DC-X's aero-shell.

After the tour of Burt's planes, Marc, Peter, and Burt headed into a conference room. Before the Angel Technologies presentation began, Peter talked about his lifelong goal of getting launched into orbit. Burt listened and thought, *This guy is a true space geek.*

"Diamandis," Burt said abruptly, "since you're such a space nut, tell me the rockets used for the first four manned space programs."

Burt's tone had shifted—from playful to needling—faster than the sand moved on the dusty Mojave runway. But Peter began ticking off the answer: "Mercury, Gemini, Apollo . . ."

"No, no! Wrong!" Burt said, seeming to delight in Peter's error. "Gagarin was first with Vostok." Peter knew that, of course, but thought Burt had been talking about the U.S. space program.

Burt asked, "What was next?"

"Mercury," Peter replied.

Burt shook his head, almost triumphantly. A heavy silence filled the room.

"Redstone! Redstone 3 was the second one," Burt exclaimed. "I'm not talking about the capsules! I'm talking about the rockets! What was the third?"

Peter hadn't felt this intimidated since transferring from Hamilton to MIT.

"Gemini," Peter said.

"No, the third was Atlas, for John Glenn's flight!"

Peter looked at Marc, who kept quiet. Burt was clearly enjoying this. Peter knew the space program as well as anyone, but Burt was looking for *his* answer.

"I could be in a room with so-called NASA historians and no one could answer this," Burt said. "What was the fourth one?"

Peter wondered whether he should even try to respond.

Burt answered for him: "The fourth one was the X-15. Fifth was the Titan II for Gemini. Sixth was the Russian Soyuz. Seventh was Saturn 1B (Apollo 7). Eighth was Saturn V." When the impromptu exam was over—and none too soon for Peter—it was time for the three men to address the challenge at hand: getting an unmanned solar-powered plane to the mesosphere.

Peter and Marc felt like evangelists for something potentially great, something that could democratize access to information—and maybe even teach an industry how to do its job better. Marc took the lead, going through a slide presentation on Angel Technologies. Marc itemized what he and Peter were looking for in a high-altitude plane: It would need to carry 1,800 pounds and have an eighteen-foot-diameter antenna pod facing downward. It would need a level payload antenna while the plane was in a seventeen-degree bank, and require liquid payload cooling. Under Marc and Peter's plan, the solar-powered plane would transmit high-speed Internet service—phone, video, information—to customers with cone-shaped antennas on their roofs, creating what they called a "cone of commerce."

Peter listened to Marc and watched Burt, who was either sketching an idea, doodling, or taking notes—Peter could not be sure which. The three then discussed unmanned aerial flight and solar-powered flight. The discussion turned to the engineer Paul MacCready, the father of solar-powered flight who had earlier won several Kremer prizes for innovations in human-powered flight.* MacCready's winning planes included the lightweight *Gossamer Condor,* which became the first human-powered aircraft to do a figure eight over a closed course; the *Gossamer Albatross,* flown human-powered from England to France; and the "bionic bat," which set a human-powered air-speed record. MacCready's early breakthroughs in human-powered flight came in part from studying the low-energy mechanics of birds in flight, measuring time, bank angle, and turning radius for the fun of it. Burt and MacCready were friends, attended many meetings together, but had never formally worked on a project together. MacCready had a concept for a plane to go after the around-the-world record if the *Voyager* failed, and had sent a photo of himself with no hair and the note: "If Rutan Flies Around the World, I'll Shave My Head!"†

After the *Voyager* succeeded, MacCready showed Burt drawings of his plane design. He said there was no value in building it now that the around-the-world milestone had been reached. Instead, MacCready's company, AeroVironment, focused on building the remote-controlled, solar-powered *Pathfinder,* being readied to fly to 50,000 feet, and the

*Named for British industrialist Henry Kremer, who was instrumental in the commercial development of construction materials, including plywood, chipboard, and fiberglass.

†The photo showed MacCready, smiling, with no hair. The hair loss, though, was due to chemotherapy treatments.

Helios, a 247-foot-long solar electric wing under development for NASA. Peter and Marc were banking on the idea that similar solar-powered technology could benefit their project.

In the conference room, Burt was excited by their high-altitude broadband idea, even if he doubted whether a solar-powered vehicle would work for this purpose. But Burt didn't express these misgivings, and he told Peter and Marc that he would begin sketching designs.

Talk turned to Burt's service on the U.S. Air Force Scientific Advisory Board, a short-term position that gave the home builder from Mojave a Pentagon badge and the power to make recommendations on what the Air Force should do following the cancellation of a program called the Orient Express. Announced by President Reagan in his 1986 State of the Union address, the program promised a "new Orient Express"—the X-30 space plane—that could fly into low-Earth orbit at 17,000 miles per hour. But the space plane was shelved as technologically unfeasible. By way of response—and out of his own frustration with government-run space programs—Burt had started a file he called "Logic Gone Amok." It said, in essence, "Okay, you don't have the guts to do the Orient Express. I propose for the next ten to twelve years that NASA should cut its budget in half," and reserve the other half—seven billion dollars—for a single prize to be given to anyone who can design, build, and fly a spaceship that had Orient Express capabilities. Burt explained, "My logic was, if this was really impossible, the government spends nothing. If it is possible, it's a win-win."

Peter loved the story. He told Burt an abbreviated version of his loving-NASA leaving-NASA tale. Peter said that when he started realizing that NASA was not going to be the one to get him to space, he began to think of his medical degree in a different light. He wanted to better understand human longevity. He believed he needed to invent something in the field of life extension to get beyond the 122-year maximum for a human. He had hoped that Harvard Medical School might give him just the advantage to figure out how to live long enough to reach space.

After Peter and Marc left, Burt scratched his head. *This space geek earned two degrees from MIT and graduated from Harvard Medical School, with no intention of practicing medicine?* Burt wondered, *Who would go through all the shit of medical school and not become a doctor?*

On the flight home, Marc and Peter couldn't help but talk about the charismatic and enigmatic Burt Rutan. They wondered whether he'd even been listening. Peter had a sense that Burt was above all an artist, a modern-day da Vinci. Burt was the type who would get involved *if* he found a project challenging and *if* he could do something entirely new. Marc and Peter both wondered what they would see when they returned to Mojave for the concept presentation.

Marc and Peter also discussed the Kremer prize; they hadn't known that MacCready had gone after it for practical reasons: he owed $100,000 to the bank after guaranteeing a loan for a relative's business that failed. Burt said it dawned on MacCready that the Kremer prize, at 50,000 pounds, came out almost exactly to $100,000. While the winning of the prize was important to MacCready—allowing him to pay off his debt—the innovations that followed proved even more significant.

Marc told Peter that he and a friend had put up $250,000 to start the Feynman Prize in Nanotechnology. The prize was run through the Foresight Institute, and was an incentive award to be given to the first person who could design and build two nanotechnology devices, a nanoscale robotic arm and a computing device that could demonstrate the feasibility of building a nanotechnology computer. Marc said the catalyst for the prize came from reading the book *Engines of Creation: The Coming Era of Nanotechnology,* by Eric Drexler.

Peter's head was swimming with ideas, impressions, and memories. As Marc talked about *Engines of Creation,* Peter remembered how it was Drexler who stood at the back of the student center at MIT in 1980 when he launched Students for the Exploration and Development of Space. Drexler spoke up at that meeting to say SEDS should remain independent from the L5 space organization, despite some urging to the contrary. Peter hoped to one day thank Drexler for his vote of confidence in SEDS.

As they continued the flight back east, Peter talked of Gerry O'Neill's passing. He had died at age sixty-five on April 27, 1992, seven years after being diagnosed with leukemia. Peter attended O'Neill's memorial on May 26, with Gregg Maryniak and the "GKO extended family," as followers of Gerry K. O'Neill were known. Peter viewed O'Neill's life as both inspiration and cautionary tale. Just as there was incredible work accomplished—and so many people inspired—there were revolutionary scientific projects that O'Neill dreamed of that were never realized.

Peter now had two ambitious companies running—International Microspace and Angel Technologies—and a third under development called Zero Gravity Corporation (ZERO-G). He was working with Byron Lichtenberg and ISU alumnus Ray Cronise on a plan to give civilians the experience of weightlessness using a modified Boeing 727–200 airplane. Still, Peter hadn't been sleeping much, feeling increasingly antsy that he wasn't getting enough done. He created a two-page contract with himself, listing his strengths, weaknesses, and goals. He wrote about ways he needed to improve himself: consolidate his entrepreneurial efforts, strengthen his piloting skills, put a stop to being a nail biter, commit himself to a regular workout routine, and find a soul mate. He was thirty-two years old: his life clock was ticking.

10

An Out-of-This-World Idea

Visiting his parents' retirement home in Boca Raton, Florida, for the Christmas holidays, Peter sat down and began reading the dog-eared copy of *The Spirit of St. Louis* that Gregg Maryniak had given him the year before. The book was filled with revelations. Peter had always assumed that Charles Lindbergh crossed the Atlantic in 1927 as a stunt, or maybe a dare. He'd had no idea that Lindbergh had made the first-ever flight from New York to Paris *to win a prize.*

As it turned out, Lindbergh was one of nine pilots who competed for the $25,000 Orteig Prize, named after its benefactor Raymond Orteig. The competition was riddled with drama, casualties, and deaths, as some of the world's top pilots and newest planes were lost to the cold expanse of the North Atlantic. *The New York Times* labeled it "the greatest sporting event of the age," and the public took to calling it "the world's greatest air derby." The barriers were both psychological and technical: The distance of 3,600 miles from Paris to New York was almost twice the distance that had been previously covered by an airplane on a single flight.

The successful flight not only made airmail pilot Charles Lindbergh famous, but it also created a global perception that flight was safe and available to the common man. Lindbergh was, after all, an Everyman-turned-Superman story: he dropped out of college to pursue his dream of aviation and then used his own engineering skills to get off the ground. He embodied the belief that adventure is essential to civilization, and that risk reaps rewards.

In between feasts of Greek food at his parents' house, Peter underlined passages and wrote notes in the margins of the book. Raymond Orteig, who grew up a shepherd in Louvie-Juzon, France, on the slopes of the Pyrenees, had immigrated to America as an adolescent and found work as a busboy in the restaurant of the Hotel Martin in midtown Manhattan. Within a decade, he was café manager, then hotel manager, and eventually, with money saved, he purchased the hotel and later another. In the years following World War I, French airmen often stayed at Orteig's hotels. He loved to hear the stories of aerial combat and developed a passion for aviation and a deep respect for the airmen. On May 22, 1919, Orteig sent a letter to president Alan Hawley of the Aero Club of America in New York City:

> *Gentlemen, as a stimulus to courageous aviators, I desire to offer, through the auspices and regulations of the Aero Club of America, a prize of $25,000 to the first aviator of any Allied country crossing the Atlantic in one flight from Paris to New York or New York to Paris, all other details in your care.*

The prize would be offered for a period of five years, which came and went without anyone claiming it. Orteig was unfazed and renewed the offer for another five years. Peter calculated that the nine teams competing for $25,000 had spent around $400,000—sixteen times the value of the cash prize. "Orteig didn't spend one cent backing the losers," an amazed Peter wrote in the margin of his book. "By using incentives, he automatically backed the winner... great return on his money. Sixteen times the purse prize."

When Peter finished the book in December 1993, he saw something that he felt had been staring him in the face for a long time. *A space prize.*

The idea of prizes and competitions was not new to Peter. He had talked with Gregg Maryniak and Pete Worden about the potential of prizes and had studied other incentive prize competitions. He had read about the longitude prize of 1714, when the British Parliament offered 20,000 pounds for the discovery of a way for ships to ascertain longitude. Nearly a century later, Napoleon and his ministers, searching for a way to end scurvy and feed troops, had offered a prize of 12,000 francs for a simple invention that could preserve food. In addition to the Orteig Prize, there were dozens of other early aviation prizes, given for everything from staying aloft for fifteen minutes straight to being the first pilot to cross the English Channel.

While the notion of a space prize seemed like the natural next step, Peter didn't know what such an award would look like or how much the winnings would be. But he was already thinking that the goal of the competition would be suborbital rather than orbital, because orbital was so much more difficult to achieve. It would be a big first step.

Peter girded himself in his seat as the small turboprop plane rattled and jolted its way from Stapleton Airport in Denver to the town of Montrose about an hour away. The wind gusted at 35 miles per hour, and visibility was less than five miles. A thick mixture of fog and snow churned in the darkness on the other side of the plane's cold windows. Peter had organized a three-night, four-day "build a rocket" brainstorming retreat for rocket scientists and space lovers at a friend's private home near

Montrose in February 1994. The log "cabin" where they were meeting had 16,000 square feet, eight bedrooms, 7.5 baths, an indoor pool, and meeting rooms with high ceilings, arched wooden beams, chandeliers, and large stone fireplaces. The home was not far from Telluride and—more important—close to Ouray, Colorado, also known as "Galt's Gulch," the mystical and secluded valley of *Atlas Shrugged,* where John Galt and the "men of the mind" went on strike to safeguard rational self-interest.

The goal of the Montrose gathering was to see whether a dozen men and women meeting over a weekend could hatch a new breed of rocket, finding inspiration in the improbable beginnings of many great companies: Harley-Davidson, Walt Disney, Hewlett-Packard, Apple, and Microsoft were conceived in garages or sheds. Likewise, engineers and pilots, not governments, had started the aviation industry. A space lover named Jeff Bezos—who had served as president of the SEDS chapter at Princeton while a student there—was reportedly starting a book company from his house to capitalize on the next big thing—the Internet.

David and Myra Wine, who owned the outsized shack in the woods in Montrose, hoped that whatever prototype was dreamed up could be built in their backyard. David and Myra had met in Daytona Beach in 1969 during the Apollo days, when Myra was working for a NASA program. David was a founder of Geostar, the satellite company based on the invention patented by Gerry O'Neill. David had been weeks away from flying three of his Geostar satellites on the space shuttle when the *Challenger* disaster happened, grounding all flying. David and Myra always told their friends, "We are spacey people."

As the shaky turboprop made its descent into Montrose, Peter tried to ignore the turbulence and focus on the meeting ahead. His goal going into the weekend was nothing less than to come up with a new rocket design that would take paying passengers to space. But he had another idea, too, one he had been studying in stealth mode.

On day one of the meeting, after a big breakfast made by Myra, after travel war stories were exchanged, the group gathered in the conference room. Peter, wearing black pants and a thick black turtleneck, wrote on the whiteboard: "SMALL TEAMS CAN DO BIG THINGS."

It was something he needed to believe in now more than ever. Like a projectile veering off course, Peter had landed in unfamiliar terrain. He had managed to sell his beleaguered International Microspace to CTA of Rockville, Maryland, which designed and manufactured satellites and made software and hardware for ground- and space-based systems. CTA bought International Microspace because of the deal that Peter forged with Pete Worden. Although the Defense Department deal was valued at $100 million for ten launches, relatively little cash was handed over up front. The money would be paid when satellites were ready to launch, something that was appearing less and less likely. Peter had scrambled to keep the company one step ahead of bankruptcy, but he now found himself doing something he never wanted—climbing a corporate ladder as a midlevel executive at CTA. He was going on thirty-three and in a job that wasn't for him. He needed to get back to his scrappy roots—of launching SEDS and ISU, of designing, building, and testing out gravity-defying ideas in the Man Vehicle Lab.

The Montrose attendees began discussing how to build a new rocket for space travel. Peter's friend Byron Lichtenberg, the astronaut, said he thought it was time to "prove that this work can be done by someone other than NASA." The year before, he, Peter, and Ray Cronise had started ZERO-G, with the bold goal of taking paying passengers on parabolic flights using the modified Boeing airplane. They had raised a few hundred thousand dollars from two adventure-investors, Mike McDowell, who had run tourist operations to the North and South poles, and Richard Garriott, better known as "Lord British" in the video game world, and son of Skylab and space shuttle astronaut Owen Garriott. If approved by the Federal Aviation Administration, ZERO-G would be the first private company to offer nonastronauts the chance to experience weightlessness.

Peter, in introducing everyone, noted that Byron had now flown more than three hundred orbits, logging 468 hours—nearly twenty days—in space. Colette Bevis, seated nearby, had headed marketing for Society Expeditions, a company trying to break into commercial space. Across the table was Gary Hudson, a college dropout who had taught launch vehicle design at Stanford, and was an entrepreneur who had been pushing for private spaceflight development since 1969, when he was nineteen. He just wanted to build and ride in a reusable spaceship, preferably one that did both vertical takeoffs and vertical landings. David Wine had known Peter since the early ISU days and was an investor in International Microspace. Wine had been in talks with Burt Rutan about moving Scaled Composites from the Mojave Desert to Montrose. Engineer Dan DeLong worked full time for Boeing and had been a subcontractor for NASA, doing the air and water recycling systems for the space station. DeLong was still in high school when he built his first submarine, electric bicycle, and a tape recorder out of a first-generation computer. Since then, he had a habit of walking away from perfectly good high-paying jobs to join experimental space start-ups. His problem with NASA was that it spent $17 billion a year and "didn't do much." He was working on a NASA contract in January 1986 when the *Challenger* broke up. Within an hour, he knew what the problem was, because he had designed thousands of O-rings and seals. He learned that on the night before the launch, as well as early the next morning, engineers had urged NASA not to launch in temperatures below 53 degrees, and were overruled. "Ten good engineers are better than one hundred," DeLong had come to believe. He was certain that the private space industry could improve on NASA's record of one catastrophic vehicle destruction for every one hundred flights.* Looking around the table, DeLong chuckled to himself. He was sur-

*Richard Feynman wrote in the Rogers Commission report: "It appears that there are enormous differences of opinion as to the probability of a failure with loss of vehicle and of human life. The estimates range from roughly 1 in 100 to 1 in 100,000. The higher figures come from the working engineers, and the very low figures from management."

rounded by people, like himself, best described in incongruent terms: hard-nosed idealists.

The participants presented ideas for vehicles ranging from a modified Learjet to multistage rockets. Gary Hudson did drawings of the holy grail of suborbital spaceflight, the SSTO, the Single Stage to Orbit vehicle. There was talk of propulsion, considered the key to space exploration. Formulas learned long before were now scribbled on the whiteboard. Tsiolkovsky's rocket equation (also called the "ideal rocket equation") calculated how much speed you gain from a rocket engine:

$$\Delta^{v} = v_{e} \ln(m_0/m_1)^{*}$$

There was mention of the teachings of Maxwell Hunter—Gary Hudson's mentor—who helped design the Thor, Nike, and other missiles during the Cold War and wrote *Thrust into Space.* There was animated discussion around speed and how to achieve it. The year before, in 1993, American sprinter Michael Johnson set a world record by running the 400-meter race in 43.18 seconds, traveling an astounding *30 feet per second.* An arrow launched from a bow can go about 350 feet per second. The average bullet travels 2,500 feet per second. To get to the inner line of outer space—62.5 miles—requires a velocity of about 5,800 feet per second, or nearly 4,000 miles per hour. Reaching orbital velocity requires 30,000 feet per second, or more than 20,000 miles per hour.

"Going to orbit is so much harder than suborbital flight, and there's a good market for suborbital missions," Byron said. "There's a lot of

---

*$\Delta v = v_e \ln(m_0/m_1)$ where: Δv = change in rocket velocity, v_e = exhaust gas speed, m_0 = initial mass of rocket, m_1 = final mass of rocket, after the burn, and ln(..) is the natural logarithm—roughly speaking ln(N) = twice number of digits in the number N. The formula describes how much speed you gain from a rocket engine. It depends on the exhaust speed of the gas and the net change in mass of the rocket as fuel is expended. The rule is that the change in velocity ΔV = (exhaust gas speed) × (logarithm of the initial/final mass). For example, if the rocket burns enough fuel to decrease the total mass by a factor of 3, then the velocity increase approximately equals the exhaust gas speed. If the mass decreases by a factor of 9, then the velocity increase approximately equals twice the exhaust gas speed. The faster you burn, the lighter the rocket gets, and the easier it is to change the velocity.

scientific work that can be done seven to eight minutes out of the atmosphere."

Discussion turned to liquid engines, hybrid motors, and solid motors. Standing at the whiteboard, Peter wrote under "Propulsion Options":

—Liquid air/jet engine
—Hybrid rocket
—LOX/kerosene
—RL-10
—$H_2 0_2$/kerosene

Peter scribbled a dozen formulas on the board, and shared his sketch of a small space plane. It had a bullet-shaped fuselage with passengers up front, wings with slats, ailerons and flaps, and triangular horizontal stabilizers.

Engineer Bevin McKinney, who had built a prototype commercial satellite launcher called Dolphin (which had a hybrid motor and was sea-launched in 1984), believed that everything being discussed was possible. The challenge was money. As Gordon Cooper said to Gus Grissom in *The Right Stuff*: "You boys know what makes this bird go up? FUNDING makes this bird go up." Grissom replied, "He's right. No bucks, no Buck Rogers." In John Clark's 1972 book *Ignition!*, a technician running a rocket engine test, and using an exotic and expensive boron-based propellant, remarked that every time he pushed the "run" button, he felt the price of a Cadillac go down the pipes.

Both of McKinney's private rocket companies had breakthroughs, but were driven out of business by government-backed competitors. McKinney's American Rocket Company had spent years developing its hybrid rocket motor technology, only to have NASA fund a competitor to duplicate its work. Gary Hudson's Pacific American Launch Systems was outmaneuvered by a program funded by the Defense Department.

On Sunday afternoon, with less than a day remaining in Montrose, Peter decided to share with the group his under-the-radar idea: the space prize. Standing at the whiteboard, Peter wrote, "PRIZES WORK."

"Prizes help to focus energy," Peter began, as if thinking aloud. "They provide a spirit of competition which has been one of the most important driving forces during the entire history of humanity." The more he talked, the more animated he became. "Space needs prizes. Space needs a return to small, well-articulated goals. Goals which involve and excite the general public."

Peter handed out a paper he had drafted titled "Spaceflight Prize Strategy." The copies were marked *proprietary and confidential.* The paper read in part:

> *There is a strong technology available which helps humans in achieving difficult, sometimes seemingly impossible feats, this technology is a forcing function which helps to focus the whole of human ingenuity at the same well articulated goal. . . . This concept, the forcing function, this technology, is the competitive "Prize." Not prizes for spelling bees or prizes for a lifetime achievement, but prizes which lay out impossible goals and tempt man to take great strides forward. Prizes such as those which were set out to the aeronautical world for speed, distance, endurance, etc. Prizes which brought forward adventurers, dreamers, and doers. Prizes such as the $25,000 Orteig Prize. Where no government filled the need and no immediate profit could fill the bill, the Orteig Prize stimulated multiple different attempts. Where $25,000 was offered, nearly $400,000 was spent to win the prize—because it was there to be won.*

Peter wanted to do for space what Orteig—through Lindbergh—did for aviation. He now had everyone's attention.

Peter told the cautionary story of Richard Feynman giving a Caltech lecture to the American Physical Society entitled "There's Plenty of

Room at the Bottom," where he spoke of building atomic- and molecular-size machinery. To promote the idea, Feynman offered $1,000 to the first person who could build a working electric motor no larger than one sixty-fourth of an inch on each side. He envisioned that novel technologies would have to be developed to permit the manipulation of individual atoms to win such a prize. A month later, one of Feynman's graduate students asked him to step up to the microscope and take a look inside. The student with a steady hand had used a very fine jeweler's forceps, patience, and ample magnification to construct a conventional, microscopic electric motor that met the prize rules. Dismayed, Feynman paid the winner the $1,000. Feynman later told a friend of Peter's that the "enemy" of the incentive prize was the "smart aleck grad student" who met the conditions of the prize without achieving the breakthrough spirit of the prize.

With that in mind, Peter emphasized the importance of clearly articulated and logical rules for the award: The prize must involve a human feat with a level of danger and drama that would capture the interest of the public. The prize must involve a feat in which the public could someday imagine themselves participating. The prize must involve competitors racing against time and each other. The prize must be sufficiently lucrative to entice a number of competitors and must be well advertised.

It was not long before the conferees came up with a flurry of questions. Would a prize for a spaceship require the vehicle to fly more than once? What would be the turnaround time, given that the goal was to introduce to the market a reusable, privately funded spacecraft? How many people would be in each vehicle? Max speed? Air launch versus ground launch? What about hybrids for the mission?

Peter, who had thought out this plan to the nth degree, was quick with his replies: The spaceship must be built privately for a cost and using a method that can be repeated. The spaceship must be reusable. The flight must return the crew and spaceship safely to Earth. The entrant must demonstrate the ability to refurbish the spaceship within seven days for a repeat of the flight. And the spaceship cannot be a surplus vehicle from any government program.

The discussion then moved on to a key criterion: the definition of the start of space. Americans consider 50 miles out from the Earth's surface "space," whereas Europeans draw the line at 62.5 miles (100 kilometers). The U.S. military and NASA had awarded astronaut wings to pilots who had flown above 50 miles.*

Dan DeLong said the goal for contenders should be to reach the Karman line—62.5 miles. "A lot of international organizations and agencies respect this as the definition of the start of space," he said. "This should be an international competition. The other reason is that's about the peak altitude you can go to and still do a pullout from vertical to horizontal with a reasonable load on the plane and pilot."

Next came literally the million-dollar question: What would be the size of the prize? Peter responded in a way that only a space geek could:

"If we use the Orteig Prize ($25,000) from sixty-seven years ago and adjust it at a compound rate of 6 percent inflation, this would yield a current value of $1,240,000 (1994 dollars)." He went on, "This value is probably at the lowest end of the right ballpark for our suborbital spaceflight prize. A prize somewhere in the neighborhood of $1.5 million to $10 million should attract a large number of entries yielding on the order of $30 million to $60 million of invested effort into getting man off the planet in reusable spaceships."

The funding of the prize would come from two primary sources, Peter said. The first would be individuals who want to create a "living monument" to honor someone. The prize would become a "shining light so strong, so compelling, that it will cause all individuals of dreams to aspire,

*Russian cosmonaut Yuri Gagarin was the first person to reach space, in April 1961. The first American in space was Alan Shepard aboard Mercury 7 in May 1961.

to think, and to strive toward greatness." The second source of funds might involve "pledges from the space advocate population," through phone-a-thons and direct mail campaigns.

As the group discussed the plan, Peter excused himself and slipped out of the room. He took the stairs to his room two at a time. Sitting down at his desk, with his view looking south to the snow-capped Ouray Mountains, he began to type:

CHARTER

"The John Galt"

The challenge of the spaces between the worlds is a stupendous one; but, if we fail to meet it, the story of our race will be drawing to a close.

—Arthur C. Clarke

There are few chances in one's life to aspire to greatness. When such an opportunity comes along, the most difficult job is to recognize it, the second most difficult job is to take the risk of acting upon it. The project described in this Charter is one such opportunity. It cannot be ignored—it must be acted upon. All those who join The John Galt Team and sign the last page of this document recognize the challenge, the sacrifice, the fun, and the need to accomplish this goal.

Throughout all of history, the greatest accomplishments of the human race have been instigated and acted upon by the individual or the small group—never have the masses brought about innovation. We have the accomplishments of Charles Lindbergh and the Rutan/Yeager Voyager team as our guiding stars, and every NASA program since Apollo as our incentive to bring about change.

Peter envisioned both a prize and a rocket. The rocket would be called the John Galt and would be built in Montrose. It would either become a part of the competition or be a separate space taxi. He completed the mission statement, the rationale for the project, and the time line. Above the signature lines, he included a list of required reading to give members of the team a current base of experience and an understanding of similar successful undertakings in the past. Peter's required reading list included *Atlas Shrugged*; *The Spirit of St. Louis*; *Voyager,* by Dick Rutan and Jeana Yeager; and *The Man Who Sold the Moon,* by Robert Heinlein.

Hours later, Peter delivered the charter to the group. It was to remain confidential, known for now only to this group. One by one, signatures were gathered.

Dan DeLong was already sketching ideas for a spaceship that he believed could win the prize. Gary Hudson had a vehicle in mind. David Wine said they needed to get Burt Rutan involved. Byron Lichtenberg said it would be crucial to find a group of backers as strong as the ones Charles Lindbergh had. Lindbergh had written: "My greatest asset lies in the character of my partners in St. Louis."

Back in his room, Peter grabbed his leather-bound journal and wrote, "This is the story of men and machines and the dreams that entwine their lives. It is, perhaps, our oldest fable: the attempt to touch the heavens."

Since childhood, Peter had continued to move inexorably toward space, in the same way Cézanne kept painting the same apples; his latest idea for his canvas of space was his most ambitious yet. Peter left Montrose suffused with the feeling of certainty and promise, the way he'd felt as a child when he watched Apollo 11.

Part Two

THE ART OF THE IMPOSSIBLE

11

Eyes on the Prize

Peter stood between *Friendship 7,* the titanium-skinned capsule that took the first American, John Glenn Jr., into orbit, and a display of a lunar rock sample said to be four billion years old. Not far away were the *Wright Flyer* from 1903 and the *Voyager,* the spindly winged Rutan flying machine that made it around the world in 1986 without stopping or refueling. Next to Peter was the one and only Apollo 11 command module, *Columbia,* and directly above was the international orange Bell X-1, the first plane to exceed the speed of sound. Suspended next to it was Charles Lindbergh's *Spirit of St. Louis,* the single-engine, single-seat monoplane with its skin of treated cotton fabric and dappled aluminum nose cone. Lindbergh was all of twenty-five when he furnished the plane with a stiff wicker chair and charted his route by placing a string on a large globe and dividing the journey into segments.

Peter had been to the Smithsonian National Air and Space Museum a dozen times before. He came to the Milestones of Flight gallery to sit and think, watch and listen. He was inspired and humbled by the genius, imagination, and perseverance on display. This was his Fenway Park, Ganges River, and Mount Kilimanjaro rolled into one.

But on this evening of May 25, 1994, Peter was looking for connections—not milestones. Dressed in a tuxedo and holding a glass of wine, he was there to do what most business advisers caution against: crash a party, furtively elbow your way to the host, and pitch your idea. The host tonight was Reeve Lindbergh, the second daughter and youngest child of Charles and Anne Morrow Lindbergh. The event was the Lindbergh Award gala. Peter hoped to get the Lindberghs to endorse his concept of a spaceflight prize.

Scanning the gala program, Peter read that Reeve was a director of the Lindbergh Foundation, and that the award was established in 1978 to recognize leaders who preserved the environment while making technological strides. This year's award recipient was Samuel Johnson Jr., who had been called "corporate America's leading environmentalist."

Peter searched the room and spotted Reeve, with her sandy hair, wire-rimmed glasses, and engaging smile. Heading her way, he rehearsed his summary and hoped he wouldn't get tossed out for bluffing his way in. He waited for an opening in the conversation before introducing himself and launching into his spiel. Talking quickly, he explained his vision for a space prize modeled after the Orteig Prize, the award that had galvanized Reeve's father to make his historic flight. Peter painted a picture of teams of rocket hobbyists and established engineers building spaceships in backyards, garages, deserts, and machine shops—daring to do what only a few governments had done before. He wrapped up his pitch by dropping the names of astronauts who had already pledged their support for his prize, and said that the prize would be offered to the first team that could reach the suborbital altitude of 100 kilometers. The winning team would change the world—just as her father had done in 1927.

Reeve was impressed by Peter's enthusiasm but had no idea what he was talking about. What she got was that he wanted to do something very new modeled on something very old that had appealed to her father decades earlier. A children's book author, Reeve was accustomed to people mythologizing her parents. Her mother was a Smith College graduate.

Anne Morrow Lindbergh's mother had been a writer and poet who had served as president of Smith College. Anne's father was a partner at J. P. Morgan, a U.S. senator, and a U.S. ambassador to Mexico when Anne and Charles first met, in 1927, during the famous aviator's goodwill tour. When the two married, Anne, who loved books more than anything, adopted Charles's physically demanding life. She learned to fly and use Morse code, and with her husband set records with flying adventures around the globe. She was the first woman in America to earn a glider pilot's license. Of Anne and Charles's children, Reeve was the most comfortable dealing with the adulation and controversy that came with the Lindbergh name. The legacy was more than a story for Reeve and her siblings: their older brother was the Lindbergh baby who was kidnapped and killed. Reeve and her siblings were raised to live quietly and modestly—buy used cars, never give an address or list a phone number—so as not to draw attention to themselves. But in a family of recluses, Reeve, born in 1945, was venturing out. The former second-grade teacher was working on a memoir about her upbringing in Connecticut after the war, about her loving, perfectionist, list-making father and her mother, who needed to write as she needed to breathe. In a similar way, writing felt inevitable to Reeve. Her mother wrote beautifully, and her father won the Pulitzer Prize for his book *The Spirit of St. Louis.* Reeve was at the point in life where she needed to set down her memories and deal in her own way with what some in the family called "Lindberghophobia."

At first, public events had been overwhelming to Reeve, until she learned to listen, nod, and smile and let people say what they desperately wanted to say. She recognized the glow in people's eyes when they learned her parentage. Now, this energetic man before her—who wasn't wearing a name tag—was talking about launching an "Orteig Prize for space." He said it was his "mission and moral imperative" to open the space frontier.

Peter, realizing his time with Reeve was up, said, "Would you consider being on our advisory board?"

Reeve thought for a moment and replied, "You should talk to the pilot in our family, Erik. He's the flying Lindbergh."

It was months later when Peter finally tracked down the flying Lindbergh, and additional months before a meeting was scheduled. Peter and Byron Lichtenberg were in a restaurant near Seattle when they saw a man with long graying hair and a white pallor headed their way. He used a cane to walk. When he introduced himself as Erik Lindbergh, Peter did his best to hide his surprise. Erik was four years younger than Peter, but looked considerably older. Erik was living in a yurt on a ten-acre organic farm on a small island near Seattle. He was not exactly the flier and adventurer that Peter had envisioned. He seemed artistic and bohemian; not the guy who was going to help get Peter's space prize off the ground.

Peter, Byron, and Erik settled into their table at the Yarrow Bay Grill in Kirkland on Lake Washington. Byron had brought his usual trove of signed astronaut photos and memorabilia.

Erik took to Byron right away, but found Peter fidgety. Peter chewed his fingernails when he wasn't talking, and kept looking around, as if expecting someone else. He reminded Erik of an antsy kid. But Peter's résumé was lacking nothing. Erik learned that Peter and Byron had met at MIT, where Peter earned two degrees before graduating from medical school. Peter had started a national student space group, founded an international space university, and built a satellite launch company. Now he wanted to create an international competition to get rocket enthusiasts to build their own vehicles to fly civilians to space. Erik smiled. No wonder his first impression of Peter was *odd*. This was not an everyday undertaking.

Byron talked about his path to the Astronaut Corps. As a kid, he was a member of the science fiction book-of-the-month club, and devoured the works of Isaac Asimov, Robert Heinlein, and A. E. van Vogt. He was thirteen years old when he heard John F. Kennedy describe the nation's space mission as "the most hazardous and dangerous and greatest adventure on which man has ever embarked." Byron's dad was in the Army during World War II, and later worked as a dairy equipment salesman. His mom ran a dress shop in their small hometown of Stroudsburg, Pennsylvania. After Byron learned that America's first astronauts, the Mercury 7 crew, were all military test pilots, he became a fighter pilot. He earned a doctorate in biomedical engineering to enhance his chances of getting to space. The strategy paid off: he had become what NASA billed as a "new breed of space travelers," more scientist than career astronaut. He had experienced the magic of NASA, was close to the missteps, and had lived through the tragedies.

Byron shared with Erik the story of where he was on January 28, 1986, the day the space shuttle *Challenger* exploded seventy-three seconds after liftoff. He had given three talks earlier that morning to six hundred high school kids in eastern Connecticut. It was the first time that a teacher was being sent to space, and Christa McAuliffe planned to give lessons from orbit. On his drive back to the airport, Byron turned the car radio on. That's when he heard the news. Tears filled his eyes, forcing him to pull off Interstate 91 and roll to a stop. There on the side of the road, he began to cry. The astronauts were his friends. He flashed to the six hundred students who had been captivated by the mission. He imagined the millions of students across the country who were watching the event live. He wondered what would happen to the shuttle program. He was seven months out from his next scheduled mission. But it would be six years before he would fly again, he told Erik.

When the conversation shifted to his experiences in space, Byron became joyful. He said that in the weeks leading up to a launch, he feared every cough, twitch and headache. Before going jogging or working out, he reminded himself to take it easy and watch his every step. Preflight astronauts got into this "Howard Hughes mode," he said, and wanted to go into a bubble. "So when you're finally strapped in, it's 'Wow, I've been training for this for five years. This is the real thing. Let's go!'"

Sadly, there were only a limited number of opportunities for people to fly to space. "Out of thousands of applicants who apply to NASA every few years, only a handful are chosen for the Astronaut Corps," Byron said. "Of those, a few will get to fly." Byron believed that NASA's future should involve buying seats to low-Earth orbit from commercial providers and spending its research money on exploration missions.

Peter nodded. "The government is not going to get us there," he said. "It is not in the business of taking risks." The three talked about the flight of Erik's grandfather, and the risks he took as the chief pilot on the St. Louis–Chicago airmail route. Twice in the latter part of 1926, the year before the transatlantic flight, Charles narrowly escaped death when his plane's motor failed and forced him to jump out and make a parachute descent. Thirty-one out of forty of his fellow airmail pilots perished in crashes.

It was Wernher von Braun, Peter noted, who had talked about the parallels between the *Spirit of St. Louis* flight and the Apollo voyage. Both missions had a shared goal of capturing the public's imagination and proving that outsized dreams were attainable. Von Braun said, "I do not think that anyone believed that his [Lindbergh's] sole purpose was simply to get to Paris. In the Apollo program, the Moon is our Paris." For Peter, Paris was getting to the start of space in a privately built spaceship. He told Erik what he had told Reeve: He was inspired by the *Spirit of St. Louis* and struck by the impact of the Orteig Prize. His new prize was to be called the XPRIZE. The *X* would be replaced with the eventual benefactor's name; *X* was the Roman numeral for ten and also stood for "experimental." The $10 million prize would be awarded to the first nongovernmental team that could build and fly a three-person rocket to the start of space twice within two weeks.

Erik listened to Peter and Byron talk in more detail about the rules of the XPRIZE: contenders were required to give sixty days' notice of their intent to fly; the flight would have one pilot and a ballast of 396 pounds to simulate two passengers; the spaceship would reach 100 kilometers twice under the dictated two-week time line; and the pilot had to stay alive for at least seven days after the winning flight.

Erik shifted in his seat to try to feel less discomfort. All he could think of was, *We've got enough problems here on Earth. Why do we need to spend $10 million to get to space?* He had flown Estes rockets as a kid and dreamed at one point of flying into space. He had heard talk of space at home, and knew his grandfather was a champion of the work of rocket pioneer Robert Goddard and had been convinced that Goddard's work might one day lead to a trip to the Moon. His grandfather even persuaded philanthropist Daniel Guggenheim to give Goddard $100,000 so he could continue his work. When Apollo 8 became the first manned space mission to orbit the Moon in 1968, Charles Lindbergh sent the astronauts a message saying, "You have turned into reality the dream of Robert Goddard."

Until recently, Erik had been a dreamer and risk taker. Even after his climb to the summit of Mount Rainier, after the mysterious and troubling symptoms of mirrored pain and swelling persisted and abated only to return again, he believed that things would be okay. But as the months passed, his denial gave way to concern. He felt worse for longer periods. He reluctantly returned to his family doctor, where he got the diagnosis: rheumatoid arthritis. The disease was punishing, effectively turning one's immune system against itself, bringing pain, swelling, and deterioration of the joints. If tissue atrophied, the bone structure would change, causing severe deviation in the fingers and wrists until they were gnarled like trees in winter. Erik felt betrayed by his best friend—his body, the body that climbed mountains, laid tracks in fresh powder, did running flips across gymnastic mats. No doctor would tell him how bad it could get. As his condition worsened, Erik tried everything, from prescribed prednisone and methotrexate to homeopathic remedies. Nothing worked. Methotrexate made his mouth painfully dry and left him with a metallic taste. He tried vodka-soaked white raisins and an array of extreme foods and diets. When he learned that beekeepers rarely had rheumatoid arthritis, he subjected himself to bee stings. He eventually found a doctor to give him syringes of bee venom, which he shot under his skin. He swelled from rheumatoid arthritis *and* bee venom. He exercised his limbs and focused on trying to prevent his fingers from curling. The only thing he hadn't tried yet was something considered a last line of defense, the injection of gold salts. The decades-old treatment was known to turn skin mauve and gray. Erik, who had never been depressed for more than a day or two at a time, now struggled with profound sadness. Standing for five minutes was sometimes difficult. Sitting hurt. Talking was no reprieve. But then, with no warning, he would have days where the symptoms blessedly subsided.

At a friend's relentless urging, Erik had earned his pilot's license. It was the obvious choice he had avoided, but it was also an interest that persisted. He immediately loved the physical part of flying—takeoffs, landings, and dealing with crosswinds—but found the memorization of rules and regulations and all of the math calculations difficult. Still, the boy who was bored in high school, maintaining a C-minus average, earned a 4.0 in flight school. He soloed for the first time on October 31, 1989, got his private license to fly the next spring, earned his commercial pilot's license in July 1991, and got his flight instructor rating in September 1991. His first job was at the Bremerton Airport in Bremerton, Washington, before he moved to Port Townsend and worked for a company called Ludlow Aviation. Erik did his best to keep his family name to himself. When the local newspaper ran a small story about a certain flight instructor with a famous last name, Erik was nervous and his boss was happy. He made $12 an hour as a flight instructor, earning money only when the tachometer was turning.

In his first years of flying, he binged on books by Ernest Gann, from

Twilight of the Gods and *Fate Is the Hunter* to *Soldier of Fortune*. It was Gann who wrote, "Flying is hypnotic and all pilots are willing victims to the spell." And it was Gann's words that resonated with Erik now: "It's when things are going just right that you'd better be suspicious. There you are, fat as can be. The whole world is yours and you're the answer to the Wright brothers' prayers. You say to yourself, nothing can go wrong . . . all my trespasses are forgiven. Best you not believe it." It was similar to the warning that Charles Lindbergh gave his children: "It's the unforeseen. It's always the unforeseen."

Erik earned money doing woodworking when he could, but it was hard on his hands. He had married in 1988, and his wife, Mara, was a massage therapist. The monthly rent on their yurt—located on a farm owned by a friend—was $50.

Finally Erik looked at Peter and Byron and said what was on his mind. "I can think of a lot of ways to use ten million dollars right here on Earth."

Byron nodded. He could see that Erik was in pain. He told him that "up in space, a change comes over you." He explained, "In an hour and a half, you orbit Earth. You see everything that holds civilization. You look down and see this beautiful Earth and look into the void of space with its blackness and white dots and you have to admire what we have. You realize that we're all on Earth together. The Earth's atmosphere looks like this one-inch-thick line and you think, 'That is what is keeping us alive.'" He didn't know of an astronaut who wasn't changed by the view, who didn't want to take better care of Earth after looking at it from afar. "When you see it with your own eyes, it's not the same as looking at pictures. I've been very blessed to go twice."

They talked about *Earthrise,* the photo of Earth shared from orbit on Christmas Eve 1968 by Apollo 8 astronauts William Anders, Frank Borman, and Jim Lovell. It was the first time people had seen home from space. The photo, called "the picture that changed the world," showed a perfect blue and white marble surrounded by a black sea. The astronauts read from the book of Genesis, and Lovell said of the image, "The vast loneliness is awe-inspiring and it makes you realize just what you have back there on Earth." The image would play a role in doomsday books by the likes of Paul Ehrlich and bleak scenarios by the *Limits to Growth* researchers. It inspired the modern environmental movement.

What Byron said resonated with Erik. He understood that on one critical level, the XPRIZE was about giving more people that view of Earth. He could see how the goals of the XPRIZE were aligned with the Lindbergh Foundation's mission of using technology to enhance life and preserve the environment. Erik brightened, temporarily forgetting the pain of sitting for too long. After being hit with waves of bad health news, he felt different in this meeting with Peter and Byron. He felt hopeful. Byron came across as part scientist, part philosopher. Peter came across as the mechanism to make this crazy idea work. Before dinner ended, Byron shared one of his favorite quotes, by Calvin Coolidge: "Nothing in the world can take the place of persistence.... The world is full of educated derelicts. The slogan 'press on' has solved and will always solve the problems of the human race." It spoke to the vision of the XPRIZE, Byron said. Erik thought it spoke to his life. He needed to press on, to find a way to a better life.

As the three stood to leave, Erik did his best to walk normally. He could still drive, though it was painful to use the manual transmission, and he couldn't afford a new car. He had a handicapped parking sticker, but hated to use it. Outside the restaurant, Peter and Byron asked Erik whether he would take part in an event being planned in St. Louis to formally announce the XPRIZE. Peter said they would return to the Racquet Club where Erik's grandfather and his backers had signed their pledge to pursue the transatlantic flight. There would be city leaders, astronauts, aviation designers, and rocket makers. It would mean a great deal if they had the support of the aviator's family.

Erik looked at the two men and promised he would think about it. He had never been to St. Louis. It was Grandfather's city, with Lindbergh

Boulevard, the Lindbergh museum, the Lindbergh school district, and a replica of the *Spirit of St. Louis* hanging in the airport. On his drive home, Erik reflected on the night. Peter and Byron had come to him in hopes of connecting with the "flying Lindbergh." He could barely walk, and only occasionally fly. But the meeting reminded him of the question he was asked all the time by fans of his grandfather: "And what are you going to do with *your* life?"

Peter stood in the slow-moving line to touch the lunar rock in the Smithsonian's Milestones of Flight gallery. He was back a year after meeting Reeve Lindbergh here, and a few months after meeting Erik in Seattle. He exchanged sad smiles with his friends Bob Richards, Gregg Maryniak, and others. One by one, Peter and the group inched forward, pausing to touch a piece of the Moon and to reflect on the meaning of the day. It was supposed to be a celebration of life, but it felt more like one of the sky's brightest stars had blinked out. Their friend, leader, and coconspirator Todd Hawley had died from AIDS. He was thirty-four.

Many of the men wore suits and ties, while others wore T-shirts from the inaugural ISU summer session held eight years before. Lapels were adorned with laminated buttons with a close-up photo of Todd's smiling face. There were family members, professors, and ISU loyalists. Peter, wearing suit and tie, touched the moonstone and closed his eyes. He had seen Todd twice in the months before his passing: for an ISU founders' reunion in April, where he, Bob, and Todd had their last photo taken together here in the Milestones of Flight gallery, and later in San Francisco, where Todd was living with his partner. The founders' reunion had been held in part to draw up an "ISU Credo," crafted and signed by Bob, Todd, and Peter on April 12—the day before Todd's thirty-fourth birthday and on the anniversary of Yuri Gagarin's 108-minute orbital flight. More important, the reunion had brought together the Gang of Three, *Peterbobtodd.*

For nearly two years, Todd had isolated himself from Peter and Bob. He had shut himself off from the space community. He broke the silence in early 1995, saying he wanted to see his friends and work together again. Writing the credo in Washington, D.C., reminded them of the adrenalized early days of ISU. They had always joked that they were "triaxially stabilized," strongest as a team of three. Their credo, which had been articulated hundreds of times but never formally written down, set forth the founding principles and goals of the space university. Comprising six paragraphs on one page, it began: "International Space University is an institution founded on the vision of a peaceful, prosperous and boundless future through the study, exploration and development of Space for the benefit of all humanity." It concluded, "This, then, is the credo of ISU. For all who join ISU, we welcome you to a new and growing family. It is hoped that each of you, as leaders of industry, academia and government will work together to fulfill the goals set forth herein. Together, we shall aspire to the Stars with wisdom, vision and effort."

Around the same time, Todd learned that he was being awarded the Tsiolkovsky Medal, given to individuals and organizations that popularize the ideas of Konstantin Tsiolkovsky. Todd was humbled by the award.

Shortly after the reunion, Todd suffered a collapsed lung and was hospitalized. Peter's last visit with the man he called his closest brother was in San Francisco's South of Market neighborhood. Even under heavy sweatpants, Todd looked gaunt. The two went for a walk and Todd had a nasal cannula and pulled a small oxygen tank. The experimental treatments in Russia had failed. Peter and Todd, only a month apart in age, had spent countless hours walking together, talking about everything from space governance structures to the development of new economic systems. On this foggy day in San Francisco, they walked slowly. Todd, a Francophile and a student of history, loved Peter's idea and inspiration for the XPRIZE. They talked about the Frenchman Raymond Orteig, and about great explorers, from Meriwether Lewis and William Clark to Ferdinand Magellan. They laughed at the term they used for their own space endeavors: "benign conspiracy." It was their goal to change the world for the better. But Todd's once-buoyant stride had turned to a shuffle. The man who always seemed to be looking up at the sky with a smile was now focused on making it one more step. When Todd tired after a few short blocks, they turned back. More than once, Peter was forced to look away so Todd wouldn't see his eyes filling with tears. As they walked, Peter remembered the consecutive nights that Todd spent in his antigravity sleeper, smiling through it, never once complaining, sleeping as if he were already in space.

In line at the space museum to touch the lunar rock was Todd's partner, Riq Hospodar, who in the final months of Todd's life changed his name to Yuri Hospodar, to honor Todd's love of space, which began with Yuri Gagarin's flight when Todd was one day away from entering the world. Yuri touched his "commitment ring" to the smooth stone. Bob Richards felt the lunar stone and thought of Todd's eternal optimism. He had talked with his friend the morning he died. Todd called to relay his final wishes. He wanted his celebration of life held in the Smithsonian's National Air and Space Museum. He wanted his ashes placed at the permanent ISU campus. Bob promised he would fulfill his wishes. He was pained as he listened to Todd struggle for breath. Hours later, Yuri called Bob to say Todd was gone.

After everyone had passed by the lunar rock, the group moved to a reserved hall nearby to share memories. Photos were shown of Todd, a blond, blue-eyed kid and with horn-rimmed glasses, next to his dog and siblings. Todd playing cowboys and Indians. An older Todd in a suit and tie, now wearing gold-rimmed glasses, giving a talk at ISU. Todd with Peter, Bob, and Uncle Arthur in Vienna. From childhood to the end of life, Todd's smile stayed the same. In a video interview, Todd talked about his "life mission." His dream, he said, was "no less than the establishment of an entire world off the Earth, inhabited by individuals including you and me." Before the tribute ended, Todd's sister took to the dais to share stories. She said half jokingly that her brother had dreamed of dying at the Air and Space Museum.

In time, the group moved to the steps of the museum. As pictures were taken, they waved and yelled in unison, "Hi, Todd!"

Born April 13, 1961, Todd died on July 11, 1995. According to his own life clock, he spent 12,507 days on Earth.

12

Cowboy Pilot

In January 1996, Mike Melvill and his wife, Sally, arrived at work at eight A.M., unlocked the door, and let themselves in. They headed through the carpeted reception area to Mike's office, which looked out onto the Mojave flight line. Their boss, Burt Rutan, was standing over his drafting table. Burt looked up, startled, squinted at them as if they were blurry, peered outside the window, and checked his watch. He muttered that his wife was *going to be mad.* He'd worked straight through another night, standing at his drafting table, nursing his umpteenth cup of black coffee, wearing his clothes of the day before.

Mike and Sally exchanged knowing looks. They had worked for Burt for nearly eighteen years, saying goodbye to him at five P.M. and returning many mornings before eight to find him where they had left him. They knew his brilliance and his quirks. He was the boss who took them on exotic vacations, only to announce a day after arriving that they were leaving because he couldn't think on a white sand beach in paradise. He was the boss who did wind tunnel testing on top of his 1966 Dodge Dart station wagon; who always said his best plane was his next plane; and who knew airplanes better than just about anyone alive.

These days, Burt was working on his thirty-first plane, the Proteus, a high-altitude research aircraft named for the Greek god who changed his shape to take on any form. The Proteus was being built for Angel Technologies to deliver broadband services from just above the stratosphere. Mike had recently attended a meeting with Burt to discuss the project, and had met Angel founders Peter Diamandis, Marc Arnold, and David Wine, and Angel chief technology officer Nick Colella.

Several times a day, Burt would wave Mike over, or appear at his drafting table with an idea or a sketch for a plane or a part that more often than not had the whimsy of a Dr. Seuss drawing. Burt was deliberately contrarian, moving engines back and wings far forward and co-opting materials once reserved for boats and surfboards.

As Burt's go-to test pilot, Mike had plenty of experience with Burt's convention-defying aircraft. He would fly anything Burt dreamed up.

Mike didn't learn to fly until he was thirty years old in 1969, but the South Africa native was drawn to danger long before his love affair with the skies began. His strong, compact frame made him a star on the local gymnastics squad, and his agile acts on the parallel bars were only the beginning.

Mike inherited a competitive spirit from his father, a world-class target shooter who routinely defeated younger rivals. A motorcycle enthusiast, Mike also got his father's auto mechanic skills. But his ability to build and fix things did not translate into confidence in the classroom. Mike found much of the teaching mind-numbing and was frustrated that the curriculum gave him no opportunity to work with his hands. He failed his senior year math final and didn't graduate from high school.

Mike met his match early when he encountered Sally, a high-energy, petite blonde with a strong will and a thirst for adventure. Sally's parents wanted her to go to finishing school and marry a wealthy sugar farmer.

Sally wanted only Mike. She defied her family and left Durban on the back of Mike's motorcycle, with her angry father in hot pursuit. Mike and Sally went to England, where Mike became a carpenter. They were married in Scotland and immigrated to the United States in 1967 to join Sally's brothers and her now friendly father. They settled in Indiana, where Sally's family had a rotary die-cutting factory called Dovey Manufacturing. Mike ran the machinery while Sally took orders from customers. His aptitude for building was apparent from his first day on the engineering-based job. He started making some of the company's tool products, and when those expensive tools failed, broke, or were incorrectly used, he was dispatched to troubleshoot.

To do that part of his job, he jumped on a plane once a week, spending long days just getting to the companies. Finally, Mike concluded that someone in the company needed to learn to fly. Most of the repair jobs he went to were near small airports. Sally's brothers had no interest in flying, so it was decided that Mike would get his pilot's license on the company dime. But Mike didn't take to flying in small planes the way he took to machining parts. He vomited every time he and his instructor were airborne. Thankfully, his instructor, Dick Darlington, told him that plenty of people were nauseated when they first learned to fly. He assured Mike that the sickness would diminish and then disappear. He was right. Mike earned his private and commercial pilot's licenses, and was soon arriving at work wishing that someone's equipment, somewhere, would break so he would have to fly off to fix it.

Flying became freedom. It was a different way of looking at the world—as Amelia Earhart described it, "You haven't seen a tree until you've seen its shadow from the sky." Instructor Darlington, aware of Mike's intense interest in flying, suggested he consider making his own plane, given that he had the skills and tools of a machinist.

Mike had no idea that people made their own planes. Darlington told him about a place in Wisconsin called Oshkosh, where the Experimental

Aircraft Association held an annual gathering of pilots and their homebuilt or modified aircraft. In the summer of 1974, Mike and Sally attended. One of the first things they saw at Oshkosh was a crazy-looking plane with a tail on the front and engine in the back. "What *is* that?" Mike asked, intrigued. Its name was the VariViggen, and it was flown by a man named Burt Rutan. The name VariViggen came from the Viggen, a highly innovative Swedish fighter aircraft built by Saab. The Viggen had "canards," essentially wings moved from the back to the front. As Mike watched Burt give people rides in the plane, doing short takeoffs, short landings, and turning on a dime, Mike said, "Now that's a plane." Burt sold kits for the VariViggen out of the back of his plane parked on the Oshkosh flight line. Mike handed over fifty-one dollars in cash—Burt didn't trust banks—returned to Indiana, and began studying the plans. He knew how to build things from blueprints, but these were not engineering plans. They were sketches and photographs and had a comic book–style narrative.

Undeterred, Mike started building. He made progress on one part, only to find himself confounded by the next part. He called the Rutan Aircraft Factory (RAF) in Mojave, and Burt walked him through the problem. At the end of three years, Mike, ever competitive, became Burt's first customer to finish building a VariViggen. Not long after finishing the plane, Mike and Sally flew the VariViggen from Indiana to a business meeting in California.

To confirm that Mike had built the plane right, they flew into Mojave. Burt was so impressed that he invited them to dinner. He told Mike and Sally that he'd left his job as director of development for Bede Aircraft in Newton, Kansas, and opened RAF in 1974. He asked Mike what he did for a living and ended up offering Mike and Sally jobs. Burt said he needed help so he could focus on designing new planes. Sally could do bookkeeping, and Mike would help him with the homebuilt airplane business. He offered the Melvills a starting salary of $22,000 a year—to be split however they wanted. They were each earning double that in Indiana.

In September 1978, Mike and Sally gave up their secure jobs in the family business and moved from central Indiana, with its bone-chilling winters, hot summers, and flat landscape, to the Mojave Desert, with its crisp winters, baking arid summers, and sandy, tumbleweed-strewn landscape. The Melvills' sons, Graham and Keith, were fifteen and twelve. Sally cried for a year, lonely for her extended family, but Mike settled in. He had found his niche. The renegade spirit suited him, and his newfound love affair with planes was the oldest story in this dusty town. Where others saw wind, sand, and Joshua trees, aviators saw a dreamland of unlimited heights and speeds. They saw a sky that was almost never obscured, a place where nearly every day brought a call from the flight tower of "severe clear." A cast of characters flew in and out of Mojave, stopping in at Burt's shop to share stories. Over time, Mike befriended legendary pilots, notably Scotty Crossfield, the first to fly twice the speed of sound, and Fitz Fulton, who set early altitude records for the military and was the first to fly the modified Boeing 747 when it carried the space shuttle out of Edwards Air Force Base.

Mike started at RAF doing whatever was needed, from sweeping shop floors to helping Burt design and flight-test new planes, including the kit-built VariEze and Long-EZ. Mike improved the instructions in the kits. Thousands of the VariEze kits sold quickly, at $54 apiece. The Long-EZ kits were snapped up for $250 apiece. Burt told Mike that the idea for the first kit for the VariViggen came in part from the Simplicity sewing patterns used by Burt's wife, Carolyn. Burt had seen Carolyn make her own dresses by pinning patterns to fabric and cutting along the dotted line, and wondered why he couldn't do the same for aircraft. Burt loved the idea that someone could build a plane in his or her garage and go and fly it.

By the early 1980s, Burt told Mike that they needed to phase out the homebuilt plane business. The money was good, but the builders needed a lot of support, and the liabilities were great. When the U.S. Air Force needed a trainer for a new fighter plane, Burt built a scaled-down replica that would give them the same flight test data. Scaled Composites opened for business in 1982. Burt set a standard for working hard, but he also evangelized the need for fun. He would stop meetings to say, "Are we having fun yet?" Employees would yell, "Yeahhhhh!" Instead of spending money on employee Christmas parties, Burt took 1 percent of the company's net profit for the year and divided it equally among employees. Burt gave himself the same bonus as the shop floor sweep. On Fridays, when the company was small, Burt announced that everyone had been working too hard. "It's time for clam chowder," he'd say. Employees would jump into planes and off they'd fly to their favorite greasy spoon in nearby California City.

In the same way a surfer studies the sea, waiting for that perfect wave—looking for a glassy surface, an offshore breeze, and the right amount of spray coming off the top of the lip—Burt was captive to the sky. One afternoon, Burt found Mike and said excitedly, "Have you looked outside?"

"Yeah," Mike answered tentatively.

"The clouds!" Burt said.

"Yeah, there are clouds."

"We need to go flying!"

Mike, Burt, and Burt's brother Dick grabbed cameras, piled into a plane, and hit the sky to fly through the rare clouds of Mojave.

Mike took advantage of any opportunity he had to become a better pilot, practicing landings and takeoffs again and again until he got it right. Burt, who had been a flight test engineer—not a test pilot—knew the maneuvers needed to get the desired telemetry, whether on directional stability or stall characteristics. Burt demonstrated a maneuver in his own twin-engine Duchess and then had Mike do the same thing. Mike learned that flight tests proceeded in incremental steps—slowly, steadily. Mike learned to take a plane that had never been flown before and start by taxiing it around the runway to make sure the brakes and steering worked and the plane cooled well. The plane was then returned to the hangar, and the team would debrief with the data and Mike's analysis. This would continue for days or weeks, until they felt the plane was flight ready. The first "flight" would be in the thin cushion of air inches above the runway.

Mike got his long-distance flying and formation training from none other than the velvet-armed Dick Rutan. The Rutan brothers agreed that Mike was one of the most instinctive stick-and-rudder pilots they had ever met. With time, Burt grew confident that Mike could do dangerous flying in a plane that had never been pushed—performing stalls and spins—and bring his baby back safely.

There were times, though, when Mike sat on the runway before a first flight test, looked at whatever unconventional contraption he was belted into, and wondered, *Am I going to be alive for long after I push the throttle?* Mike narrowly averted disaster many times, including the day a mechanic left a wrench inside a wing of the prototype Starship, flown in 1983. The controls jammed midflight. Mike tried everything he could think of before grabbing the stick and putting all of his weight on it. He was lucky; the wrench popped loose.

Sally served as Scaled's director of human resources. It wasn't easy being the wife of a test pilot of experimental planes. When someone asked what it was like, she pointed to her wrinkles. But being a pilot herself, Sally said, "It's for Mike to question whether the plane is safe." Both she and Mike had to believe that Burt would never put Mike in a plane that wasn't safe.

But Burt pushed the limits. In 1992, Mike and another Scaled pilot, Doug Shane, endured a plane that Doug called "a new and unwelcome experience," and that Mike labeled "harrowing." Burt had designed a new radio-controlled unmanned aerial vehicle (UAV), intended for forty-eight-hour flights at 65,000 feet. The drone was called the Raptor and had a wingspan of sixty-six feet. It was designed to carry a 150-pound payload, including underwing antimissiles. The fuselage was too narrow to accommodate a cockpit. The Raptor was part of a ballistic missile defense concept, engineered by Nick Colella while he was at Lawrence Livermore National Laboratory.*

Mike arrived at work one morning to find the shop guys having fun with a saddle they'd thrown over the newly built Raptor. The maintenance manager was a horse owner, and the crew had apparently been taking turns in the saddle. When the guys somewhat nervously asked Mike whether they could take his picture up in the saddle, he was game and climbed aboard, just in time for Burt's arrival. For a moment, no one said a word, fearing the boss would not be amused. But Burt studied the situation and exclaimed, "That's it! That's what we needed and I never thought of." Burt had been worried about losing the unmanned prototype on its first flight tests. That morning, again to the surprise of the crew, Burt had the shop build a fiberglass saddle with a back and shoulder support. He would give the pilots the ability to override the remote controls. *They could just ride on top of the plane!* He'd give them parachutes, too, just in case.

When it came time for the Raptor's test flights, Mike climbed warily into the fiberglass saddle on top of the plane, put his helmet on, and got his feet into the stirrups. Sally was deep inside Scaled—not about to come out to witness the love of her life riding *on top* of a plane. Project engineer Dave Ganzer controlled takeoff and landing remotely, making Mike—straddling the fuselage—feel like a pawn in someone else's nutty video game. The landings proved particularly terrifying, as the Raptor came in at nearly 100 miles per hour. There was Mike, riding on top of a plane in the open air without even a windshield. It took every bit of his strength not to reach for the controls.

A few days into the testing, Mike got airborne and very quickly

*"Raptor" stands for "Responsive Aircraft Program for Theater OpeRations," and the slightly fanciful idea was to have UAVs loitering on the edges of a battlefield where they would detect and respond to theater ballistic missile launches and intercept them in the launch phase with a hypervelocity TALON missile. It was to be a direct predecessor of the Predator and Reaper UAVs.

realized he had no rudder control. Ganzer and his crew were in a chase van on the runway, and Ganzer reported the same problem. Mike couldn't land if he couldn't line up with the runway. Mike radioed Ganzer to say he was going to fly over to the dry lake bed in Rosamond and try to land there. He didn't have the option of parachuting out, as he couldn't gain even a foot of altitude. Ganzer sped out of Mojave, following the imperiled Raptor's path. Ganzer was in Burt's old white van with the roof cut out to make way for a plastic bubble. Ganzer would stand up in the bubble holding the controls to fly the Raptor.

With the dry lake bed below, Mike considered putting one wing down to drag himself to a stop. The plane would surely break apart. Ganzer had said something to him about the Raptor's "adverse yaw." This suggested the plane would react in an opposite way to the normal push of the stick. If he pushed the stick to the left, the plane would initially yaw or turn to the right. Mike said his pilot's prayer and plunged the stick all the way to the right. The plane turned to the left initially and then had one beautiful moment of leveling out. Mike immediately put the plane on the ground and taxied to a stop in a cloud of dust. By the time Ganzer and crew came tearing in, Mike had dismounted from the death trap. He was *almost* breathing normally again when Ganzer and the crew came to a stop. Ganzer discovered there on the lake bed that a relay that controlled the rudders had locked up. The relay was replaced, and Mike was asked to fly the drone back to Mojave. His first reaction was "No way! I thought I was dead!" As the hours passed, though, it was clear that the only way to return the Raptor to Mojave was for Mike to fly it. He reluctantly got back in the saddle.

Now, in early January 1996, Mike was getting ready for the first test flight of another new plane, the Boomerang, a five-passenger twin-engine that took defiance of convention to a new level. It was intentionally asymmetrical, and looked all wrong. The wings didn't match, with the right wing coming in fifty-seven inches shorter than the left. One of the engines was mounted on the fuselage, the other on the left boom, and the right engine was more powerful than the left. The horizontal stabilizer, joining the fin on the twin tails, extended past the right fin but not past the left. The "door" for the copilot and pilot was through the windshield. Burt's goal with the forward swept-wing plane—thus the name Boomerang—was to solve the problems and dangers of engine failure and asymmetric thrust—the "P effect"*—in conventionally designed twin-engine planes. Burt assured Mike that the asymmetrical design was actually "more symmetric than a symmetric airplane" when flown. He said that the P effect slowed a symmetric plane down, requiring rudder, but that the P effect on an asymmetric plane made it symmetric at low speeds and asymmetric at high speeds, when the pilot wouldn't notice it.

Before leaving the office, Mike checked out the drawings for the Proteus, the Angel Technologies' high-altitude plane still in the drafting phase. The plane shared similarities to a Klingon warship from *Star Trek*, a praying mantis, and a dragonfly. The Proteus needed to be capable of doing small circles in the sky for up to fourteen hours at a time and carrying payloads of different size and weight. "You'll have to wear a Moon suit for this," Burt said. Mike had no doubt Burt was serious and noted happily that at least the plane had a cockpit—and no stirrups. Mike studied the drawings and realized that the temperature cycling would be extreme: the Proteus would have to be capable of taking off in the Mojave summer of 110 degrees, and at 50,000 or 60,000 feet would encounter temperatures of minus 110 degrees, a delta of 220 degrees.

As he was pondering this, another Proteus drawing on Burt's drafting

*The P effect is caused by asymmetry in the action of a propeller blade that is running into the air at an angle. Picture a spinning propeller as a disc. If the disc is facing straight into the direction of motion, then the action of the blades is completely symmetrical. But if the disc is tilted as it would be when the aircraft is climbing, then the lower edge is ahead of the upper, and the blades on the way "down" the disc are moving faster into the air than on the way up. The effect is to move the center of thrust. In a single-engine plane, this puts the center of thrust along a line parallel to but offset from the centerline of the airplane. It tends to yaw the craft, and you compensate with rudder control. Burt's asymmetric design made this effect go away: the asymmetries cancel out.

table caught Mike's attention. In this sketch, the Proteus carried a rocket underneath. Mike looked closer; there was a *cockpit in the rocket.* This couldn't be serious, he thought.

Burt smiled at him expectantly. Mike had seen that dreamy expression before. He had to wonder: What on Earth was Burt going to build next?

13

History Repeats Itself

It was the morning of Saturday, May 18, 1996, and Peter was pacing in front of his hotel in St. Louis, studying the names of confirmed participants and going through his checklist. Today was the day he would formally announce the $10 million XPRIZE and invite teams around the world to compete, and he needed everything to go perfectly. He had local and national media attending. He had commitments from more than twenty astronauts, including his childhood hero Buzz Aldrin. He had top honchos from NASA and the Federal Aviation Administration, as well as rocket designers and aviation stars like Burt Rutan. And he had Erik Lindbergh and his brother Morgan in St. Louis for the very first time—the city where Charles "Slim" Lindbergh had found the support he needed to fly.

Peter finished his cup of black coffee and watched as the superheroes of his youth, the astronauts, began to appear outside the lobby, looking like Secret Service agents in dark suits and aviator shades. As a line formed to board the van to take them to a spot near the city's Gateway Arch, Peter went over logistics with Byron Lichtenberg, who had corralled many of his fellow astronauts into coming. When he looked up from his notes, Peter saw Burt talking with Dan Goldin, the head of

NASA. Something about their body language didn't look right. While the line was moving, the two men had faced off and were not moving. Peter walked closer.

Burt loomed over Goldin like Apollo Creed over Rocky Balboa. He was chiding the space agency for its lack of innovation and declared that NASA should be pronounced "nay say." Peter bit his nails. "There is no growth. No activity. Nothing," Burt said. "Why isn't NASA doing this?" he asked, gesturing around. "Because risks don't register with NASA today."

Goldin, who grew up in the Bronx, ran marathons, and did 100-mile bicycle races for fun, was not about to let anyone attack his agency or employees. He gave it right back to Burt. Burt didn't have to live under the constraints of government rules or expectations. He didn't have to answer to the president, the Congress, or the American people. He could do his thing in the Mojave Desert, with no interference, little oversight, and without the "gotcha" media watching and waiting. Goldin respected Burt, considered him brilliant, and had been at the receiving end of his needling before. NASA was a lot of things, he said, but it was *not* risk averse.

"'Failure is not an option' was the mantra from *one human being during Apollo 13!*" Goldin said angrily. "[Gene] Kranz wasn't speaking for *all* of NASA. He was saying, 'These three people's lives are at stake. We cannot fail. We gotta bring them back.' It has been misinterpreted." Goldin said that people expected "perfection from NASA, and it's a news story when NASA fails." Goldin was the first to admit he had never wanted to run the space agency, but got what he called the "hug of life" from President George H. W. Bush. When he took the position, he pushed for a "faster, better, cheaper" approach. Four years into his tenure, he was a passionate and irascible defender of NASA and didn't tolerate anyone in any domain calling NASA "mediocre" or "risk averse"—or for that matter, "nay say."

As Peter watched the titans go head to head, he feared that his big event could end before it began. He was angry with Burt for being so antagonistic, but he had seen before how Burt could go from playful to challenging. The men moved forward, but were still sparring like prizefighters heading to the ring. Peter whispered to an aide that he wanted the two separated on the bus. Goldin told Burt that he wanted NASA to experiment with different approaches. Failure was a "way out of mediocrity," Goldin said loudly. Fear of failure "would keep America grounded," and expecting perfection was "unfair to the wonderful people at NASA."

Burt shook his head. "You've got a budget of *fourteen billion dollars*. Why don't you take the money NASA spends on *coffee* at its centers and do what the XPRIZE people are trying to do? Why don't you take one percent of that, or a half a percent—you wouldn't miss it—and just throw it out there for someone to do this stuff? Someone will have a breakthrough, and it would be the best money you would spend while you're running NASA."

Finally, Goldin thawed. He knew Burt was critical of the space agency because his life's work had been inspired by the X-planes of the forties and fifties, and NASA of the sixties. Burt was Burt in large part because of the risks taken by the likes of Chuck Yeager, Wernher von Braun, Alan Shepard, Neil Armstrong, and Buzz Aldrin. What he was saying was that he wanted NASA to keep inspiring.

"I'm here," Goldin said by way of response. "I'm clearly very receptive to wild and crazy things."

On the van heading to the Gateway Arch, Erik Lindbergh and his brother Morgan heard the last of the barbs between Goldin and Rutan. Erik was amused, intrigued, and impressed. He had figured that NASA's administrator would be an agreeable civil servant, but Goldin was the opposite. Erik liked the passion coming from both sides and thought, *Alpha dogs in the presence of other alphas will fight*. The trip to St. Louis had been memorable from the moment the Lindbergh brothers arrived, and they'd been here for less than a day.

Starting at the airport, they were treated like celebrities and surrounded by homages to their grandfather. It had been nearly seventy years since their grandfather had set out from St. Louis to win the Orteig Prize, but his presence was everywhere in this city. It was here he found his backers and believers. The city also happened to be steeped in aerospace history, as the headquarters of McDonnell Douglas, builders of the Mercury and Gemini capsules, the Skylab space station, and the new Delta Clipper.

Unfortunately, Erik was in pain just riding in the small bus. His rheumatoid arthritis was worse than when he had first met Peter and Byron in Kirkland a year earlier. He had slowly warmed to the idea of participating in the XPRIZE, though he remained wary of being a public Lindbergh. Morgan, on the other hand, had taken to the XPRIZE dream right away. He was the youngest of Jon and Barbara Lindbergh's six children and had gone through his own challenges in dealing with the complicated Lindbergh legacy. At one point, Morgan disassociated himself from the family altogether. He found his way back only after reading his grandfather's autobiography. Morgan was moved by how his grandfather's time in the air gave him powerful insights into the vastness of the universe and man's place in it. A practitioner of meditation, Morgan was searching for his own epiphanies. When he read that his grandfather had sat on a beach and studied his own hand as a sort of time travel to primitive life, Morgan's mind drifted to Apollo 13 astronaut James Lovell as he famously looked back at Earth from space, put his hand up to the window, and realized he could hide all of Earth with just his thumb. Morgan was certain that the world needed the XPRIZE; that peace and wisdom were attainable through access to space. He intended to talk onstage about the need to inspire a new generation of dreamers. Morgan had another motivation that went beyond giving a speech: he wanted to help his older brother find his passion again.

As dozens of members of the press and about one hundred invited guests filed into the staging area under the Gateway Arch, Peter took a moment to look around. He noticed Erik Lindbergh, moving slowly to his seat, relying on his cane. He saw astronauts representing Apollo missions 7, 10, and 11, Gemini missions 6, 9, and 12, and a dozen Skylab and space shuttle missions. Burt and Dan Goldin had arrived in one piece, to Peter's great relief, and now the two men were exchanging friendly banter like the best of friends.

Peter had garnered endorsements from key organizations, including the U.S. Space Foundation, the National Space Society, the Space Frontier Foundation, the Society of Experimental Test Pilots, and the Explorers Club. Byron Lichtenberg, a founding member in 1985 of the Association of Space Explorers, had gotten many of the international fliers and astronauts to attend. Peter had snagged commitments from Patti Grace Smith, associate administrator for commercial space transportation at the Federal Aviation Administration, and from a group of St. Louis civic leaders.

Two months earlier, on March 4, 1996, committee members had convened at the historic brick-façade Racquet Club in the leafy Central West End of St. Louis. Taking a page from Lindbergh's playbook, Peter and the XPRIZE backers gathered at the very same table that Lindbergh and his supporters had used to sign their intent to enter the race for the Orteig Prize. Lindbergh had found his support slowly. His first backers were insurance executive Earl Thompson; Frank and Bill Robertson of the Robertson Aircraft Corporation; and Major Albert Bond Lambert, the city's first licensed pilot and an avid balloonist. He heard enough noes for a lifetime. The fund-raising challenge surprised him. He wrote in *The Spirit of St. Louis*: "Aside from Mr. Thompson and Majors Robertson and Lambert, I've found no one willing to take part in financing a flight across the ocean. The men I've talked to who are interested don't have enough money. Those who have enough money consider the risk too great." Lindbergh thought about raising money by popular subscription. "Maybe I could get a thousand people in St. Louis to contribute ten dollars each."

His luck improved when he met Harry Knight, president of the St. Louis Flying Club, who introduced him to others, including Harold Bixby, head of the St. Louis Chamber of Commerce, and E. Lansing Ray, publisher of the *St. Louis Globe-Democrat.* Lindbergh soon had the financial fuel needed to fly. As Bixby handed the aviator a check for $15,000, he asked him, "What would you think of naming it the *Spirit of St. Louis*?"

Some of St. Louis's biggest names showed up for Peter's March 4 organizing event, invited by Al Kerth, a civic leader, senior partner at the public relations firm Fleishman-Hillard, and Peter's newfound guardian angel; Doug King, chief executive of the St. Louis Science Center; and Dick Fleming, head of the St. Louis Chamber of Commerce. Kerth's idea was to get one hundred people in St. Louis to donate $25,000 each—$25,000 was chosen because it was the amount of the Orteig Prize. Seven people agreed to donate $25,000 each, becoming the first members of the "New Spirit of St. Louis" group. The funding was enough to get the XPRIZE up and running. Ralph Korte, president of Korte Co., a construction company based in Highland, Illinois, was the first to write a check. Support also came from Dr. William Danforth of Washington University; Enterprise Holdings' Andrew Taylor and his father, Jack Taylor; Sam Fox of Harbour Group; Hugh Scott, former mayor of Clayton, Missouri; Steve Schankman of Contemporary Productions; John McDonnell of the McDonnell Douglas Corp.; and lawyer Walter Metcalfe.

Peter had also gotten help from his Angel Technologies partner, Marc Arnold, who had moved Angel to St. Louis and was connected with members of the Young Presidents' Organization. Many locals embraced the XPRIZE as a chance to revitalize the city and revisit its most glorious chapter. It came at a time of city renaissance efforts; St. Louis had committed to a multibillion-dollar renovation of its historic properties. At the end of the organizing event in March, Peter raised his glass of gin and tonic to toast Al Kerth, who raised his glass of scotch. Kerth was Peter's Harold Bixby and Harry Knight rolled into one. He had not hesitated when Peter pitched him on the XPRIZE idea, nearly jumping from his chair. "I get it! I get it!" Kerth had said. "Let's do it!" He was proving to be an indomitable force. Kerth had come up with the XPRIZE logo, created the bronze medallions for "New Spirit of St. Louis" members, and hatched the idea of unveiling the prize under the Gateway Arch.

Now, as the final guests arrived at the arch for the May 18 ceremony, Peter took one last look at his checklist. He'd set out to make this event impossible to ignore, something he described as launching "above the line of super credibility." Peter wanted this event to be heard around the world.

The show began with the luminaries of old space and new space, reluctant Lindberghs and born-again Lindberghs, St. Louis old-timers and newcomers. Buzz Aldrin paused on his way to the stage to sign autographs. The stressed steel Gateway Arch glistened in the late-morning sun. Every seat was filled.

Peter, in suit and tie, his parents in the front row—supporting him while not yet fully grasping the importance of a prize for suborbital flight—began, "The *Spirit of St. Louis* carried Charles Lindbergh from New York to Paris and into the hearts and minds of the world. Today, all eyes are on St. Louis again."

To rousing applause, he said, "The XPRIZE has been created for one major purpose, to accelerate the development of low-cost, reusable launch vehicles and thereby jump-start the creation of a space tourism industry."

Gregg Maryniak, who had plotted and planned this event with Peter, listened with pride. Peter was the most relentless person he knew. At the dais, Peter talked about the incentive prizes of the 1920s and 1930s, "the hundreds of aviation prizes that pushed the envelope of speed, distance, endurance, and safety in the fledgling aeronautical industry. In 1926 and 1927 alone, more than $100 million worth of prizes (in 1996 dollars) were offered to challenge the flying community. Today, only seventy years later, aviation is a global multibillion-dollar industry. This is the first-ever human spaceflight prize." Peter told the story of Raymond Orteig and his prize—a competition that was not without casualties. In the summer of 1926, Charles W. Clavier and Jacob Islamoff, two members of Captain René Fonck's flight crew, died when their plane, designed by Igor Sikorsky but grossly overloaded, crashed and ripped apart on takeoff from Roosevelt Field on Long Island. In spring of 1927, U.S. naval pilots Noel Davis and Stanton H. Wooster perished during a final test of their aircraft. Weeks later, on May 8, 1927, French aviators Captain Charles Nungesser and Captain François Coli flew westward into the dawning skies over Le Bourget, France, and were never seen again. While Orteig expressed sadness at these losses, he never wavered from his offering. On May 20, 1927, Charles Lindbergh departed from Roosevelt Field, flying nonstop, thirty-three hours and thirty minutes in a single-engine, single-pilot aircraft to Le Bourget Field outside of Paris. There had been others to fly across the Atlantic, but Lindbergh was the first solo pilot to fly nonstop and connect these major cities.*

Peter expressed his hope that vehicles born from the global XPRIZE competition would "bring about change in the stagnant aerospace world."

When Peter was done, Erik Lindbergh made his way to the podium. At first, he spoke softly, and then he gained confidence.

"I found some notes taken by my grandfather when he was preparing to get funding for his flight across the Atlantic," Erik said. "There are some notes here about why he wanted to make the flight: 'Make America first in the air. Promote and demonstrate the perfection of modern equipment. Advertise St. Louis as an aviation city.' I think there are some great parallels for what is going on with the XPRIZE today." Erik believed that the XPRIZE had the ability to start an industry and bring humanity together. "That's where the XPRIZE has the most potential."

---

*On June 15, 1919, in pursuit of another prize—of 10,000 British pounds offered by the *Daily Mail*—John Alcock and Arthur Brown were the first to fly nonstop across the Atlantic. They flew approximately 1,890 miles across the shortest part of the Atlantic, taking off from St. John's, Newfoundland, and landing in what was described as a "gentle crash" in a bog in Ireland.

Erik offered another insight from his grandfather, this one written as the foreword to astronaut Michael Collins's book *Carrying the Fire.* His grandfather acknowledged the "awareness" that came with scientific and technological breakthroughs, whether with his flight or the push into space. He wrote: "Alone in my survey plane, in 1928, flying over the transcontinental air route between New York and Los Angeles, I had hours for contemplation. Aviation's success was certain, with faster, bigger, and more efficient aircraft coming. But what lay beyond our conquest of the air? What did the future hold? There seemed to be nothing but space. Man had used hulls to travel over water, wheels to travel over land, wings to travel through air. Was it remotely possible that he could use rockets to travel through space?"

In closing, Erik said, "The XPRIZE is an event that has the potential to capture the world's imagination. It has the potential to shift people's interest from conflict and war to an adventurous goal."

Morgan Lindbergh also gave an impassioned talk, but with a focus on the potential and imperative of the XPRIZE to inspire young people. Buzz Aldrin, sixty-six years old, took to the microphone and lamented that close to twenty-five years had passed since man had stepped foot on the Moon with Apollo 17. He hated to see America lose its leadership role in space exploration and was pouring his energy into campaigning for new resources for space travel. "America must dream again," he said. "I am still awed by the miracle of having walked on the Moon. That sense of awe in all of us can be the engine of future achievement."

Toward the end of the ceremony, NASA chief Dan Goldin stepped up to give his endorsement of the XPRIZE. "We need to encourage the participation by as many people, by as many organizations, in this noble venture," he said, wearing an XPRIZE pin on his left lapel. "I hope that my grandson, Zachary, who is two years old, will be able to go with his children on a trip to a lunar hotel."

Gregg Maryniak, taking it all in, believed that the XPRIZE would soon find its benefactor, and the *X* would be usurped by the person's name. Attracting teams, on the other hand, might be more difficult, he thought. Gregg watched the Lindberghs and saw Erik trying his best not to show his physical discomfort. Occasionally, a grimace would make its way through. Erik demonstrated a lot of strength by being here.

As the crowd dwindled and the television trucks rolled away, Peter looked back at the 630-foot-high Gateway Arch, a monument to fur traders and explorers, to the spirit of pioneers. It was the shape of a parabola, the very trajectory he imagined a homebuilt spaceship would one day fly to win his $10 million prize.

Hours after the announcement under the arch, Peter and the XPRIZE crew were cleaned up, dressed in black tie, and at the St. Louis Science Center. The evening gala, cochaired by Buzz and Erik, was to include fog machines, an elaborate laser show, and talks by luminaries. Tickets went for $500 per person.

When it was time for the dinner to begin, Hollywood producer Bob Weiss found his seat in the tented dining area. Bob, who had met Peter the year before, found a certain poetry to the entire day, with the army of astronauts, captains of industry, and the parallels to a sixty-nine-year-old dream. Bob had produced a range of films, including *The Blues Brothers*, and had a new science fiction TV series out called *Sliders*. He was a self-professed space geek who went to space conferences and had grown tired of listening to pessimistic projections of when man would return to space. He had found Peter's XPRIZE idea brilliant: offer the right incentive, use Darwinian forces, and stimulate innovation. It was taking human nature and marshaling it for a specific purpose.

Settling in at the table, Bob soon began to wish that he could direct this event. He was happy to be across the table from Burt Rutan and wanted to hear what he had to say, but with no warning, the fog machine would come on and make half the table disappear. People said Burt's mind was in the clouds; now he was in a cloud. Once the fog dissipated, Al Kerth gave the welcoming remarks and introduced a narrated film that combined clips from Kitty Hawk, Lindbergh arriving in Paris, Neil Armstrong and Buzz Aldrin walking on the Moon, and Burt Rutan at his drafting table creating what would be the *Voyager*. Peter spoke on video, saying, "Sixty-nine years ago, Charles Lindbergh and the *Spirit of St. Louis* changed the way people think about air travel. The XPRIZE is trying to change the way people think about space travel." There were humorous clips of Jimmy Stewart as Lindbergh, talking with Robert Cornthwaite, who played Harry Knight.

Knight, looking serious, said to Lindbergh, "Slim, you understand we have to make sure we're not financing a suicide."

Lindbergh replied, "The idea of suicide never crossed my mind."

"Except you're flying over the ocean," said one of the men at the meeting.

Lindbergh responded, "But the idea is not to set it down on the water. The idea is to set it down on Le Bourget."

"Will this stimulate aviation?" another man asked. "I mean, a man went over Niagara Falls in a barrel. Did that stimulate an industry?"

"That was a stunt," Lindbergh replied. "I'm not a stuntman; I'm a flier."

Next up was a video message from Peter's longtime supporter Arthur C. Clarke:

"I'd like to send my fondest greetings to Buzz and Peter. I recently had the pleasure of having Peter here. He explained the commitment you have made to launching a new era in private space travel. Thirty years ago, Stanley Kubrick and I made this little movie, *2001: A Space Odyssey*. We predicted by that time, space tourism would begin and if you had money, anyone who wanted to could go to orbit. Sooner or later this will happen, and I hope the XPRIZE will contribute to that. I think I may need to revise my predictions to the date 2004 instead of 2001."

Clarke smiled and went on, "It's always been our nature as humans to explore our surroundings and turn frontiers into future homes. Now, space beckons. During the birth of the space age, it was the competition between the former Soviet Union and the U.S. that drove it so far and so fast, from Yuri Gagarin's flight in 1961 to landing on the Moon only eight years later. It is my belief that the XPRIZE will reintroduce in a constructive fashion this element of competition. I invite teams from every nation in the world to lay their plans and begin the competition for the prize. May the best team win. I am Arthur Clarke, signing off in Sri Lanka, to you in St. Louis—to be known one day as the gateway to the stars."

Peter grew anxious when it was time for Burt to speak. Burt had told him he planned to talk about the importance of prizes in aviation history. Peter just hoped that Burt would rouse the crowd, not roil them.

At the podium, Burt, wearing a tuxedo and a New Spirit of St. Louis medallion on a ribbon around his neck, opened by saying he wanted to share "what's in my heart" when it comes to the meaning of the XPRIZE. Peter grew even more nervous.

"Imagine something that didn't happen but could have happened," Burt began. "Back in the golden age of the development of aircraft, back when people had this fantasy to leave the Earth and fly through the atmosphere, we had XPRIZEs—we had a lot of them. Over a tiny amount of time from the Wright brothers to when you could buy a ticket to Chicago or have a private airplane to enjoy the skies. But let me imagine . . . let me ask you to imagine . . . what if in those days, we didn't have XPRIZEs? The prize for the first flight over the Alps. The prize for a flight across America. What if we had, between 1903 and 1920 or 1930, a government-owned, government-developed, government-flown program, where they are the only ones who could go into the air? You would have seen large and extremely expensive craft. You would have seen the government's airplane fly with seven pilots, and only those who had worked for fifteen years got to fly twice. That could've happened.

"I feel not just embarrassed as an American citizen, in a society that is supposed to be free, but frankly I am mad as hell that we have the kind of limitations we do for us to leave the atmosphere.

"I believe seriously that this XPRIZE is what is going to break that open. I have seen myself, personally, all of a sudden, get extremely creative in design. I have dreamed of making a homebuilt spacecraft since I've been doing homebuilt aircraft. That's since 1968, when I started on the VariViggen. But I have never, by myself, been as creative as I have been in the last couple of months, eyeballing this goddamn prize.

"I am not going to tell you what I've come up with because I want to win this thing, but I am going to tell you I'm not the only one that's going to be creative. I'm going to tell you to try to think about something entirely different from what you imagine to be a spacecraft. It's not a throwaway Atlas. It's not a space shuttle. The guys who were barnstorming in the old days, they'd fly a Jenny [an early 1900s biplane] over and land in a field and give people rides for two dollars. Could they have imagined a 747 or a Concorde or TWA's baggage system?" The crowd laughed. "Think about it: Did they have that kind of info? Could they have imagined a Bonanza or a Long-EZ? They had no idea. I'm telling this crowd tonight I have myself just got a touch of this.

"I'm looking out here at some very sharp entrepreneurs who are going to go after me like crazy, and they're going to have phenomenal breakthroughs. What's going to happen is way beyond our imagination, and it's going to happen very soon. It's going to create the best roller coaster in the world. We'll be sending people to orbit. We're going to the planets and the stars—and we're going to do that because of Peter Diamandis."

Burt got a standing ovation. Peter was stunned by Burt's compliment, and by what Burt said about the prize. Was he serious? Had Burt Rutan just announced he was a contender for the XPRIZE?

Later, when most of the dinner attendees had gone home, Peter lingered with a close group of friends and family. It was after midnight, and Peter shared the story of how he had first met Buzz Aldrin eight years earlier. He told them how Aldrin had agreed to talk at the founding summer conference of the International Space University. After his talk, Buzz had dinner with Peter and some of the students and faculty. They ended up at the MIT faculty club, where they spent five hours, captivated by Buzz's stories.

Putting his weary feet up on a table, Peter asked Gregg for his thoughts on the night. Gregg was amazed, he said, that everyone who RSVP'd yes actually showed up. He looked at his hands, with more than their share of nicks and cuts from stuffing envelopes, and said, "We're not far from our SEDS days of getting paper cuts after midnight." On the table were copies of their eight-page, full-color invitation. The cover image was of a family of space travelers standing in front of rockets shooting at different trajectories. The rockets were given names: John Galt, Byron Lichtenberg, Doug King. The biggest of the rockets was reserved for their friend Todd B. Hawley.

As Peter gathered his belongings and his strength—the man with seemingly limitless energy was finally exhausted—his mom and dad appeared with a birthday cake blazing with candles. His name and the XPRIZE were on it. Peter was about to turn thirty-five, and the XPRIZE had just come to life. He made a wish and blew out the candles. No one had to ask what he wished for.

14

The Space Derby

Thin, pale, and prone to dressing in Zombie Apocalypse T-shirts, John Carmack was still in his "larval stage" of rocketry when he first heard about the $10 million XPRIZE, news of which was quickly making its way around the globe.

The twenty-seven-year-old Carmack had long been fascinated with space and had dabbled with Estes rockets in middle school, but his interest then went dormant for years as he focused on computers, coding, and video games. Once called "a brain with legs," the multimillionaire Carmack had attracted a cult following for creating some of the hottest games of the computer underworld, including *Commander Keen, Wolfenstein, Quake,* and *Doom*. As a founder of id Software in Mesquite, Texas, he had helped pioneer the first-person shooter genre, giving players the feeling of being inside the game.*

Carmack had started designing video games when he was in the third

*First-person shooter games are made by building a perspective picture, such as of a scary basement. Then as the player moves the controls, the perspective changes, creating the illusion of turning or moving forward or backward. The problem is to store all the textures and images that compose the constantly changing scenes.

grade, when his favorite pastimes were reading comic books and *The Lord of the Rings* and playing *Dungeons & Dragons.* He liked the logic of computers. Unlike his home life—where his games were dismissed as a waste of time—and unlike the religious "myths" he was taught in parochial school, computers made sense to him. There was no magic, no mystery. He might not understand something at first, but if he put in the time, he could eventually unravel every seemingly weird and quirky behavior. He could look at a computer and understand its operating system, interconnect protocol, compiler, chip set, and peripheral hardware. The beauty of programming was in the knowledge that a computer could be controlled. The specific programming language was less important. Carmack felt less at odds with the world when he learned to command computers.

But by 1997, around the time he released the source code for *Doom 3,* Carmack felt he'd achieved all he could in the realm of video graphics and games. He had built fast games and bought fast cars—a Ferrari Testarossa and Ferrari 328, to which he added a turbo engine—and he was beginning to look around for the next hot rod to hack. So when one of his programmers at id Software, Michael Abrash, started feeding Carmack some old Robert Heinlein books—*Stranger in a Strange Land, The Moon Is a Harsh Mistress, The Man Who Sold the Moon,* and *Rocket Ship Galileo*—it was not long before Carmack's passion for rockets was reignited. Carmack loved the archetypal Heinlein hero, the fierce individualist who uses technology to solve problems. He believed to his core that building things was the best way to shape the world.

In short order, Carmack was shelling out thousands of dollars for seminal rocket textbooks, including *Rocket Propulsion Elements* by George P. Sutton, and *Modern Engineering for Design of Liquid-Propellant Rocket Engines* by Dieter K. Huzel and David H. Huang. The rocket treatises filled bookshelves and consumed his mind. Carmack pored over NASA publications from the early sixties and seventies, impressed by the nuts-and-bolts descriptions of what worked and didn't work for the Mercury, Gemini, and Apollo programs. The early NASA studies were dazzling in their details, right down to diagrams of welding procedures. NASA publications from more recent times proved the opposite: full of meta-analysis of surveys of simulations. During his embryonic phase of rocket study, Carmack also looked online for tutorials and signed up for all the right mailing lists.

When Carmack got to the point where he felt like he could hold an intelligent conversation with industry types, he began attending space conferences. At first, he wandered around asking questions without anyone knowing who he was. He was just another space geek who dreamed of the stars but looked like he never felt the warmth of the sun. Soon, though, whispers circulated that he had the magical designation of an "accredited investor," a high net worth individual who could make risky investments. Suddenly, around every corner came a new pitch.

That's when Carmack discovered the XPRIZE. He also learned about the $250,000 CATS (Cheap Access to Space) Prize, which would be given to the first private team to launch a 4.4-pound payload into space, 124 miles or higher, by November 8, 2000. Carmack wasn't sure whether to finance a team for one or both contests, but he was certain of a few things: He wasn't in this for the money—none of his work had ever been about the money. Truth be told, he relished the idea of entering a new arena where many of the folks would know a lot more than he did. He would always love computers, of course, but it was time to look beyond the screen in front of him.

In Bucharest, Romania, Dumitru Popescu had just sat down in an Internet café to browse a Web site called Astronautix when he saw something about the XPRIZE. The twenty-year-old aerospace engineering student had come to the café to research the liquid fuel used in SCUD missiles when his attention was drawn to the words *ten million dollar prize* and *suborbital flight*. By the time he got a few paragraphs in, he downed his espresso, hurried to find his phone, and called his wife.

"You need to come to the café," he told her, as they didn't have e-mail at home. "I have to show you something." His wife, Simona, arrived within the hour, took a seat, and read the story. Before she could respond, Dumitru said, "Let's try to do something on our own."

She could see he wasn't joking. "We are students," she said by way of protest. "We have no money. We don't have a thousand dollars to even *enter* the competition." Dumitru shook his head. "Let's start building something and maybe we will be able to attract sponsors."

That afternoon, Dumitru went from the Internet café to the campus of the Faculty of Aerospace Engineering at the University Politehnica of Bucharest, where he was a sophomore. He wanted to discuss the XPRIZE with his peers and professors. The reaction he got was "This is unreachable for us," and "Forget it." But he couldn't forget it. His father was a policeman, his mother an accountant. Life seemed more about drudgery than dreams. He had a degree in theology, which was his parents' idea. Now he was studying aerospace, which was his idea. He wanted to build manned rockets, and feared the Romanian space agency was making no progress in this direction.

Soon, Dumitru had persuaded his wife to help him. Together, they talked her father into letting them build their rocket in his backyard in a small town west of Bucharest.

Thirty-two-year-old Pablo de León was talking with friends in Buenos Aires about the lamentable monopoly on space by the world's largest governments when he heard of the XPRIZE. "This is right out of science fiction," he said excitedly. After researching the prize, he confided his interest to colleagues. He had founded the nonprofit, nongovernmental Argentine Association for Space Technology, and his friends there told him, "Pablo, listen, don't even think about it. This is something that just isn't possible. You'll burn your credibility."

But de León couldn't shake the idea. When he read that Burt Rutan was a possible contender, he said to himself, *Boy, I have to do that. If Rutan is thinking about it, then it's serious.* He already knew of Peter Diamandis as a founder of the International Space University. De León had wanted to attend ISU for many years, but couldn't afford both the flight and tuition. After eight years of applying, he had finally been accepted on full scholarship. He believed the stars were aligning in the direction of the XPRIZE.

De León had been captivated by space for as long as he could remember. When he was five years old in July 1969, his parents woke him up in the early morning hours to watch the lunar landing of Apollo 11. There he sat with his parents and grandparents on their farm in Cañuelas, Argentina, watching man walk on the Moon. He remembered the dim lighting in their small family room and the quiet that fell over the family when images of the astronauts appeared on-screen. They were one of the few families in town that had a TV. By the age of nine, he was launching homemade rockets in the pasture out back, scaring the cows and sheep and getting stern warnings from his parents. He went on to design and build spacesuits used at Kennedy Space Center and wrote two books about the history of the orbital efforts in Argentina, the only country in South America with a space program.* At the time he heard of the XPRIZE, he

*The Argentine space program consists of two main components: scientific satellites and the development of a small launch vehicle, the Tronador. The country also has two communication satellites in geostationary orbit, launched by European Space Agency Ariane rockets. They've done a few scientific satellites, mainly in collaboration with NASA; however, the first Argentine satellite was launched from Russia. The last Argentine scientific satellite was launched from the United States on a Delta 2 from Vandenberg Air Force Base in California. The Argentine government space program is not pursuing human spaceflight as one of its priorities.

was working from Buenos Aires as a payload manager for seven experiments to be launched on the space shuttle in early 2000.

Pablo would find a way to enter the contest. As the designer of any future craft, he would surely have to fly it, too, he told himself happily. He could picture the rocket, and he already had a name for it: *Gauchito,* for Little Cowboy.

Steve Bennett was working as a laboratory technician for the toothpaste company Colgate in Manchester, England, when a coworker showed him a small newspaper story about a prize being offered from America for anyone who could build and fly a rocket without the government's help to the start of space. Bennett had taught himself the ins and outs of rockets, from types of propulsion to ablative skins. He was thirty-three years old, but he had never forgotten how his parents had refused to wake him up for the Apollo 11 landing in 1969, when he was five. His mother had told him that he had school the next day, a Monday, and assured him there would be "plenty of rocket launches to see later." By the age of thirteen, he was on a first-name basis with workers at the local chemical supply company, buying ingredients to propel his homemade rockets. As a young adult, he watched in dismay as England's auspicious space program regressed. In its aeronautical heyday of the sixties, England had a sophisticated and impressive satellite carrier rocket called Black Arrow. The rocket was retired in 1971 by decision makers who opted to go with less expensive American-made vehicles.

Now, as Bennett toiled in the toothpaste factory, he saw rockets in the shapes of the tubes and propellant in the texture of the paste. He wanted to live and breathe rockets, but he couldn't leave Colgate until he had another source of income, however paltry. His wife wouldn't hear of it.

To Bennett, humanity's destiny was outside Earth and required bravery, exploration, and adventure. He wanted to see space, with its luminosity and darkness, milky patterns, planets, stars, galaxies, and nebulae. On breaks from work, he began sketching his rocket. It would be at least forty feet tall and have a capsule on top. Like the boy with the purple crayon, he drew himself into the cockpit.

There were many others rumored to be gearing up to make a run for the XPRIZE. Brian Feeney, an inventor from Toronto, was living in Hong Kong when he happened upon a story about the prize while browsing his local newsstand. Geoffrey Sheerin in Canada was already said to be designing the ultimate homebuilt hot rod, a fifty-four-foot-long rocket modeled after Germany's V-2 missile. A team in Russia was reportedly interested in using a solid engine to power something resembling a mini space shuttle. A former NASA propulsion specialist in Texas was building a ship in a rice field. The vehicle would launch vertically from water and land horizontally like a seaplane. In California, Peter's friends Gary Hudson and Bevin McKinney—who had been at the John Galt gathering in Montrose, Colorado—were working on a Buck Rogers–style ship, only with helicopter blades on top.

Even the grandfather of rockets, Bob Truax, who had been pushing for space tourism since before Peter Diamandis was born, was eyeing the race. The eighty-five-year-old Truax* had a forty-foot-long spaceship, two fuel tanks, and a rocket engine in storage just outside of San Diego. Having designed and built Evel Knievel's Skycycle, Truax had known both success and failure. At an age when most of his peers were

*Truax was involved with some of the highest-profile U.S. military rocket programs of the twentieth century: Thor, Viking, Polaris (submarine missile), and Sea Dragon. He was interested in space tourism, with his "Volksrocket." And he built Evel Knievel's Skycycle, a steam-powered rocket with wheels that Knievel used to try to launch across the Snake River Canyon in 1974. The parachute deployed early, and the vehicle drifted back to the canyon floor.

relaxing into retirement and a round or two of golf, Truax confessed, "I just like to go out and play with rockets."

In trying to decide which rocket team to finance, John Carmack first reached out to contenders for the CATS Prize. This was a much less ambitious contest than the XPRIZE, but the methodical Carmack had always taken measured steps—and the smaller prize was the perfect place to start.

Carmack found out that the CATS Prize was in response to the XPRIZE: the money was put up by Walt Anderson, who'd had the bitter falling out with Peter during their time together at International Microspace. The smaller prize was being run by Rick Tumlinson, president of the Space Frontier Foundation and a longtime space advocate who had resigned from the XPRIZE board because Peter had announced the award without having the $10 million. Tumlinson told Carmack that he'd said to Walt, "If you want to poke at Peter, endow this little prize."

Carmack didn't know Peter or want to get involved in any intramural fights; he was focused on looking carefully at the CATS competitors. He sent letters out introducing himself and saying he was interested in possibly sponsoring a few teams. Several teams never bothered to respond. After interviewing the others, Carmack came to a conclusion: he found that a great number of people in the space community were out of touch with reality. Several team leaders told him how they would use his money, but didn't offer any plans for testing hardware. When Carmack asked about this, he was told by one team, "We're not going to tell you how to make video games, so don't tell us how to make rockets." He repeatedly heard how easy it was going to be to win the CATS Prize, but few people were building anything. He met people with thirty years of experience in the industry who had never screwed a nut onto a bolt but insisted, "There are no technological challenges. All we need is funding." One company founder told him he needed "a million dollars to turn the lights on." Carmack talked with Patrick Bahn of TGV Rockets and was impressed by his business plan, but worried how the company could go for so long without having a rocket to show for it. Carmack marveled, too, at how space professionals attended conference after conference, presenting the same presentation decks with only minor tweaks. After months of research and interviews, Carmack drew up a list, putting teams into three categories: Loony, Unrealistic, Maybe.

He ended up funding a few groups, including XCOR Aerospace in Mojave, which he found to be the best of the bunch. He gave $10,000 to JP Aerospace, a volunteer-based, do-it-yourself effort that balloon-launched high-powered rockets—or "rockoons." Carmack went to a few launches of hardware by other companies, including one in the Black Rock Desert of Nevada, where he waited hours and hours only to see the vehicle blow up on the pad. He favored that over doing nothing. When he saw hardware built and launched—even if it failed—he was happy. Failure meant learning.

As he continued working full time at id Software and sponsoring the efforts of a handful of rocket makers, Carmack began studying the programming required for rockets. He delved into high-reliability programming, designed not to fail under stress or regular use. NASA specialized in designing systems with three duplicate computers calculating the same thing, such as navigational positioning or main engine burn. For safety, the engineers would compare the three results and use the majority result. Aircraft control systems were subjected to automatic verification of no errors. Error-correcting systems had to tolerate storage and network errors by using redundant data encodings. In a way, game code and rocket code were like the same play cast in two different ways, two apparently different objects with substantially identical underlying structures.

To Carmack, game development was far more complicated than rocket development. Games involved millions of lines of software code. Games contained more program objects. A single game might contain

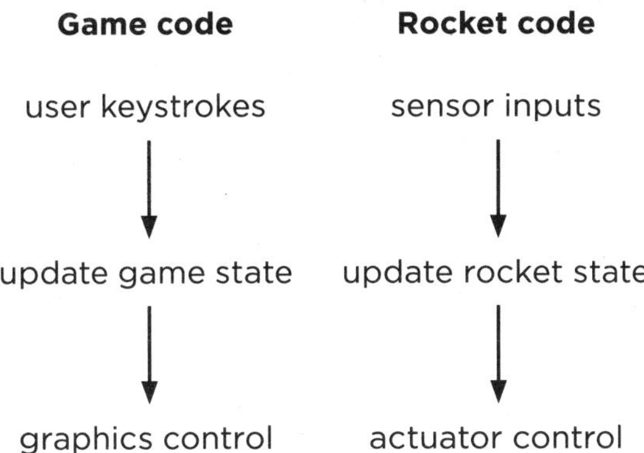

thousands of items that needed to be tracked, updated, and rendered. A rocket, by comparison, had a small set of sensor inputs and control surfaces, such as the angle of the rocket nozzle. Reliability requirements, though, were biased the other way. A game flaw had few repercussions. Nobody got hurt if something went awry. But a bug in a rocket control code could cost millions of dollars and endanger lives. In the end, the rocket code was smaller, but needed more validation.

Carmack wanted to make the programming of rockets more like software development. He didn't want to write software the way the space shuttle team did it, where everything was reviewed to death and changed every few weeks. The method could work, but it was cost ineffective and schedule ineffective. His observations and interviews also taught him what he didn't want to do if he started his own rocket company. He didn't want to work for six months to a year, make a pilgrimage to the desert, push a button, and see something go wrong. He wanted to build and test something new every week, to let problems express themselves. He wanted to be open source with rockets as he had been with video games, posting all he learned, right down to where he'd purchased the parts. The breakthroughs of the Internet, personal computer, and smart phones came from a production-efficient method in which failures were expected and iterations were the norm.

One day in his office, Carmack found himself studying the exposed ductwork on the ceiling. He thought, *Rockets should be made by spiral welding tubes. Then you wouldn't have the hoop stresses in there and you could build a pressure-stabilized vehicle like the Atlas. You could build a Saturn V out of sheet metal.* Returning home that night, he waded through boxes of rocket parts filling his garage and hallway, and through the rocket textbooks stacked on the hood of one of his Ferraris. When Carmack had first started his research, he had assumed that he should finance other rocket builders, but he had slowly realized that many of the so-called experts had no idea what they were talking about. During his computer career, Carmack had turned the video game industry inside out; he wondered if he could show the aerospace industry how to build spaceships in a faster and cheaper way. He knew it was time for him to stop watching and start doing.

Carmack reached out to the president of the Dallas Area Rocket Society to see whether anyone there would be interested in joining him in developing and building experimental rockets. He was given some names, and began to think about the team he needed. The best times at id Software involved overloaded circuits that made their basement office go dark, pizza- and Coke-fueled all-nighters, and a belief they were doing something entirely new. Over the years, that passion was replaced by schedules, output, production, and professionalism. With rockets, he would be back to tapping unexploited energy and traveling into the unknown.

15

Epiphanies in the Mojave

Dick Rutan was in the pressurized capsule of a giant hot air and helium balloon, climbing to a cruising altitude of 30,000 feet. The higher he went, the calmer he felt. It had been a year of nonstop planning and building to get to this point, day one of attempting to be the first to circumnavigate the globe in a balloon.

It was early January 1998, twelve years since Dick's historic around-the-world *Voyager* flight, which had gotten him an invitation to the White House and speaking engagements across the globe. But his newfound fame as a pilot hadn't made Dick any less hungry for his next run at the record books.

Ten minutes into the balloon flight, which had begun in New Mexico, everything felt right. Dick monitored the carbon dioxide scrubber he had built to balance the nitrogen and oxygen in their hermetically sealed, eight-foot-diameter carbon fiber sphere. His copilot, Dave Melton, controlled the helium release during ascent, creating the right amount of buoyancy.

Dick checked his Inmarsat satellite communications and radio altimeter, and continually fine-tuned the pressurization in the chamber. He removed his gold Rolex watch—a gift from a sponsor—and replaced it with his trusty Casio, which could do things the Rolex couldn't. He set the Rolex on a shelf. As the balloon reached its target altitude, the buoyant force equaled the craft's weight. Melton removed his boots, tucked them away, and put on his slippers. It was time to relax a bit—this balloon would be their home for the next month. Dick had constructed the balloon's capsule, which his brother Burt had designed, and now the craft was in the stratosphere, heading east. "Cruisin' now," Dick said. On their way.

Suddenly—*BOOM!*

The capsule floor jumped up like a trampoline. The bottom of the helium cell ruptured. Parts of the inside of the balloon hung in shreds. They were falling. Not like a freefall, but going down. Dick was on the radio: *MaydayMaydayMayday! It ruptured... we're on our way down.* He grabbed his knife to cut the survival gear free, but stopped himself. *I'm going to use this parachute,* he thought. *I better not cut it with a knife.* The sounds were not comforting—ripping, tearing, rushes of air. A rip to the very top and the helium would release. They would zigzag to the ground like a small party balloon with helium let out. The three-story-high silver balloon would be a flag marker for the point of impact.

This is the end of a nice day, Dick thought drily.

"Can you continue to fly it?" Mission Control asked. Dick and Melton depressurized the cabin and started throwing things out to slow the descent. They pulled the emergency helium release valve and used the long rope to tie it off and keep it open. The two men and the mission controllers went back and forth on what to do, until Dick made the call: "We have an airborne structural failure... it's deteriorating... we're going to bail out." He had bailed out of burning planes in Vietnam and parachuted into hostile Viet Cong territory. He could get out of a balloon above the New Mexico desert. Dick helped Melton into his boots and parachute. Melton peppered him with questions: *When do I pull the rip cord?... How much time do I have to wait?* Dick grew mad as hell: Melton had told him earlier that he had done thirty-five parachute jumps. Dick had asked because he wanted to make sure that Melton had the training—in the unlikely case they had to bail out. *His experience appears to be slim to none,* Dick thought as they stood at the edge of the capsule.*

Melton said repeatedly, "Don't hit me, don't hit me, don't hit me."

"What the hell do you mean?" Dick asked. "You think I'm going to coldcock you with my fist?"

"Don't fall on me when you jump," Melton said.

Dick shook his head. *He doesn't know jack shit about this. The guy has no concept of separation in skydiving. This is not good.* The winds were at least 40 miles per hour on the ground, dangerously high for a parachute landing. He looked up: the aluminized Mylar was shredding on the inside, with strips pulling like tentacles. Melton asked more questions. Dick realized he just had to get him out and under a parachute.

Dick told him: "Keep your head down. Arms in. Jump. Wait a handful of seconds. Pull the rip cord down by your crotch. It should open."

Finally, Dick said, "Go!" Melton jumped, pulling the cord almost as his feet left the capsule. Dick didn't have time to tell him to land backward in high winds and always protect the arms. *I don't have time for Parachuting 101,* he thought.

Now about six thousand feet above the ground, Dick got himself ready to jump. There was no time to retrieve his $5,000 Rolex. He saw the pilot Clay Lacy and business magnate Barron Hilton in Hilton's Learjet. Hilton was sponsoring Dick and Melton's flight, and a half dozen other teams in various locations were vying to be the first to race around the world in a balloon. A film crew was following along. Dick stood ready to jump. *I'm going to wait until they come close, so as I jump I'll be right in frame for the camera,* he thought. *You gotta make the best of a bad day.*

---

*In Melton's defense, he was a last-minute replacement for the flight. Dick had worked for over a year with balloonist Richard Abruzzo, who quit the race only weeks before the scheduled departure. Melton was a talented and experienced balloonist, but he and Dick had little time to prepare.

Dick had taught people how to jump out of balloons and planes. But he was distracted—by the film crew, by the shredding balloon. He kicked off when he jumped, just as he told students not to do, and started falling and rolling. He tried to grab air. As he plummeted, he yelled at himself: *Don't flip over! Don't flip over! Don't do it!* He flipped over. *I'm pissed.* Then he got himself into a freefall position. He turned and maneuvered, and loved the feeling. He was caught up in the reverie of flying, free and unencumbered, until he realized he had flown to terminal velocity of 125 miles per hour and couldn't pull back. *Crap, this is really going to hurt,* he thought, pulling the chute. He had an emergency parachute, designed to open fast. He shouted: *Holy bananas!* He was sure he'd have crotch burns and raspberries for life.

Dick surveyed the terrain of eastern New Mexico: patches of snow, desert, scrub brush, sloping hills, roads, power lines, cows, a few pastures. There were high winds on the ground. He had not had time to tell Melton that if you land forward going forty miles an hour, your feet hit, you break your toes and kneecaps, you slam down hard, you hit your face. If you put your hands down—your natural instinct—you'll injure your arms and fingers and you won't be able to disconnect from the canopy. You need both your hands. If you can't land backward, there's a chance you'll be dragged to death through the desert.

It was quiet as Dick neared landing. He could hear the ripple of air going through his parachute. He looked between his legs and saw the ground going by in a blur. He turned backward—an uncomfortable feeling. *Resist the urge to turn around! Don't turn around! Low enough now. Elbows in. Ground close, no turning now. Grab the four risers! Hold them.* Bam! *Feet hit. Pull the risers across your chest!* His back slammed to the ground. Then his head. *Did my helmet crack?*

Dick lay motionless. He looked up and around. *I'm in a goddamn cactus patch,* he muttered. *Cholla cactus. The most hated cacti around. A cactus that grabs you, that has a straight needle that turns and hooks under your skin. Cholla all over my face.*

He'd crashed his motorcycle enough to know that seconds could pass before the pain really kicked in. He waited. The cholla on his face and hands was one thing. *There isn't any horrible pain elsewhere,* he thought. Ops check time: *Neck works, okay. Hands work, okay. Knees work, okay.* He looked to his feet—the big test. *Can I move my feet? Yes! The central nervous system is working.* His right hand was full of needles, which left him in a quandary. *Do I take the hand already full of needles to get needles off my face, or do I use the good hand, and get needles in that, too? Maybe I can find a knife.*

Seconds later, a television news helicopter landed. Soon, a cameraman—with a huge camera on his shoulders—walked around him in a big circle. He didn't say, "Hey, Dick, are you okay?" He filmed everything and said nothing. Dick soon heard the helicopter shut down and told himself the pilot would come to his aid. *Okay, a fellow aviator is going to get out and help me.* The helicopter pilot appeared and looked at Dick, wearing his jumpsuit and helmet, and said, "You okay?" Dick, still supine, responded, "Yeah. How about getting this fucking cactus off my face?"

Wearing gloves, the pilot carefully extracted the cactus. Dick's nose and cheeks looked like they'd been in a fight with a tiger. The pilot took Dick's hands—needles now removed—and carefully, slowly, pulled him to his feet. Dick was able to walk away. Melton wasn't so lucky. When he landed, his femur went through his hip socket, and he went end over end. But Melton was alive. He would be okay.

Their balloon eventually hit a power line in north Texas that cut the envelope loose. The envelope landed in a cow pasture. The capsule, infused with 100 percent oxygen and propane, caught on fire and burned like a cauldron when it crashed down. Dick and his crew chief, Bruce Evans—who had been on the *Voyager* team—went to collect it. There was no sign of the Rolex; nothing at the crash site stood higher than a pair of shoelaces.

Erik Lindbergh sat at a table in a restaurant in Mojave and listened in awe as Dick Rutan told the jaw-dropping story of his attempt to fly the Barron Hilton balloon around the world. It was one of the most incredible adventure stories that Erik had ever heard, one that got better and better as it got worse and worse.

Dick was recounting his balloon misadventure in mesmerizing detail for a small XPRIZE event that had drawn board members, local aviators, and rocket makers to Mojave. Erik was happy to be there; the XPRIZE was a fresh, daring venture. Attending events like this one helped him forget his own misery. Erik repositioned himself in his chair. He was recovering from a fusion of the talonavicular joint in his right foot, a surgery that involved screws and a bone graft. Somehow, Dick's colorful storytelling made Erik's pain more bearable.

Dick told the gathering that if anyone completed the around-the-world balloon journey, a $1 million prize would be offered by Anheuser-Busch, the beer company. But he emphasized that money was never a motivation for him.

"I did this for a milestone," Dick said. "There are a certain amount of milestones in aviation. We looked at the *Voyager* as the 'last first.' People set all sorts of *records* for speed, altitude, and distance. But there are only certain events that happen that are *milestones*." He ticked off a few: the first Moon landing; John Alcock and Arthur Brown's first nonstop transatlantic flight in 1919, crossing from Newfoundland to Ireland; Leigh Wade's flight around the world (in segments and with four planes) in 1924; Lindbergh's New York–to–Paris flight in 1927; British captain Charles Kingsford Smith's first flight across the Pacific Ocean in 1928; his *Voyager* flight. "These are events that are landmarks, that mark a change in development, in what is possible. It's more than a record, where, okay, you get your name in a record book."

Other participants in the balloon race had included Steve Fossett in his *Solo Spirit*; Richard Branson, the billionaire head of the Virgin Group, in his *Virgin Global Challenger*; Swiss psychiatrist Bertrand Piccard and crew in the *Breitling Orbiter*; and Kevin Uliassi in the *J. Renee*. So far, no balloonist had ever come close to achieving such a distance—more than 25,000 miles—and several had died trying.

"I didn't know anything about ballooning when I started," Dick told the group. "I started asking questions, and talked to some people and said, 'We oughta fly a balloon around the world. I know something about flying around the world. Why not?'" Dick said the problem with the Barron Hilton balloon was a manufacturer's defect that caused the helium cell to rupture. He was already working on building a second capsule for his next attempt. He had a new constrained volume helium lifting system, and a new name—*World Quest*.*

Erik stretched his legs. *This guy has courage,* he thought. He spends a year building the capsule by hand in Mojave, planning, getting ready, lifting off. He would fly over hostile areas with uncertain clearance: Russia, China, Afghanistan, Iraq. Then, about twelve minutes after hitting the stratosphere, just as he thinks they're off and cruising—the balloon explodes. And it intensifies from there, right down to landing in a massive cactus patch. But he makes the most of it. *You gotta make the best of a bad day,* Dick had said.

Erik's rheumatoid arthritis brought a lot of bad days. He was still trying to figure out how to operate in a world that he couldn't attack physically, as he was used to. He was figuring out how to make a living when he didn't know whether he would wake up able to move or not. He'd had

*In 1999, the team of Bertrand Piccard and Brian Jones became the first balloonists to circumnavigate the globe with a nonstop, nonrefueled flight. At that point, once they succeeded, Dick Rutan was no longer interested in making the flight.

the one foot operated on, and was going to have to have his left foot fused as well. Dick ended his talk by sharing a story about his brother.

"There are two things I always say about Burt's designs: no way is this going to work, and no way can we get it done that fast. Then Burt comes back and uses his favorite saying, 'Gray today, white tonight.' The plane is gray now, but you can paint it white tonight." Dick laughed. "We'd be like, 'There's no frickin' way. It's twenty percent done!' But Burt would stay with his 'white tonight' mantra, and more often than not, miracles happened."

The talk of miracles stayed with Erik. Change *was* always possible, always right around the corner. Erik told himself, *Gray today, white tonight.*

A few months later, on July 25, 1998, Mike Melvill walked out onto runway 30 at the Mojave airport to take his first flight in the Proteus, the Angel Technologies high-altitude experimental plane developed to deliver broadband services. The plane's design had evolved from solar-powered to twin-engine early in development. Burt knew that a solar-powered plane wouldn't be strong enough or reliable enough to carry and power the payload. It also had to fly at night, when there is no solar power. If all went according to plan, the completed Proteus would fly higher than Mike had ever gotten close to flying. The all-composite plane with graphite-epoxy construction was designed to fly to above 60,000 feet and carry a large, downward-pointing antenna.

The plane was beauty and beast; big yet delicate. After getting into the aircraft, Mike and flight test engineer Pete Siebold taxied down the runway. The Proteus flexed, twisted, and bent. When Mike tested the brakes, even the wide-stance landing gear seemed to flex back and forth. He took his time, taxiing it around the Mojave runways, becoming familiar with the plane's behavior on the ground. When he was comfortable, he moved on to do the plane's first high-speed run along runway 30. He carefully lifted the nose wheel off the runway, reduced the power, and let it run the length of the runway on the main wheels. He got a sense of how the plane would look when he touched down his main wheels for the first landing. Approaching the end of the runway, he lowered the nose wheel to the runway and applied the brakes. He turned the Proteus around and used more power, rapidly accelerating to Burt's predicted liftoff speed, then reduced power to maintain the speed and pulled back on the side stick control, lifting the Proteus just a couple of feet into the air.

Mike smiled. As he flew toward the end of runway 12, still only a few feet in the air, the plane felt remarkably controllable. He gingerly tried small inputs of all three axes of control, and was happy with the handling qualities in ground effect. In time, he reduced the power to idle and gently landed. Burt, chasing him in a company truck, looked thrilled with the two high-speed runs. They taxied back to Scaled, where the ground crew would go over the plane, like a groom with a prized racehorse. They would prepare the Proteus for its first flight up high the next morning.

Months before, Mike and Siebold had traveled to Beale Air Force Base in northern California to learn about high-altitude flying and get trained in the use of pressure suits. When Mike first stepped into the fitted suit, he wasn't sure he was cut out for the Astronaut Corps. He felt terribly claustrophobic. He spent two days at Beale. Enduring rapid decompression tests on the second day, the chamber crew, including a doctor, monitored him from the other side of windows. The atmosphere was raised to 70,000 feet from sea level in less than two seconds. The experience was terrifying, even with the classroom training he'd had the day before. Instantly, the Gore-Tex pressure suit went from a soft fabric to one of baseball material, making it hard for Mike to move his legs and arms, and nearly impossible to simulate flying. The efforts were exhausting. Mike saw a bowl of water on a windowsill inside the chamber. As he passed

63,000 feet, the water boiled furiously until there was none left. This was what would happen to his blood if he were not wearing a functioning pressure suit.* *This space stuff is scary,* he thought to himself.

Early on the morning of July 26, on a beautiful, clear, windless day, Mike and Siebold—a talented young engineer and naturally gifted pilot—suited up, donned their parachutes, and climbed into the Proteus. It was time for the plane's first real flight. Mike taxied out to runway 30 and applied maximum power to the two jet engines mounted on the aft fuselage. The Proteus accelerated rapidly down the runway, rotated, and lifted into the sky for the first time. The rate of climb was smooth as could be. Mike reached 12,000 feet, circling the airport to remain within gliding range. He went through the test card, including approaches to the stall, which he found benign. Roll forces were higher than on most planes he had flown, but appropriate for an aircraft of this size. Pitch and yaw forces were light and just about perfect, while control authority in all three axes was outstanding.† Because the plane had a canard—a forward wing—it flew in some ways like the Long-EZ, but had a flexible airframe and an unsettling ride in turbulence. The main wing was long and narrow, and bent more than any aircraft he had flown, almost like a car with soft suspension. As soon as Mike got used to the flexing, though, the Proteus proved a comfortable ride.

With the first flight tests behind them, Burt and the Scaled crew began thinking about late September as a time for the first public flight of

*A metaphor describes what was happening with the boiling water: The water molecules are trying to jailbreak from the liquid phase (their prison) through the surface (the steel bars) into the open (the vapor phase). Either you give them a more powerful drill (that is, more heat) or you make the bars lighter by reducing the external pressure. At the Armstrong limit (63,000 feet), the bars all but vanish, and the molecules escape en masse, free at last. It's water below the line and water vapor above the line. The molecules bounce around and eject off the surface into the vapor layer above the bowl of water. The atmosphere is bearing down on the surface, making it harder for the molecules to get out. You can either give them more energy by heating the water or make their life easier by reducing the atmospheric pressure. At the Armstrong limit, the pressure is so low that water at human body temperature will boil away.

†During yaw the nose of the aircraft moves from side to side. During pitch the nose moves up or down. During roll the nose rotates around the direction of flight like a spinning football. The rudder creates yaw force around the aircraft center of gravity, the elevators create pitch force, and the ailerons on the wings create rolling motion.

the Proteus, which would also be the flight attended by their clients, Marc Arnold, David Wine, and Peter Diamandis. Burt and Mike also talked excitedly about the possibility of setting national and international world altitude records in the Proteus in its weight class. The plane was impressive, and attracting some major attention.

Burt was in his office on the Mojave flight line when he looked out the window and saw a Boeing 757 Business Jet taxiing to a stop. It wasn't every day that a 757—measuring 155 feet long with a wingspan of 124 feet—arrived in off-the-grid Mojave, but Burt was expecting a visit from Microsoft cofounder Paul Allen and Vern Raburn, who handled Allen's technology investments.

Burt gazed at the 757, sitting way up high, as it came to a stop in front of Scaled. He frowned. Mojave had no commercial infrastructure, and was not the kind of airport with a mobile stairway to use for the next billionaire who rolled up. Burt grabbed a couple of his guys and headed outside. The 757 was the same plane used by the vice president for Air Force Two. Suddenly, the door to the 757 opened. Out flipped, in a sort of seamless triple flip, an elegant air stair. Burt looked up at Paul Allen and thought, *God is here.*

Allen and Raburn were in Mojave because Allen was investing heavily in broadband, buying cable systems, Web portals, wireless modems, and fiber builders. He was staging an initial public offering to raise billions for his cable firm, Charter Communications. He was looking at all areas of infrastructure of the Internet, and flew from Seattle to Mojave to find out what the Proteus could do. He was intrigued by the high-altitude plane, which was designed to stay at altitude for twelve hours, spit out broadband, and be replaced with the next plane for another twelve-hour shift.

Burt told Raburn and Allen that he had done the Proteus's preliminary design work between November 1994 and May 1996. The second phase, involving a more detailed design and the building of the prototype—the idea was to have a fleet of these planes—began in December 1996. The plane's general mission capabilities included commercial telecommunications, communication and data relay, atmospheric sciences, reconnaissance, and microsatellite launch. The midfuselage area was the dedicated payload component, and Burt had made the wing and canard tips extendable to adapt to the aerodynamics of a wide range of payloads. As a mechanism for broadband delivery, the Proteus was less expensive than broadband delivered by satellite, could go for twelve hours with 12,500 pounds at takeoff, and circle at altitudes of between 52,000 and 64,000 feet.

Burt went over the other figures for atmospheric science, reconnaissance, and microsatellite missions. Then he got to a final possible use for the Proteus, one he hadn't discussed publicly: space tourism. Burt broached the possibility that the Proteus, or some version of it, could be employed to launch a spaceship from the air. The spaceship would then rocket out of the atmosphere, giving the crew about four minutes of zero gravity and about the same view of Earth that you would see from orbit, and then fall back for landing.

As soon as he uttered this idea, Burt knew he had the quiet billionaire's full attention.

Allen, born in 1953, was ten years younger than Burt. He was a classic space geek: he loved science fiction, grew up knowing the names of the Mercury 7 astronauts, and followed every NASA launch. Like Burt, he built balsa wood model planes and made and launched model rockets. In 1969, the tenth grader who loved music and machinery in equal parts had a banner year. In May he went to his first rock concert—Jimi Hendrix—and in July, watched Apollo 11 land on the Moon. In more recent times, he had commissioned architect Frank Gehry to build a rock-and-roll and science fiction museum in Seattle. Allen was the world's third richest man (after Bill Gates and Warren Buffett) with a net worth of $22 billion.

He owned the Portland Trail Blazers basketball team and the Seattle Seahawks football franchise. He had a yacht the size of the White House, and the Boeing 757 was but one of the planes in his stable.

Burt told Allen that he wasn't entirely convinced that the Proteus was the right vehicle for suborbital spaceflight, but that the Proteus was inspiring sketches for another space plane, possibly an air-launched craft modeled after the X-15, which had been carried aloft by a modified B-52. His idea was to design something "safer and cheaper, something you could sell tickets for." Burt made it clear he wasn't looking for money, and told Allen and Raburn, "I don't know if I can do this, it's just something I've been thinking about. It's something that might happen." Allen made it clear that he was interested. If Burt got to a point where he believed his design would fly to space, Allen wanted to be the first to hear about it.

A short time later, Allen and Raburn boarded the 757. The elegant air stair retracted with the ease of a red carpet being rolled up. Off they flew into the cloudless Mojave sky.

After Allen and Raburn had left, Burt stood in his office thinking about the XPRIZE. He had been approached early on by Peter to help develop and refine the rules and requirements. Burt told him, "I will not help you. I may want to compete and win this, and it wouldn't be a good idea if I was involved in writing the rules." Looking at some of the sketches he'd done for the Proteus, he felt it would be tough to launch something into space with three seats—as was required by the XPRIZE. He had all sorts of ideas, including launching a capsule with parachute recovery, like Mercury and Gemini. He considered helicopter recovery, knowing he could use his neighbor's Huey helicopter to attempt in-flight pickup of a capsule, where the helicopter grabbed the top of the capsule's parachute and set it carefully on the ground right in front of Scaled.

But there were tough obstacles. Even the world's largest governments hadn't succeeded in building a fully reusable manned space vehicle.* And Peter didn't have the $10 million prize money; Orteig had put up the $25,000 right away, just like Kremer for human-powered flight.

As Burt worked throughout the day, he thought of his meeting with Paul Allen and brainstormed about the most vexing part of human spaceflight, the holy grail of manned missions: the return to Earth. He sat down and began to draw. When he looked up again, everyone had gone home.

*The space shuttle came the closest. The fifteen-story external tank of the space shuttle was the only component not reused.

16

Peter's Pitches

The XPRIZE gala dinner at the St. Louis Planetarium was sold out. Once again, Peter had astronauts. He had renowned space scientists. He had military brass. He had captains of industry and a who's who of endorsing organizations. He had members of competing XPRIZE teams. He had artwork and models of the teams' designs. He had attention from the media. He even had an elaborate trophy for the eventual winner. It was what he didn't have that weighed on him.

After issuing a clarion call to rocket makers, investors, and entrepreneurs, Peter had contenders with ideas for vehicles in all shapes and sizes, from the familiar to the flying saucer. He had assurances from teams that funding was secure or imminent. But at the end of the day, no one appeared to be building anything. In Peter's mind, the competition would be real only when the hardware was real.

In the same way Peter was jonesing for hardware, he was hustling for dollars. He was pitching his heart out, and here he was, in the middle of a dot-com bonanza, when wildly speculative companies including Pets.com and Webvan were taking in hundreds of millions of dollars in investment capital, and he was coming up empty handed, for something that could make history and launch an industry.

To make matters worse, his keynote speaker for the XPRIZE gala had canceled at the last minute. Buzz Aldrin had phoned a few days before the event to say he couldn't make it. Adding even more to Peter's headache was the reality that the one possible contender with actual hardware was Burt Rutan. Burt still hadn't officially registered for the XPRIZE competition, but he was making moves in that direction with the Proteus aircraft. Aware of the major conflict of interest, Peter sent out a letter to the teams disclosing how funding for the Proteus had come in part from his company Angel. He really didn't want the Proteus to be Burt's solution to win the XPRIZE competition.

Al Kerth, emcee of the St. Louis gala dinner, opened the evening by noting, "Attendance tonight has grown by more than thirty percent over last year. This is a sellout crowd. Only the stock market has shown more growth recently. I was thinking the only other difference between the stock market and the XPRIZE is we know the stock market is going to crash." He paused for the laughter. "Okay, bad joke."

After welcoming representatives of the XPRIZE teams, Kerth talked about exciting developments in the realm of space, including a half dozen shuttle missions; the Clementine unmanned mission, which revealed water ice at the lunar south pole; the Hale-Bopp comet paying Earth a visit; the Pathfinder landing on Mars; and the Global Surveyor entering the orbit around Mars. He noted that aviator, balloonist, and adventurer Steve Fossett was in attendance, as was Pete Worden, DC-X engineer Bill Gaubatz, and Clementine deputy program manager Stu Nozette. Peter's parents, Harry and Tula, were there, and had donated $25,000 for a New Spirit of St. Louis membership.

At the dais, Kerth was handed a note. Reading it, he looked up and scanned the crowd. He said he had an announcement to make. The surprise news was that someone in the audience had just bought not one, not two, not three, but *four* New Spirit memberships, valued at $100,000. Kerth said, "The donation comes from a man of surprises. You know him already as a great author—of books including *The Hunt for Red October, Red Storm Rising, Patriot Games, Clear and Present Danger, Without Remorse.* But how many of you knew he started as a humble insurance agent? He's also an investor in one of the XPRIZE competitors, Rotary Rocket. He's part owner of the Baltimore Orioles. His personal military contacts rival those of most nations. His job tonight is to lay to rest the rumor that the XPRIZE is really a secret conspiracy propagated by aliens stranded here." He paused for effect. "His name? Tom Clancy."

Clancy, drink in hand, made his way to the stage. At the podium, he said, "Let me tell you a story. July 20, 1969. Apollo 11. It was the night of the watermelons. I was driving home. There were all of these watermelons everywhere. I finally catch this tractor. I see watermelons falling out the back. July 20 was a great day to be an American. That was when NASA really meant something. The people at NASA are good people and smart people, but they are working in a system that doesn't reward achievement. The government doesn't really work.

"When the government wants to do something intelligent, they need us. Who invented the personal computer? The government? No. IBM? No. Two college dropouts named Jobs and Woz in a garage! Okay, it takes a big garage to build a rocket, but that's the American way. The difference between private industry and the government is we have to be efficient. If we don't make money, we go out of business. We need to create a private industry around space."

He continued, "America has brought democracy to the world. Freedom. Liberty. It happened here. Now, our job as Americans is to get the hell back out there in space. Progress depends on the unreasonable man."

He concluded, "Let's have some fun. I just pledged $100,000 for this—because it's fun! How often do you get a chance to make history? How often do you get a chance to see something cool happen and say, 'I had a piece of that'? To tell your grandchildren, 'I kind of helped.' Our next legacy will be to start human expansion into the next dimension.

"We do impossible things. That's why we're here. Look at things that never happen and say, 'Impossible?' Impossible means we don't know how to do it today. We'll figure it out. The future is something we will build." The evening ended with enthusiasm, encomiums, and more Tom Clancy dazzle. Funds were raised, and the night was a success. What still eluded Peter, though, was a big-name sponsor.

Peter's very first pitch, after the XPRIZE announcement under the arch, had been to St. Louis civic leader Bill Maritz, who ran a billion-dollar business, Maritz Inc., structuring in-house sales incentive programs for companies including General Motors. Peter and Gregg Maryniak believed that Maritz would be a perfect title sponsor, given his awareness of the XPRIZE and the company's focus on incentive-based competitions.

Gregg had flown in from New Jersey. He and Colette Bevis—now doing marketing for the XPRIZE—met Peter at a Kinko's in downtown St. Louis at around nine P.M. the night before the meeting. They had copying, printing, and binding to do. They wanted to make high-resolution, full-color prints from nearly sixty slides. Hours into the project, copies done, they used a 3M adhesive spray for the backing of the eleven-by-seventeen-inch color images. They managed to get as much glue on themselves as on the heavy cardboard stock. They didn't leave Kinko's until three A.M., returned to their hotel at four A.M., and were at the meeting with Maritz at nine A.M.

There were about ten people on the Maritz side. Peter did most of the pitching, starting out with an overview of the XPRIZE goals. "We had no money and no teams when we kicked off the XPRIZE," Peter said. "Our goal is to incentivize a twenty-first-century Charles Lindbergh." He went on for several minutes.

Al Kerth, who was present at the meeting, discussed the impact that the prize would have on St. Louis. A local sponsorship would change people's view of St. Louis from a "has been" to a "futuristic" city; create a new industry in St. Louis; excite youth about St. Louis's image—"space is sexy"; and offer "huge revenue potential for the city and its surroundings." Kerth had dreams of the prize being waged and won in St. Louis and recreating the romance and pride of the 1904 St. Louis World's Fair, which attracted visitors from near and far.

Maritz and his team wanted to know why NASA wasn't doing what the XPRIZE was attempting. Maritz asked about the risks to the sponsor if something went wrong. He wanted to know the likelihood of anyone pulling this off. While there was a flurry of questions, there was also considerable warmth and encouragement. When the meeting ended and Peter and crew walked out, Gregg surmised they had a "fifty-fifty chance."

"They seemed to really get it," Gregg enthused. Peter thought the odds were even better. That night, Peter and Gregg let themselves imagine what it would be like to land their title sponsor on their very first try.

A few days later, Peter got a call from Maritz's office. The idea was great, Peter was told, but it didn't "align with who we are."

That was the beginning of passionate pitches and succinct noes. Peter pitched to the founders of Enterprise Rent-A-Car, selling it as EnterPRIZE. He and Gregg and crew did the same with just about every major company in St. Louis. Then they moved on to Boeing, Cadillac, Champ Car, Charter Cable, Cisco, DHL, DuPont, EchoStar, Emerson, E*TRADE, Gateway Computers, JetBlue, Hilton, Lexus, Mars Inc., Miramax, Orbitz, Red Bull, Sprint, Wendy's, and more. Every time, there were the same concerns: "Why isn't NASA doing this? Can any small team really do it? Isn't it too dangerous? What if someone dies?"

A few months of failed pitches later, in late 1998, Peter, who was living part time in St. Louis and part time in Rockville, Maryland, had a meeting in London with the man he was sure held his winning lottery ticket. This was their guy. Peter could feel it. He was scheduled to meet with none other than Richard Branson of the Virgin Group. Branson, the rebel billionaire. Branson, the adventurer. Branson, the space lover.

The Virgin Group, started in the sixties with a magazine called *Student* and a record shop that evolved into a label called Virgin Records, was now made up of hundreds of companies, including Virgin Atlantic Airways. Branson lived on his privately owned Necker Island, which was, naturally, located in the Virgin Islands. He exuded gusto, with his mane of blond hair, year-round tan, and open-collar shirts. He had an international best-selling book out, *Losing My Virginity*. He was a humanitarian, and he was on a quest to break records—in balloons, boats, and amphibious vehicles. Peter wanted to convince Branson that it was time to trade in his dangerous balloon adventures for something more sedate, like spaceships.

Peter met up with his friend and business partner Eric Anderson at the airport in Newark, New Jersey, to fly on Virgin Atlantic to Heathrow Airport in London. Eric was twenty-three, fresh out of college, and working for a software company in Philadelphia. Eric and Peter had met several years earlier, in the mid-1990s, when Anderson was an aerospace student at the University of Virginia. Eric, a member of UVA's SEDS chapter, interned for Peter and spent the summer living in his basement, helping him in the early days of both the XPRIZE and ZERO-G Corp. He and Peter had just started a new company called Space Adventures with Mike McDowell, who had founded a polar expedition company called Quark Expeditions. Space Adventures was an umbrella organization for all things space, from private tours of rocket facilities with astronauts to classes on propulsion systems to zero gravity experiences. They wanted to broker deals between viable rocket providers—including in Russia—and wealthy citizens interested in the ultimate joyride.

Peter and Eric arrived at Heathrow before seven A.M. London time. Peter told Eric they would be met at passenger pickup by a Virgin Limobikes, courtesy of Branson. Eric had never heard of a motorcycle limo, but was too tired to give it much thought. It was freezing outside: drizzly, gray, snow seemingly imminent. They met their motorcyclists, were given leathers and helmets with microphones, and watched as their wardrobe bags were strapped to the back of the bikes. They took off in a blur, and held on as the red Virgin motorcycles darted in and out of morning rush-hour traffic. Eric's mind was on strong hot coffee. Peter was having a great time. They arrived at Branson's three-story town house in Holland Park, and were directed to a den off the living room. The first thing the two noticed was a model of Vela Technology's Space Cruiser displayed on the mantel. Peter and Eric fretted that the Vela folks had gotten to Branson before they had. Eric wondered if Branson would milk them for information on Vela, which was doing its own thing and wasn't involved in the XPRIZE. The company, based in Virginia, was trying to develop a fully reusable, two-stage rocket to carry at least six passengers for $80,000 each on suborbital flights. Eric knew the company was trying to raise more than $100 million.

Fifteen minutes later, Branson walked down the stairs, smiling warmly as he greeted them. Branson was in his trademark khakis and white shirt. He appeared relaxed, and asked about the motorcycle limo service. Peter, who had been anxious about the meeting—Branson was someone he deeply admired—began his pitch, talking up the power of the "Virgin XPRIZE," and detailing some of the contenders and likely entrants.

Branson said he had been to Mojave years before to meet Burt Rutan to get a sense of his thinking on pressurized balloon capsules. Peter, who had brought mockups of Virgin XPRIZE logos, told Branson about

Charles Lindbergh and the history of the Orteig Prize. Branson was intrigued, and said he loved the idea. He had been drawn to space as a teenager and was nineteen when Apollo 11 landed on the Moon. Branson had a number of favorite space quotes, including one by Carl Sagan about the richness of the cosmos: "The total number of stars in the universe is larger than all of the grains of sand on all of the beaches on planet Earth."

As Eric talked about Space Adventures' mission of broadening access to space by offering zero-gravity experiences, suborbital expeditions, and orbital flights on Russian hardware, Branson asked a question that took Eric by surprise. The billionaire wanted to know whether there was really a market for space tourism. No private citizens had ever flown on a commercial rocket to space. Private space as an industry did not exist.

"Is there really a market for people going up and floating around?" Branson asked. "I mean, I would do it, but I'm kind of crazy."

Peter listened, sure Branson already knew the answer and was playing devil's advocate. Eric thought Branson seriously doubted that a market existed. Peter and Eric talked about different events that showed the public appetite for private access to space. In 1958, Bantam Books published an "extraterrestrial travel reservation form" in the back of selected science fiction paperbacks. It asked readers to "Reserve a future trip to the planets. Your name and destination request will be kept on record until the technology is available to take you there." More than 250,000 responses poured in. Ten years later, during the height of the Apollo program, Pan Am Airlines offered reservations for a trip to the Moon. Ninety-three thousand people signed up. In 1985, hundreds of people put down $5,000 deposits for a private flight to orbit, this one offered by adventure travel agency Society Expeditions. Major checks from wealthy individuals soon followed. When NASA refused to sell shuttle seats to Society Expeditions, Gary Hudson was hired to design and build a private spaceship. That pursuit ended when the *Challenger* exploded.

Branson listened and nodded. He was a dreamer; these guys were

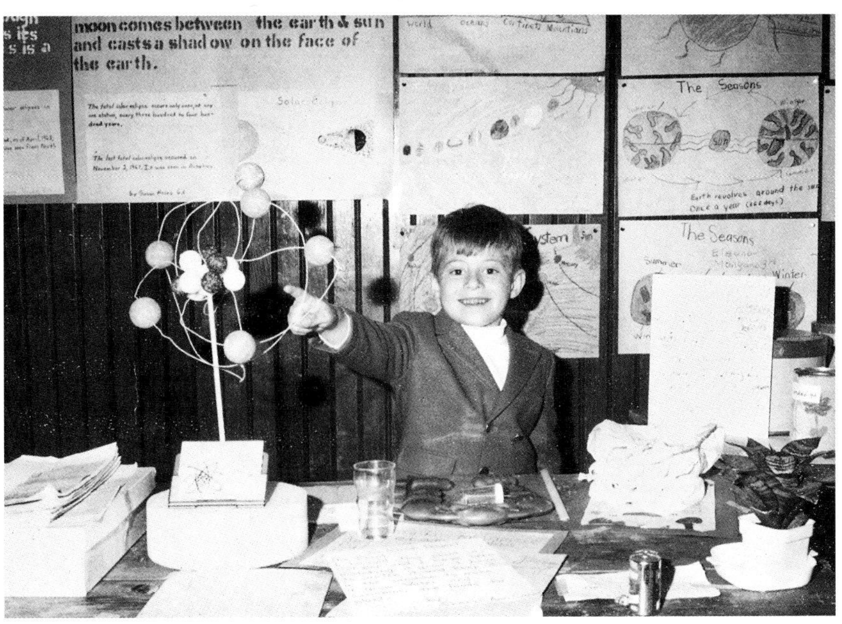

Peter Diamandis displaying the model of an atom he made for his first grade science fair. He was upset he took only second place.

Peter Diamandis

Peter with his dad, Harry, mom, Tula, and sister, Marcelle.

Peter Diamandis

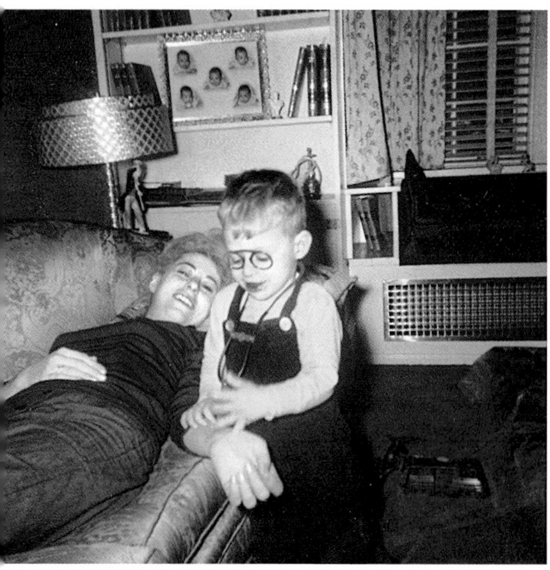

Six-year-old Peter Diamandis playing doctor with a kit his parents gave him. He checks his mother's vital signs.
Peter Diamandis

Peter when he first met Arthur C. Clarke at the United Nations' 1982 Unispace conference in Vienna, Austria.
Peter Diamandis

Peter (right) and his rocket-making friend, Billy Greenberg, with their homemade Mongo rocket.
Peter Diamandis

Peter (right) and Todd Hawley clasping hands when Peter turned over the SEDS chairmanship to Todd at George Washington University.

Peter Diamandis

Peter and his father upon Peter's graduation from Harvard Medical School in 1989.

Peter Diamandis

At a gathering (the "John Galt meeting") of rocket makers and space enthusiasts in Montrose, Colorado, Peter and others shared ideas for getting to space without NASA's help. This was where the idea for the XPRIZE began to take shape.
Peter Diamandis

The "build a rocket" brainstorming session in Montrose drew a half dozen commercial space enthusiasts to the home of David and Myra Wine.
Peter Diamandis

(Left to right) Peter, Todd Hawley, and Bob Richards in the Smithsonian Air and Space Museum in 1995. This was the last picture taken of the three of them.
Peter Diamandis

The $10 million XPRIZE is announced in St. Louis in 1996. St. Louis was chosen as the city for the announcement because that was where the young aviator Charles Lindbergh found his backers to make his transatlantic flight. *Peter Diamandis*

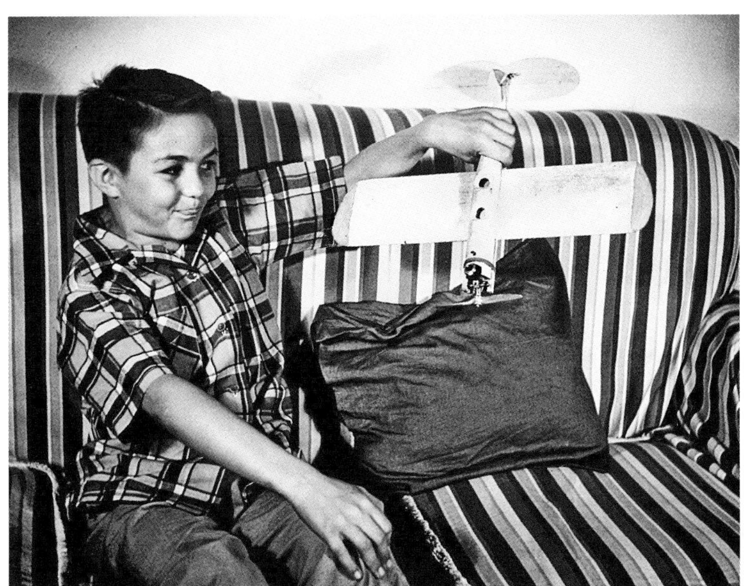

Burt Rutan at around age six at home in Dinuba, California, with a model plane he built. Burt never built models through kits, preferring to make his own designs using mixed pieces of balsa wood. *Burt Rutan*

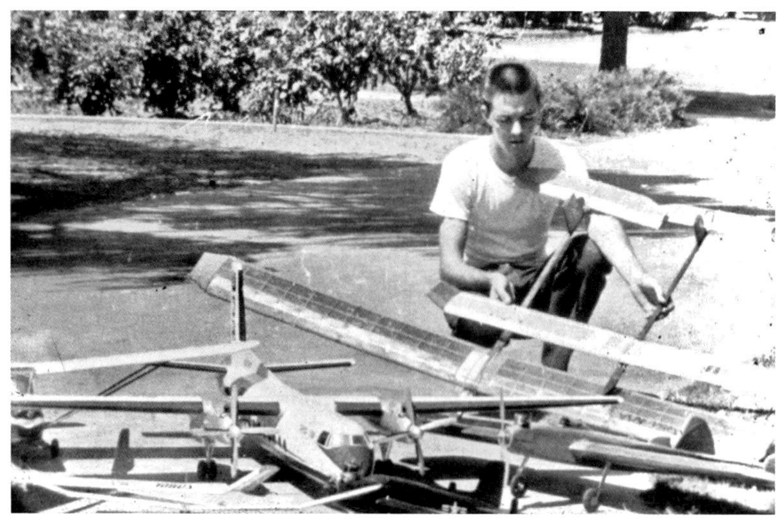

Burt Rutan, sixteen, with model planes of his own making assembled for entry in the 1960 Academy of Model Aeronautics national competition.
Burt Rutan

The *Voyager*'s historic flight by Dick Rutan and Jeana Yeager in 1986, trailed here on the final leg home by brother Burt Rutan and friends Mike and Sally Melvill in their Duchess chase plane. *Mark Greenberg*

Scaled Composites engineer Steve Losey works on the cockpit of what will become *SpaceShipOne*. *Dave Moore*

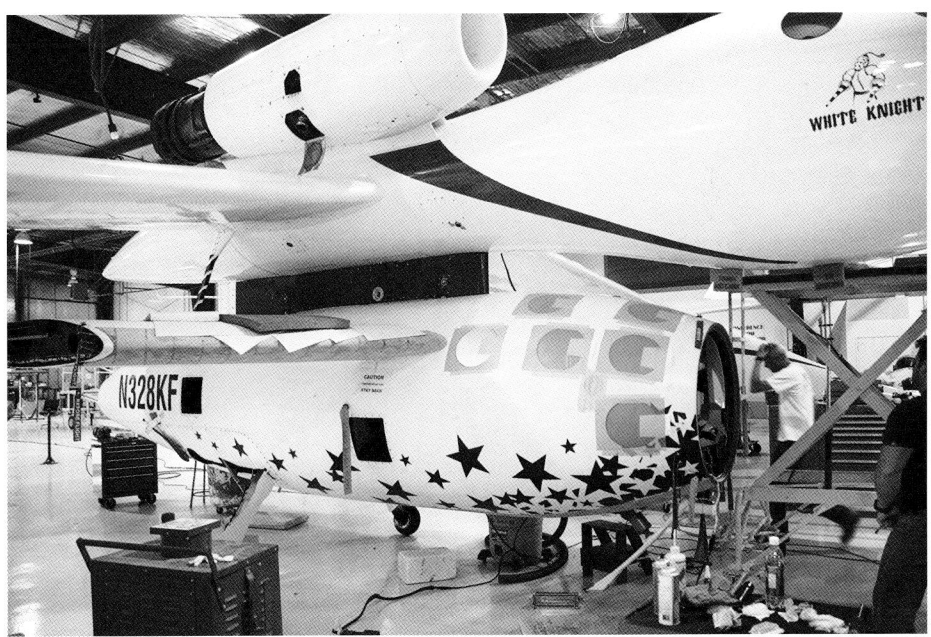

The construction of Burt Rutan and Paul Allen's secret spaceship program, *SpaceShipOne*—the rocket "mated" with the *White Knight*—at Scaled Composites in a hangar in the Mojave Desert, California. *Dave Moore*

Scaled Composites unveiled its spaceship program to the world in April 2003. Here, Burt (left) with Mercury and shuttle spacecraft designer Max Faget (center) and astronaut Buzz Aldrin. *Bradley Waits*

LEFT: Also at the April 2003 rollout of the *White Knight* and *SpaceShipOne*, Scaled's pilots (left to right) Doug Shane, Mike Melvill, Pete Siebold, and Brian Binnie. *Bradley Waits*

RIGHT: Steve Bennett, a Brit, left a secure job in a Colgate factory in England to start his own company and build rockets. He was the first XPRIZE contender to launch a vehicle. Here he is with his rocket in Morecambe Bay, England, in 2001. *Starchaser Industries Ltd.*

Argentinian space scientist Pablo de León and his XPRIZE capsule in Argentina. His colleagues warned him he would lose his credibility if he tried to win the XPRIZE.

Pablo de León

Romanian Dumitru Popescu dropped out of aerospace engineering school in Bucharest to build a rocket that could win the XPRIZE. Here he is with his XPRIZE contender in September 2004.

Dumitru Popescu

Erik Lindbergh, the grandson of Charles and Anne Morrow Lindbergh, after double total knee replacement surgery. He fought to get his life back after being diagnosed with rheumatoid arthritis.
Barbara Robbins

A triumphant Erik Lindbergh lands his Lancair at Le Bourget Field in Paris in 2002, seventy-five years after his grandfather made his historic flight. *The Lancair Company*

Burt Rutan (seated left) hosted Microsoft cofounder Paul Allen (seated center) and Virgin Group founder Richard Branson (seated right) at his Mojave home on June 20, 2004—the night before Scaled Composites' attempt to make history with the world's frist private manned space program. The "vision summit," as Burt called it, focused on goals for exploring and inhabiting space.
Tonya Rutan

ABOVE: Cars pour into the Mojave Desert in California in the early morning of June 21, 2004, in hope of witnessing history.
Mark Greenberg

LEFT: Burt Rutan's *White Knight* carries *SpaceShipOne* to altitude. After being released from the *White Knight* at 48,000 feet, spaceship pilot Mike Melvill lights the rocket motor and tries to make it to the Karman line—sixty-two miles up.
Mark Greenberg

Experimental test pilot Mike Melvill is surprised by his celebrity status following his June 21, 2004, suborbital flight, where he took off and landed on a Mojave runway, just a few dozen feet from cheering spectators. Here, Melvill signs a woman's back as others ask for his autograph. *Mark Greenberg*

Burt Rutan turns and smiles as pilot Mike Melvill (center) and *SpaceShipOne* backer Paul Allen greet photographers and others in Mojave following the historic June 21, 2004, suborbital flight. *Mark Greenberg*

Following the flight, a triumphant Mike Melvill holds a sign plucked from the crowds by his boss and friend Burt Rutan. Burt and Paul Allen (in cap) ride on the back of a truck as the rocket is towed back to its hangar in Mojave. *Mark Greenberg*

Mike and Sally Melvill fell in love as teenagers in South Africa and ran away together to get married. Here they are in Scotland the day before their wedding on October 6, 1961. *Mike and Sally Melvill*

Mike and Sally Melvill share a private moment before the first XPRIZE flight on September 29, 2004. Sally was acutely aware of the danger of flying the spaceship and feared losing the love of her life. *Mark Greenberg*

On October 4, 2004, the morning of the "money flight," Mike Melvill (left) tells Brian Binnie that he'll do great. Binnie is out to win the $10 million XPRIZE and to restore his reputation as an ace pilot. *Mark Greenberg*

A special anguish is shared by the wives of experimental test pilots. Here, Brian's wife (left), "Bub" Binnie, and Sally Melvill on the runway right before the start of the October 4, 2004, flight.
Dave Moore

After a lifetime of dreaming, Peter Diamandis (right) and his father, Harry, watch the winner-takes-all flight of *SpaceShipOne* on October 4, 2004. October 4 was chosen by Burt Rutan to commemorate the anniversary of the launch of Sputnik (1957), the world's first artificial satellite to orbit Earth. *Peter Diamandis*

The *White Knight*, carrying *SpaceShipOne*, takes off for its XPRIZE flight.
Mark Greenberg

As thousands of spectators cheer, *SpaceShipOne* pilot Brian Binnie comes in for a landing to try to win the $10 million XPRIZE.
Mark Greenberg

Richard Branson (in white shirt) hugs Burt Rutan after *SpaceShipOne* wins the XPRIZE. For both men, the day heralded a new era for commercial spaceflight.
Mark Greenberg

An exultant Peter addresses the crowd after the XPRIZE is won, October 4, 2004.

Peiwei Wei

The XPRIZE is won. (Right to left) Richard Branson, Brian Binnie, Burt Rutan, Paul Allen, Peter Diamandis, Amir Ansari, and Anousheh Ansari, October 4, 2004.

Dave Moore

dreamers. His motivation, though, was not necessarily to create another company because he saw a market for it. He started companies because of simple urges: to listen to better music, to have a better experience flying, to get a better drink at a juice bar, to get to London from Heathrow faster. If *he* had the desire to do something, he figured a lot of other people would, too. He definitely had an urge to go to space. He looked at Peter and thought, *Everything tells me I should say yes.*

Days later, when Peter was back in the States, he got word from Branson's people. Dr. Yes, as he was known, had said no. Peter sat in shock. He wondered: "If Richard Branson says no, who will say yes?"

In early 1999, Peter landed another major meeting, with another major billionaire, who also seemed *the perfect fit* for the XPRIZE. This time, he was heading to Seattle to meet with Jeff Bezos, head of the impossibly fast-growing Amazon.com. The e-commerce company, not yet five years old, had a stock market value of more than $30 billion. The stock had risen 1,000 percent in a year, making the Seattle-based company more valuable than blue-chip giants like Texaco. Bezos, thirty-five years old, was worth at least $9 billion—and he was a space lover. Peter and Bezos had an e-mail exchange in which Bezos agreed to the meeting but cautioned, "I'm so busy I'm trying to optimize my tooth brushing time."

Peter couldn't help but get his hopes up again. As a kid, Bezos watched *Star Trek* religiously and spent recess playing Spock or Captain Kirk. When he graduated as valedictorian of his high school class, he spoke of building colonies for millions of people in orbit to "preserve the Earth." Before graduating summa cum laude from Princeton in 1986 with degrees in electrical engineering and computer science, Bezos served as the chapter head of the campus SEDS group. Peter and Bezos had been in the same circles but had never met.

The two got together for breakfast at a diner in downtown Seattle.

Bezos, *Time* magazine's person of the year, was dressed in jeans and wore a watch that updated itself thirty-six times a day from the atomic clock. His laugh was quirky and unabashed. They talked about their shared interest in space, and about the SEDS days at Princeton. The SEDS chapter hosted movie nights, showing 16 mm films ordered from NASA about the history of the Apollo program. They had Friday night showings of James Bond films. They would charge $2 or $3 for tickets and sometimes make a few hundred dollars. The money paid for field trips to air shows, the Smithsonian Air and Space Museum, or nearby Air Force bases. They sent students to symposiums on Robert Goddard and conducted campus-wide polls asking students how they felt about the U.S. space program.

Like SEDS chapters elsewhere, SEDS at Princeton was all about networking with like-minded students, learning more, and plotting a more active future in space. Up to two dozen students showed up for regular meetings, held in hallways or open rooms of the Princeton student union. Their biggest meeting followed President Reagan's speech about diverting nuclear weapons by building a peace shield. An advocacy group formed, and soon a meeting was held on the subject, drawing hundreds of people and high-ranking military leaders.

Peter and Bezos were seated in a booth in the dark-wood diner. Peter, three years older than Bezos, explained the XPRIZE idea. He talked about the teams and showed him vehicle drawings. He went over specifications and propulsion systems. Bezos listened closely and asked highly technical questions. The two talked about Gerry O'Neill, who was at Princeton while Bezos was a student, and about their shared long-term vision of going to the Moon and mining space for resources. By the end of the meeting, Peter was clear on one thing: Bezos wasn't going to fund the XPRIZE. As an engineer, Bezos didn't want to just sponsor spaceflight; he wanted to be involved in making this dream come true himself. He wanted to get a smart group of people together to design and create his own star chaser. Bezos told Peter that Amazon was his means to make money to get to space. The more money Amazon made, the better his chances were of opening space. Amazon was his focus at present. As the two stood to leave, Bezos picked up the check. Peter noticed that Bezos ripped up the receipt and left it there. He was well beyond the point of expensing a meal.

The two parted ways but agreed to stay in touch. Another space-loving billionaire had turned Peter down. Peter was disappointed again, but also felt that if the XPRIZE didn't succeed, Bezos had the wealth and vision to one day pull it off himself. As Peter trudged up the hilly streets of Seattle, the drizzle soon turned to a downpour. He didn't have an umbrella.

Filmmaker Bob Weiss watched Peter talk to anyone and everyone about the XPRIZE. He was the evangelist going door to door to win converts. When the two first met in New York years earlier, Peter had excitedly told Bob about a screenplay he'd written that he expected to sell in Hollywood to fund the XPRIZE. Peter believed the script would garner millions up front and millions more once the movie was made. Bob listened, moved his glasses lower on the bridge of his nose, and began telling Peter the reality of Hollywood. He was sorry to have to disabuse him of his "fantasy notions of the economics of movies."

Bob told Peter that even if he did manage to sell the movie rights, the chances were remote that he would get any real money. Bob found Peter exceedingly bright, and Peter had in his soul what Bob called "the extraterrestrial imperative." He was driven to get off the planet. Bob had the same imperative and wanted to help Peter make his dream come true. It was science fiction come to life. Bob came from Hollywood, where his job was to create alternate realities. This story of Peter Diamandis and the XPRIZE had a similar mission. Since their first meeting, Bob had worked as much as he could on XPRIZE matters, until his wife reminded him they had children in private school. Then he'd go back to moviemaking to make some money before returning to the XPRIZE.

It was now nearly three years after the 1996 announcement under the arch, and the XPRIZE was pulling in just enough money to stay afloat. Peter had done at least fifty pitches for sponsorships, meeting with executives at Sony, Chrysler, Anheuser-Busch, Rolex, Breitling, Ford, FedEx, Airbus, Northrop Grumman, AOL, Discovery, Enterprise, Nissan, and Xerox, to name a few. All turned him down. The New Spirit of St. Louis memberships had kept the XPRIZE alive, and Tom Clancy's support had come at just the right time. But the memberships were increasingly scarce. The XPRIZE was getting $25,000 here, $25,000 there. Peter was like a gambler hooked on intermittent rewards. He was either going to win the jackpot or go broke trying. There was nothing in between.

Peter passed by the XPRIZE office in St. Louis near the St. Charles Airport on his way home from another trip and checked messages left by his assistant. Flipping through the notes, he saw one reading, "First USA called about donation." Peter laughed and thought, *Someone is asking* us *for a donation?* He started to crumple up the note to throw it away, but decided at the last minute to have his assistant call the following day to get more information.

The next day, his assistant came back with promising news. First USA Bank was not *seeking* a donation, but wanted to *make* a donation to the XPRIZE. The bank executives based in Boston had apparently read a story in *The Christian Science Monitor* about the XPRIZE gala event where Tom Clancy donated $100,000, and they were interested in meeting with Peter for a possible deal.

Within a week, a meeting was scheduled at the St. Louis Science Center. Peter met Gregg, Al Kerth, and Doug King there. The four were standing curbside when four First USA executives clad in black suits stepped out of black limos, prompting Peter to quietly refer to them as the Men in Black. Once inside, the executives told Peter, Gregg, Al, and Doug that they were interested in creating an XPRIZE credit card. They explained how they made specialty credit cards for colleges, alumni associations, airlines, and the like. They believed the space community was sizable enough to attract new cardholders. They would want access to the XPRIZE mailing lists, to pilot and general aviation lists, as well as mailing lists for a number of space clubs and organizations. First USA XPRIZE cardholders would get a chance to win various space-related flights, including eventual ZERO-G flights, and flights on Russian MiG jet fighters. Cardholders would also be able to donate directly to the XPRIZE.

A deal came together quickly—and it was big. First USA Bank, anticipating significant revenues through the XPRIZE cards, agreed to fund half of the purse prize. But the offer came with a caveat: the $5 million would be awarded only if the prize could be won by a certain date—December 17, 2003, the one hundredth anniversary of the Wright brothers' historic flight. And there would be no prize unless Peter could come up with the other $5 million.

Back from his trip to St. Louis, Peter got on his running shoes to head to the gym. He was filled with mixed emotions. On the one hand, he could announce to the teams that he had secured half of the $10 million. On the other, his title sponsor remained elusive. Still, in the face of a torrent of noes, the reality was that someone had actually said yes.

17

A Lindbergh Sculpts a Dream

Erik Lindbergh sat with a doctor considered brilliant in the field of rheumatology. He was at the Mayo Clinic in Rochester, Minnesota, and had endured day two of being processed like meat, moved from test to test he didn't know he needed, being shuffled along in lines wearing flimsy paper gowns ignominiously open at the back. Erik had tried anything to get better, meeting with alternative healers and osteopaths, adopting the latest in experimental treatments. Nothing worked. Finally, his dad told him, "You have to talk to somebody good," and offered to pay for him to go to the Mayo Clinic. So here Erik sat, across from an esteemed rheumatologist. The doctor studied his X-rays and records, swiveled in his chair and said matter-of-factly, "It's obvious. You need knee replacements. Both knees."

Erik didn't hear much after that. He knew his knees were shot, but no doctor had been this blunt. The doctors he had seen before didn't put expectations on him; they treated him where he was rather than looking too far ahead. Erik left the hospital in a stupor. He was by himself, relying on his cane to navigate the icy walkways and make it down the steps.

Instead of staying at a hotel connected to the Mayo Clinic campus by heated and covered walkways, he had tried to save money by booking a room in a motel across town. He kept hearing the words: *total bilateral knee replacement.*

Arriving back in his room, he sat down on the bed and didn't bother to take off his coat. There was a jagged line of brown liquid creeping down one wall. The light was dim and flickering and the radiator sputtered. The thick shag rug was the color of mustard. Someone outside yelled to the manager that there was a rat in the pool. He moved his hands along his long legs. Strong, agile, ready for adventure. Fly down a trail on a mountain bike. Carve glassy water with a slalom ski, huge spray, controlled, aggressive cuts through the wake. Aerial cartwheels, boxes of trophies. That was then. Now his legs were thin, brittle. Stuck like rusted metal. The room, the sounds, the doctor's words, the feeling of his legs. It hit him. He curled up on the bed and cried. Hours passed. The light in the room was tobacco brown. He slowly got up and forced himself to look in the mirror. Really look. For most of a decade, he looked past his reflection. He shied away from having his picture taken. He even avoided his shadow. He was young and fit. Erik the gifted athlete. Not frail and hobbled. His smile would conceal his pain; no one would notice the cane. He was still the same Erik. This is what he told himself. The disease, with its pattern of attack and retreat, pain and easing off, was complicit in his denial. It enabled him to put off acknowledging that he had a serious and chronic disease. He studied himself in the mirror. He felt his beard. He was not yet thirty-five, but he was an old man.

The good thing, Erik told himself later, was that there was nowhere to go but up from the motel in Rochester. Six months later, he had new knees. He kept his old knees—chunks of bone—in a jar in the refrigerator until his wife, Mara, said the knees had to go. He had wanted to hold on to them for as long as he could, maybe even have a burial. He reluctantly moved them out of the fridge and put them in a wooden box outside, where they were eventually eaten by mice. The recovery was slow and painful. He had eight-inch vertical scars over both knees. But the pain of surgery and recovery was better than the pain of rheumatoid arthritis. He spent his recovery at the small beach home of his mother, Barbara, on Bainbridge Island. He needed constant care. His friends came to visit, and physical therapists were in and out. Not long after surgery, with his knees still huge from swelling, Erik was up and walking around. Barbara returned home one day to find Erik had somehow made it down the stairs and was out wading into the salt water. She just watched him—her happy-go-lucky boy who rarely complained about anything. It wasn't long before he could walk steadily and without a cane. Barbara worried, though, about what was ahead. She feared that the new knees were like new tires for an old car. His wrists were shot from using the cane so hard. He still needed his right foot fused like the left. And the surgeon who replaced his knees cautioned that he would need to go through the same surgery again in a decade or two.

In his drive to get better and have a normal life again, Erik learned of a new medicine called Enbrel. The drug was a "biologic," made in living cell cultures rather than through a manufactured chemical process. The cultures were made from genetically modified Chinese hamster ovary cells. The hamster's ovary cells were said to produce the proteins needed to combat rheumatoid arthritis by stopping joint damage and inflammation. Erik's rheumatologist was cautiously optimistic about Enbrel, but told him that his health insurance wouldn't cover the cost of the drug. Fortuitously, his doctor was involved in a double-blind efficacy study on Enbrel. He got Erik into the trial and told him that if he showed improvement while on the drug, his insurer would then probably cover the costs of the medication. His doctor warned him that the treatment required suppressing his immune system, which in turn could make the body vulnerable to everyday infections or more serious problems. He also said the long-term side effects were unknown, and that it "may cause cancer."

Erik listened to all the warnings, but didn't hesitate. Parts of his body were worn out, bone on bone. Even with his new knees, his quality of life was so low that he was willing to take the chance. He was given a kit with a sterile powder and solvent to mix at home. He would inject the drug once a week just below the skin of his thigh. After about three weeks, he began to feel something, like coming back from a debilitating and prolonged flu. The active inflammation, caused by his immune system turning against itself, was stopped. Gradually, everything began to hurt less. He felt more stable and could sleep through most of the night again. As the months went by, Erik put on needed weight, even adding muscle mass. He felt less brittle. Bumping into things no longer brought searing pain. New damage was being halted. He was told that if improvements continued, he wouldn't need the right foot fused like the left.

With his improved mobility, Erik began taking long walks on the beaches around Bainbridge Island, a ferry ride from Seattle. As he walked, he picked up pieces of driftwood. He had made furniture for years, but only intermittently because of the flare-ups. The first piece of furniture he made was a bench inspired by a gorgeous piece of maple driftwood. The wood looked like a wishbone, which he split into two to form the legs. He used a large chunk of redwood for the seat, and had another perfect piece for the back of the bench.

He knew of cancer patients who planted rows of baby trees, determined to be a part of something living and growing. Erik needed to carve and build, to take something unwanted, strangely formed or riddled with knots, burls, and twisted grain, and make it beautiful, strong, and useful. He pulled wood destined for the furnace or fireplace and gave it permanence. He needed to carve and build almost as he needed to move. He found two pieces of beautiful wood that formed oddly complementary lines of an *X* and brought the sculpture to an XPRIZE event. He made a lamp of twisted juniper and madrone wood, carved a butterfly from juniper and dogwood, and made a "Felix the Chair" out of holly and cherry wood. He brought his furniture to the local farmers' market, and talked about the shapes and types of wood. He got to know some of the regulars, including a man who asked him whether he would consider making a sculpture of the *Spirit of St. Louis*. Erik demurred, telling him he could buy a model from an aviation magazine. But the would-be client persisted, saying he had become a pilot after reading *The Spirit of St. Louis* and loved Erik's bohemian style. Erik returned to the woodshop, thinking about the request. He sifted through pieces of driftwood, studying the shapes, colors, and grain. He went out in search of wood, looking in neighbors' wood piles, walking through forested areas, and finding old and worn branches. He hauled pieces into the shop. Then he went back to the shoreline, to spots he knew were laden with sea-scrubbed driftwood in all shapes and sizes. The pieces were knotted or satin smooth, weathered by sand, sun, rain, and sea.

At the same time he searched for wood for his grandfather's plane sculpture, these walks on the beach connected him to his grandmother, Anne Morrow Lindbergh. As he collected driftwood, he thought of Grandmother out collecting seashells. Her book *Gift from the Sea* was about lessons learned from the shells she collected on Captiva Island, on Florida's Gulf Coast. First published in 1955, the book had by now sold millions of copies. His grandmother was easier to think about than his grandfather. She was a softer presence. She was a candle, he was a spotlight. Grandmother was forty-nine years old when *Gift from the Sea* was released. She organized the book as a collection of meditative essays on how to find inner peace, outer harmony, and meaning, particularly as a woman. Each chapter focused on the shape, purpose, and meaning of one kind of shell, whether an oyster or a channeled whelk. The shells spoke to simplicity and paring back, to the beauty of solitude, to different needs at different times of life.

Erik's grandmother was now ninety-two and living with his aunt Reeve. On the surface, Grandmother was a delicate flower. Inside, she had the strength of a redwood. She raised five children after having her first son kidnapped and killed. She married Charles when he was the most famous man on Earth and became a decorated pilot herself. She lived through her husband's adulation and endured his castigation. His isolationist views during World War II and his praise of Nazi Germany's military and aviation programs turned him from hero to villain. Grandfather had died in 1974 at the age of seventy-two, when Erik was nine. Grandmother was soft-spoken and eloquent and called Erik "nature's nobleman" because of his love of the outdoors and his easygoing mien. He marveled that she could see through his growing-up-in-a-pack-of-boys energy.

Erik often returned to his favorite passages from *Gift from the Sea*. He loved what his grandmother wrote about serendipitous finds: "One never knows what chance treasures these easy unconscious rollers may toss up, on the smooth white sand of the conscious mind; what perfectly rounded stone, what rare shell from the ocean floor. Perhaps a channeled whelk, a Moon shell, or even an argonaut." She wrote: "The sea does not reward those who are too anxious, too greedy, or too impatient. To dig for treasures shows not only impatience and greed, but lack of faith. Patience, patience, patience, is what the sea teaches. Patience and faith. One should lie empty, open, choiceless as a beach—waiting for a gift from the sea."

That was the perfect way to walk along the beach—though in the Pacific Northwest, driftwood was more abundant than shells. While embracing his grandmother's call to accept what came his way, Erik felt happy to be in a greedy phase. He had a fever to create. Building was more than a meditative act. There was the commerce of it—he needed to make a living—and there was the catharsis.

Back in the shop, he was surrounded by the smell of wood and the tools of the trade, the grinders, sanders, and drills. He arranged a dozen pieces of wood on the table and hoisted a heavy piece of lumber. He'd spotted the beauty in his neighbor's woodpile—destined for the fireplace—and asked whether he could have it to make art. He loved wood weathered by salt; it was like seeing smile lines on a beautiful face.

He looked at the pieces of wood laid before him. He was reminded of the stories of the astronauts he'd met through the XPRIZE. They talked about the unparalleled force of the rocket engine, the black sky, the g-forces, the weightlessness, and leaving the bonds of Earth. The rockets that launched them were ballistic capsules and missiles, flying like supersonic bottles to space, skyscrapers launched to the heavens. Erik hadn't known what to do with these oddly shaped pieces. Now he saw it: *That is a smoke trail for a rocket ship!* He picked up a branch he'd found in eastern Washington. *That's a parabolic arc!* Another piece was Jupiter with its ring. The wood from his neighbor's pile would make amazing rocket ships. Images of traditional Buck Rogers–style rockets, abstract rockets, and even a few of the Estes models he'd built as a youth played like a slideshow. He didn't need to search for pictures; he had it all in his mind. If he could do spaceships, he could certainly do the *Spirit of St. Louis*.

He needed the right wood. The *Spirit of St. Louis* was not the most stable plane. Grandfather had designed it to bump and rattle and keep him awake. And it was not the best-looking plane. Astronaut Neil Armstrong once called it a bird only its mother could love. Erik couldn't sculpt a replica; the plane was too iconic. It was entirely Grandfather's.

He could make his version of *Spirit* aerodynamically improbable, if he wanted. He could make it abstract beyond obvious recognition. He could make it as *he* saw it. It was like walking in his grandfather's footsteps without feeling obligated to fill his shoes.

He remembered a story about his grandfather's friend Jilin ole "John" Konchellah, a Masai warrior. Konchellah had to slay a lion with a spear to become a man. Erik's grandfather had to cross the unforgiving sea. Erik's father had to dive to the ocean's uncharted depths. His grandmother removed herself from everything familiar, finding lessons in collected shells. Erik was reduced to nothing by a foreboding room in Rochester, Minnesota. Maybe he had to break before the fixing could begin. He studied a piece of wood and saw what he had been searching for: swells of the Atlantic in the waves of the grain. He held it up and started flying the piece around the shop. For the first time, he imagined himself in the cockpit. He thought of what it must have been like for his grandfather to fly the plane for more than thirty-three hours nonstop. *What was it like to take off in relative anonymity and land in global fame?* The flight changed his grandfather's life—and it changed the world.

Erik held the wood up to the light. He flew the piece through a ray of sun flecked with sawdust and imagined his grandfather flying through the night to get to a golden dawn. It no longer made sense for Erik to keep running away from being a Lindbergh. He had done that all of his life, and it had gotten him nowhere. Like old lumber rescued from the fire pit, he had been given a second chance.

18

Peter Blasts Off

The first thing Peter noticed when he pulled up to Idealab in southern California was the show of cars: limos and Porsches were double-parked out front. Inside the office—all open space, exposed ductwork, doors used for desks—employees didn't walk, they ran. This was not Pacific Standard Time; it was Internet time.

Peter had sold his Rockville, Maryland, home in one day, put his furniture and all of his belongings into storage, and moved west to run a new Idealab company called Blastoff. The mission of Blastoff was to be the first private company to land a rover on the Moon and transmit images back to Earth. Peter had made a deal that once the Blastoff lunar mission was fully funded, the XPRIZE would get funded next. It was his riskiest venture yet, but all Peter could think was: *I was offered the Moon, the fucking Moon, a funded mission to the Moon.*

It was early 2000, and the tech-heavy NASDAQ had doubled in little over a year, though it had started to drop. The beleaguered "old economy" Dow was in retreat. Equity trumped cash and e-commerce elbowed out brick and mortar. Netscape had gone public five years earlier, Google had begun operating in a garage in Menlo Park two years earlier, and eToys had a value of $7.8 billion on its first day public in 1999.

Idealab, founded by a small, wiry, constantly-in-motion engineer turned entrepreneur named Bill Gross, was worth $9 billion and comprised more than forty dot-com companies, including eToys, Pets.com, Friendster, NetZero, and CarsDirect. An inventor from an early age, Bill Gross had found inspiration for Blastoff from a new company called eBay. Bill had gone to the online auction site to try to buy a Moon rock, which he had dreamed about owning since he was eleven years old and transfixed by the Apollo 11 lunar landing. Bill told his brother Larry, "Someone must have it at some price." But their eBay search of lunar souvenirs left them empty-handed. They learned that all Moon rocks were exclusively owned by the government, locked up in hurricane-proof vaults in Houston or loaned to teaching institutions like the Smithsonian. Even lunar dust was supposed to be the property of the U.S. government. At that point, Bill decided to form a private company that would send a human to the Moon to collect and return lunar samples to sell on eBay. The second iteration of Blastoff replaced the astronaut with a machine—an unmanned robotic rover, equipped with a camera, would land on the Moon and collect the samples. The mission would be covered on television, beamed through the Internet, and financed by sponsors. Bill disliked the idea of plastering ads on a rocket and a rover, but he had even more disdain for the government's hold on space.

Jim Cameron, a pioneer of special effects and director of some of the world's top-grossing films, including *Titanic* and *Terminator*, was interested in filming the lunar spectacle for Blastoff, and filmmaker Steven Spielberg was an investor. The technical head of Blastoff's space missions was the white-haired, Einstein-like Tony Spear, project manager for the NASA Pathfinder mission, which landed a rover on Mars on July 4, 1997. Bill and his brother Larry had taken notice when the Pathfinder landing drew more than eighty million Internet hits a day in the first few days, including nearly fifty million hits to the Jet Propulsion Laboratory Web site, where cameras were trained on Spear and his team as they watched, waited, and celebrated.

A few of Bill's aerospace friends told him that he could probably land something simple and efficient on the Moon for less than $10 million. But as the project grew, so did expenses. No matter. Bill was starting companies the way the Treasury printed money. An IPO for Idealab was planned for later in 2000, an event that would make the forty-one-year-old a decabillionaire. Besides, if the online grocer Webvan—another of his companies—could raise more than $800 million just so people could have groceries delivered to their door, the cost of his little Moon project would be pocket change. The fact that there hadn't been any private space missions didn't matter, either; in the halcyon days of the Internet, when things like profit were irrelevant, anything was possible. Bill could lure the best rocket scientists, the best filmmakers, and the best space entrepreneurs, including Peter Diamandis.

When Bill, who had recently raised more than $1 billion in private equity, recruited Peter as the CEO of Blastoff, he told him that he had set aside up to $60 million for the Moon exploration company, but wanted Peter to bring in outside funds where possible. Funding for the XPRIZE would kick in once Peter was able to raise outside financing for Blastoff.

Peter, who had grown tired of friends' telling him they were making a mint on Internet companies, was the eighteenth employee at Blastoff, based in Pasadena in Los Angeles County, and down the street from Caltech and the Jet Propulsion Lab. Opting for a lower salary and more stock, Peter had a base pay of $145,000 and 1.3 million options. Peter believed that Blastoff would be the rising tide to carry the XPRIZE and his other space ventures. He still had Gregg Maryniak and the team in St. Louis working on the XPRIZE, and he planned to continue managing it from the West Coast.

Working at Internet hyperspeed, Blastoff had a mission date of summer 2001 to get to the Moon, just in time for a fall 2001 IPO. As Bill saw it, the private lunar landing would be an Internet phenomenon and another billion-dollar business. As Peter saw it, Blastoff was a way to make all of his space dreams come true.

Peter and Larry Gross, who was on Blastoff's board of directors and involved in day-to-day oversight, met with Jim Cameron at his Santa Monica office, where the Terminator T2000 robot stood just outside the door. In a corner of Cameron's office was the ship's wheel from the *Titanic* film. Cameron, known for his love of the oceans, had an equally strong fascination with space, and grew up on a steady diet of science fiction. For many years, faced with a long bus ride to school, he read a book a day, devouring works by Arthur C. Clarke, A. E. van Vogt, Harlan Ellison, and Larry Niven. He was fascinated by aliens and interstellar travel and liked the blurring of lines between reality and fantasy, science and art.

During the two-hour meeting, Larry and Peter offered the director 300,000 shares in Blastoff, discussed his level of involvement, and tossed out ideas for the title of the lunar film. Peter suggested that Cameron serve as Blastoff's "mission producer," but Cameron had a similar title with Fox. The three decided on "imaging supervisor."

The biggest thing to come out of the meeting was Cameron's insistence that they needed not one camera-equipped rover, but two. "You need one that is going this way and another that sees it, so you have this point of view," Cameron said, mapping out the camera angles on a whiteboard. The two cameras playing off each other would promote stronger audience involvement.

Meetings like this prompted Peter to start an audio journal that he hoped would document history in the making. In his tape-recorded observations about the Cameron confab, Peter said that he found the director extremely smart, friendly, and generous with his time—but Cameron's handlers were a different story.

"The big issue now," Peter said in his audio journal, "is dealing with his lawyer, Bert Fields. Cameron and Spielberg play super nice on their side, but then you have to deal with their accountants and lawyers who are basically recalcitrant. I was impressed today by how Jim gave us all the time we needed. His organization, on the other hand, is frenetic trying to take care of him."

Peter also noted in his audio journal—which he often recorded in his car—that although it was exciting to have Cameron on board, the move to two rovers instead of one would double the cost and workload of the program and send the engineers into overdrive. Peter wondered how these strange bedfellows—hardwired aerospace, frenetic Internet, and fantastical Hollywood—would work together. On the passenger seat next to Peter was Heinlein's *The Man Who Sold the Moon,* written in 1949 about a businessman obsessed with the idea of being the first to reach, control, and monetize the Moon, selling naming rights to craters, offering publicity stunts, and returning diamonds from the lunar surface.

It was a novella that Peter assigned to the entire Blastoff team to read.

One of Peter's first staff meetings at Blastoff lasted for seven hours. Peter and the team went over all of the systems requirements. Peter wanted the company to have the "culture and spirit of the early days of Apple Computer," when a pirate flag flew over the building.* The team went over particulars, from the landing site and new hires to the rovers and the delivery rocket that would get both lander and rover to the Moon. Buzz Aldrin had ruled out Bill Gross's original idea of landing near the Apollo 11 site, where the flag planted in 1969 was no longer standing. Bill's

*The team behind the early Macintosh hoisted the pirate flag over their office as a testament to Steve Jobs's statement that it's more fun to be a pirate than to join the navy.

vision was to move the rovers slowly toward the iconic site—cameras rolling—and have a rover pick up the flag and replant it. Aldrin nixed the idea, saying they couldn't "traipse all over the historic site," make tracks and "go over our footprints." The footprints should remain for millennia.

The Blastoff team consulted with Harrison Schmitt, a geologist and Apollo 17 astronaut who was the twelfth man to set foot on the Moon and the second to last to leave (right before Gene Cernan). The team considered the Taurus-Littrow valley and highlands, the landing area for Apollo 17, but settled on the Apollo 15 site at the foot of the Apennine mountain range. Schmitt told team members that the photographs of the Moon released to the public decades earlier were not entirely accurate. The original photos, taken with Hasselblad cameras, were put into a vault and copies were made into prints. Schmitt said the Moon was not white and colorless as portrayed, but had a range of colors, mostly in shades of brown and brick-red.

Discussion turned to landing times that coincided with spaceflight milestones, lunar cycles, and IPO targets. Bill wanted the craft to land on the Moon, American flag in tow, on July 4, 2001. The engineers were looking at the end of 2001. The new Moon for December 2001 would fall on December 17, with the full Moon on or around December 28. Students across the globe could begin studying the mission and its physics and technology in the fall. One of the engineers pointed out that December 25 was Isaac Newton's birthday.

Rex Ridenoure, Blastoff's chief mission architect and number-two hire (after a Czech-born engineer named Tomas Svitek), went over some of the systems challenges, including transmitting video, images, and data from both lunar rovers back to Earth. The plan, he explained, was to transmit video and data from the rovers independently to the lander—the mother ship—via the same sort of radio links (a 2.4 GHz microwave radio band) used by remote-controlled cars and toy drones. The data streams would be merged on board the mother ship. All of this could be done live or in record/playback mode. To ensure transmission to Earth,* the mother ship would have to be parked in a certain orientation on the lunar surface. The command link on the lander could be active at any time, even when the mother ship was on the move. Commands from Mission Control to the rovers would be routed through the link with the mother ship.

Ridenoure, who like the Gross brothers was a graduate of the California Institute of Technology, had been recruited to Blastoff from SpaceDev, one of the first commercial space exploration companies. He had met Peter around 1992, when he was working at the Jet Propulsion Laboratory, managed for NASA by Caltech. Peter had come to JPL to talk about the potential launch services of International Microspace. Ridenoure believed in the Blastoff mission and thought they had a good shot at becoming the first to launch a private venture to the Moon.

The group discussed the types of rockets that could be used to carry the mother ship to the Moon, including Orbital Sciences' Taurus XL rocket, a Boeing-designed Delta II rocket, and the Russian Dnepr launch vehicle, a converted SS-18 Satan ICBM. As they brainstormed design ideas, tension began to emerge between the engineering side and the marketing side. The engineers wanted function; the marketers wanted form. They wanted a cute "family" of rovers, where the mother ship would dispatch two childlike rovers to "play and explore." The rovers would need to be anthropomorphized, with squat white bodies, thin necks, big eyes, and maybe even some sort of cap or eyebrow features.

The engineers debated the best approach to reach the Moon. They could go by "direct injection," à la Apollo 11, where the rocket would take three to four days to reach the Moon. The challenge with this approach was to time the launch at just the right period when the Earth-Moon

*The Moon-to-Earth link would be established through a higher-frequency X-band radio link from a fan-shaped medium-gain radio antenna on the rover to a global network of space-mission tracking antennas and dishes operated by the U.S. company Universal Space Networks (USN). The USN antennas would also be used to send commands to the rovers from Blastoff's Mission Control. The USN dishes were capable of transmitting commands to the mother ship at relatively high power levels.

geometry was lined up. Direct injection gave engineers at most one try a month. By contrast, a "phased orbit" would put the hardware in a low- or medium-Earth orbit, require trips around the Earth (of varying times based on low or medium Earth orbit), and allow for multiple slots per month to inject to the Moon.* The Blastoff team went with the phased orbit strategy.

Peter left the office late, and exhausted. Walking to his car, he looked at the Moon in the midnight sky and said, "That's where we're going."

For Peter, months after arriving in sunny Pasadena, the euphoria of Moon missions and Internet time began to dim. Peter was feeling unsettled by the seemingly infinite iterations of Blastoff. The mission statement changed faster than cars pulling in and out of Idealab. During pitches to potential investors, Bill Gross touted Blastoff as an education company. Then it was an entertainment company. Next, it was a sports venture, with rovers racing on the Moon. The latest pitch by Bill to venture capitalists had Blastoff as the "broadband Napster." Peter listened and thought, *This has zero validity.* As much as he tried to accept the new way of doing things, Peter felt whiplashed by the pop-up business ideas. Larry Gross said they needed to create a "level of pizzazz," a vision that would excite people. Peter was dispatched to Brazil twice to work on the "Olympics on the Moon" concept, with corporate sponsorships of national flags. The idea was to have the mother ship carry six rovers—now six!—representing six nations. Children from different countries would man the joysticks and control the rovers as they planted their national flags and raced on the lunar surface.

The engineering team remained more rooted in reality. They believed they were eighteen months from launch. They had invited a serious

*With the Moon's twenty-eight-day trip time around the Earth and a Moon-bound spacecraft orbiting the Earth once every two days, say, you may get up to fourteen times per month to inject to the Moon.

top-to-bottom technical peer review of the entire project, with sixteen experienced space project managers. The reviewers included a former JPL director, JPL deputy space mission project managers, former senior Department of Defense space managers, expert rocket engineers, and more. The consensus was that their designs were solid and the goal was attainable, if the money kept coming in.

That was the problem. So far, Bill had allocated only $12 million for Blastoff, and Peter was told he needed to personally raise another $10 million before Idealab would kick in what would be a final $10 million. The budget would be capped at about $30 million, even though $60 million was supposed to have been set aside in the first place. "I didn't sign up to raise money," Peter said in one of his late-night audio recordings. "I signed up to make this company work. The capital promised—$60 million—is not committed. Is this the Internet way, to lure someone in and then yank the rug out from under them?"

Peter scheduled a meeting with Bill and Bill's wife, Marcia Goodstein, Idealab's chief operating officer. He loved meeting with Bill, who was the consummate optimist and brainstormer. He dreaded meeting with Marcia, who was steely, by-the-numbers, and made every meeting personal and unnerving. At the meeting with the two, Peter was told he needed to "make it work" with what he had. He was told they wanted to support him, but the softness in the market was delaying the Idealab IPO and making things generally "squeamish." Some pundits and economists believed that a market correction was under way, while others, including Bill, said the fundamentals remained solid and the boom would continue. Still, as summer turned to fall and the markets continued to drop, money was becoming impossible to raise.

Unwelcome surprises presented themselves at every corner. One engineer resigned, fed up with what he saw as unsubstantiated promises made to potential investors. Blastoff management was trying to hawk a product that engineers found illegitimate. Dot-com companies, meanwhile, were running out of cash or liquidating. Idealab's Pets.com, with its popular sock puppet, had closed its doors and filed for bankruptcy.

During a meeting in early November, Peter asked his team for support, as part of his feverish bid to save Blastoff from the same fate. But Peter got no sympathy. Blastoff's avionics systems head, Doug Caldwell, told Peter in the meeting that he didn't have the backing of engineers. Caldwell said point-blank that engineers wouldn't support him because they didn't respect him. What was being told to investors was not reality. Peter listened and felt a mix of anger and betrayal. He felt that many of the engineers—while brilliant—were naïve or ignorant when it came to marketing and raising the capital needed to save their jobs. They were upset that the mockups he was showing to venture capitalists didn't reflect actual hardware. As Peter let the engineers vent, he thought to himself, *How the hell did I get stuck in this situation?* He had sold his house for far less than he could have made had he taken his time, and he'd gotten pennies to the dollar when he'd brought his things out of storage and held a one-day estate sale. Now he was trying to inspire engineers while fighting for the company's survival. He was selling what he still believed in his heart was possible, but it was hard when no one seemed to believe in *him*.

Peter felt even worse because he had personally recruited a number of people away from JPL, including a brilliant young engineer named Chris Lewicki, who grew up in the dairy country in northern Wisconsin and was a leader of the SEDS chapter at the University of Arizona. Lewicki and many others had left stable, high-paying jobs to join this mission to the Moon. Peter had moved into a house in the hills of Pasadena, and two of his talented recruits—inventor Dezso Molnar and space entrepreneur George Whitesides—were his roommates. Peter had also brought in Bob Weiss as his number-two man and the head of marketing. It took Peter months to persuade Weiss to take the job. Weiss thought that Blastoff sounded like a project, not a business. But Peter was unrelenting, and would track Weiss down on vacation to woo him with some new angle. Weiss would retreat with phone in hand to a quiet room or even a closet so his wife wouldn't hear his latest idea about leaving moviemaking—to go to the Moon. Eventually, Peter sold him on the idea.

The public relations team at Blastoff, led by Diane Murphy, who had boundless energy and great contacts, had a massive media rollout in the works, including a multifaceted Web site and a polished video that began, "What if we told you that for the past year, an amazing team of scientists has been working on a mission to the Moon? . . . What if we told you this was not a part of a secret government program?"

But with each passing day, the dreams were turning to desperation.

Peter was in save-the-ship mode, working around the clock, sleeping a few hours here and there. He and Larry Gross were in talks to buy an Athena II rocket built for the Navy by Lockheed Martin. The government had canceled the project after spending $40 million, leaving Lockheed with a brand-new rocket. Lockheed said Blastoff could have it for $20 million.

Dezso Molnar, who had been involved in private space for more than fifteen years—he started working with rocket maker Bob Truax in 1984—knew that a bloodbath was coming. He had been hired at Blastoff after running the land-speed racing team for Craig Breedlove in the Black Hawk Desert, where the goal was to build a car to break the sound barrier. At Blastoff, Molnar told everyone to go to the doctor or dentist while they still had health coverage. "George, have you been to the dentist?" he asked George Whitesides. "I'm calling to make you an appointment. You're about to lose your job. I've seen this movie before."

By the end of 2000, Blastoff was splashing down. A decision was made to keep people on through the holidays. In early January, Peter and Bill called a meeting. The mood was grim, like standing on the deck of the *Titanic*. "Guys, I hate to say it," Peter began, "but this mission is being put on hold." Bill apologized that Idealab wasn't able to support the mission in the way it had intended. There were talks of resurrecting Blastoff when the economy improved, of Hail Mary passes, of investor meetings still scheduled.

Bob Weiss watched as employees packed up their things. An engineer carrying a box with belongings stopped at his desk on the way out the door. The engineer offered a joke.

"Hey, Bob, what did one aerospace engineer say to the other?"

"I give up," Bob said.

"You want fries with that burger?"

It took Bob a few moments to get it.

The pace had slowed to a stop at Blastoff. The parking lot was now mostly empty. Peter stayed at the company after the last employee left. He needed to circle back on contracts, including the deal with Lockheed, which tied Blastoff to monthly payments of $1 million. He needed to figure out what to do with the impressive hardware and software developed by Blastoff.

Bob Weiss had been in talks with a Japanese company that wanted to invest in Blastoff and deploy some sort of inflatable house on the Moon. Peter had a meeting with Jeff Berg, head of the talent agency ICM, who had leads on potential investors. Filmmakers Ron Howard and Brian Grazer had been about to come on board when the market tanked.

Early in 2001, Peter was forced to do some more personal downsizing, moving out of the gorgeous house in the hills above Pasadena and into a small two-bedroom apartment in Santa Monica. The address was 2408 Third Street, an address Peter remembered as: 2 (to the first power); 2 (squared); and 2 (cubed). Before leaving the Pasadena house, he sat by the pool and remembered how he, George, and Dezso had envisioned great pool parties. He had gone swimming maybe once.

Now it was time to refocus on the XPRIZE. He couldn't let the failure of Blastoff end his most important dream. Peter used the second bedroom of the Santa Monica apartment as his office. It had been nearly five years since the XPRIZE was announced under the arch in St. Louis. Gregg Maryniak and his team had kept things going—but barely. Peter had spent some time almost every day on the XPRIZE—connecting with teams, talking with board members and benefactors—but his focus had been on Blastoff, on this crazy moonshot, on the promise that Blastoff would generate millions for his real baby, the XPRIZE. He knew that teams were building actual hardware. Insiders told him Burt Rutan was working on something predictably counterintuitive and super cool. The XPRIZE had all of the ingredients it needed, except the prize money. He had $5 million promised and needed $5 million more. This was becoming a familiar refrain for Peter. In joining Blastoff, Peter had taken the biggest gamble of his life. And he had lost.

19

Elon's Inspiration

Peter's part-time assistant, Angel Panlasigui, opened the door to Peter's Santa Monica apartment late in the morning and saw what was becoming an increasingly familiar sight in early 2001: Peter still in his bathrobe, hair disheveled. The place was dark, the shades down. Peter had yet to emerge from the rubble of Blastoff.

Things hit a new low in early March, when Peter sat with a copy of *Fortune* magazine featuring Bill Gross on the cover. The title of the story by Joe Nocera was "Why Is He Still Smiling? Bill Gross Blew Through $800 Million in 8 Months (and he's got nothing to show for it)." It was painful to read the piece, which made no mention of Blastoff. Peter also had nothing to show for his last ten months. Unlike Gross, Peter had put everything on the line.

Then, in May 2001, when Peter was whiling away another afternoon in his apartment, Angel said there was a call from Larry Gross.

"Peter," Larry said, "I've got a fortieth birthday present for you."

Peter arrived at the Skybar, a rooftop watering hole on Sunset Boulevard, a week after Larry Gross told him he knew who was going to fund the continuation of Blastoff. It was two men who had made a fortune on the Internet and who were interested in space. Peter had never heard of the men, so he wrote down their names: Adeo Ressi and Elon Musk.

Peter usually approached pitch meetings with great enthusiasm, but tonight he felt subdued. He spotted Adeo by the Skybar pool, smoking a cigarette and looking out at the gold and glimmering Los Angeles sunset. He was tall and thin, a Giacometti walking man figure, and immediately affable. Adeo said Elon was running late but on his way. Elon was working on getting his pilot's license and was flying down from San Jose with his instructor. He had a new plane being built.

It was Sunday, June 3, 2001, and Adeo, Elon, and Peter were scheduled to have dinner at Asia de Cuba, adjacent to the Skybar in the Mondrian Hotel. Peter took in the beautiful women in filmy tops and short skirts, and felt overdressed in his suit and mock turtleneck. Adeo was in casual slacks and a shirt open at the collar. The music was pulsating, the lychee up martinis flowing, and the entire hotel was bathed in white, with minimalist accents of Hermès orange. Even the matches were stylish, with lime-green tips. Peter had taken notice of the white-clad valet team when he pulled up to the white-façade hotel. The valet attendants all clasped their hands in exactly the same way.

Peter joined Adeo for a drink near the pool. Adeo, who knew Larry Gross from a two-month stint as CEO of Idealab in New York, had just sold his Web development firm Methodfive and was working on turnarounds of lagging public companies. He said he and Elon had been housemates at the University of Pennsylvania. Elon was South African and had founded Zip2, a mapping and business services company, and cofounded PayPal, the online payment service being bought by eBay.

Their shared interest in space came to light during a late-night car ride the weekend before Memorial Day. As they drove back to New York City from Long Island on a cloudy night, talk turned to what they wanted to do next. As a joke, one of them said, "Why don't we do something in space?" When the laughter died down, Elon said, "Well, why *can't* we do something in space?" The debate went back and forth: Space was too expensive. *Why was it so expensive?* Space takes a lot of infrastructure. *Why does it take so much infrastructure?* Space is controlled by governments and strict regulation. *What happens if it is taken out of the government's hands*? Finally, they asked each other, *Why do we even think space is interesting?* This led to a discussion of where they would go if they could go to space. By the end of the car ride, they had their answer about what to do next. They knew exactly where they wanted to go.

Before Adeo could continue, Elon arrived and apologized for being late. The three men moved with drinks in hand from the Skybar to the restaurant and ordered a feast: pan-seared ahi tuna, miso grilled salmon, the Asian noodle box, and more. Tracks from the Buddha Bar collection played in the background. Peter found Elon immediately likable: good chemistry from the start, soft-spoken, polite, his words well chosen. Adeo was also great, but more of an extrovert who seemed to enjoy playing devil's advocate. Peter knew very little about the two coming into the night's dinner. He'd had a quick conference call the week before with Elon and Adeo, thought they said all the right things, and took Elon's accent to be British.

Adeo made Peter laugh knowingly when he said, "I think every geek is a bit of a space buff." Peter talked about Blastoff, the XPRIZE, ZERO-G, and Space Adventures, his company with Eric Anderson, which had brokered the final part of the deal to send the world's first space tourist, Dennis Tito, to the International Space Station aboard a Russian Soyuz spacecraft for $20 million. Peter talked briefly about how NASA had tried to stop Tito from flying, but Tito had launched on April 28 and landed safely in Kazakhstan on May 6. It was big news in space circles that Tito, an American, had to fly with Russian cosmonauts and was not allowed on the U.S. side of the space station. Peter had met Tito when he came to California to interview with Bill and Larry Gross for the Blastoff position. Peter pitched Tito on an XPRIZE sponsorship, but Tito said he didn't want to sponsor a prize; he may want to fund a team and compete himself.

As Peter talked about what the "ultimate space company" would look like, putting on Moon missions and suborbital and ZERO-G flights, Elon and Adeo said they had set their sights on something different, something even more difficult. They wanted to reach the Red Planet. Their mission, decided that night on Long Island, was to put humanity on Mars. They wanted to spend money to "shame, embarrass, or prod" the government into doing a human mission to Mars.

Peter cautioned that he had seen a lot of great missions fail because one wealthy backer or another expected other wealthy individuals to support his vision. "But every wealthy person has his own vision," Peter said. To make such a mission work, Peter now believed, one very wealthy and determined person would need to be willing to pay for everything, something he had yet to encounter. As Elon listened to Peter talk about Blastoff and his other companies, he thought Peter's heart was in the right place, and it was obvious that he cared deeply about the future of space travel. But the Blastoff plan didn't make sense to him. He didn't think sending a rover back to the Moon was going to reignite space travel.

Returning to the subject of Mars, Elon said, "We want to do something that's significant enough but does something for a reasonable budget—for a couple million." Adeo added that they had $10 million to $15 million to spend, but wanted to start with a $1 million or $2 million project.

Peter was stunned to hear such a low figure of "a couple million," but knew that in aerospace, a couple million often led to many more millions. He listened to them with interest, but made sure not to get his hopes up.

Still, at the very least, even if Elon and Adeo did nothing, Peter had met some smart guys who would be friends. Elon was a major Trekkie. He had watched all of the episodes as a kid in South Africa, dreamed of spaceships, and read Heinlein, Asimov, and Douglas Adams. He said his successes in Silicon Valley had paved the way for his future in space—not unlike what Jeff Bezos had told him.

Adeo, Elon, and Peter shared an interest in using small teams to accomplish what only the government had done before, though Elon remarked that he saw the government as "a corporation—the biggest corporation." And like Peter, Adeo and Elon didn't see NASA as the bad guy, but instead saw the public's expectation of perfection as an unnecessary speed limit on innovation. The expectation that everything needed to go right caused NASA to be overly cautious.

Elon talked about how he had been trying to understand why the world had not made more progress in sending people to the Moon or Mars. "There was a lot of excitement with the Apollo program and the dream of space travel," Elon said. "It was ignited, and somehow that dream died or was put into stasis." He said he was "trying to figure out if there is anything we can do to bring back the dream of Apollo. Maybe even a philanthropic mission."

Peter could see that Elon—a logician and engineer above all else—needed to understand the physical and psychological limitations of why rockets hadn't improved since the sixties. Peter knew that Elon and Adeo were in research mode, talking to a mix of major players and fringe players in the world of aerospace. Peter told them he thought Mars was a great place to set up a future colony, but "the Moon was economical. The Moon is a place where you can go to gain access to resources and you are close enough to Earth that you can build on it."

But Elon was not interested in the Moon. "Maybe we do a mini greenhouse to Mars," he said. "Maybe mice to Mars. Maybe we grow samples of food crops." He said it had been obvious to him since childhood, when the Moon was already reached, that "Mars is next." Also, Mars was even more mythical, more unattainable. The Moon was 240,000 miles away from Earth. Mars was about 34 million miles away when their orbits were together on the same side of the sun, but as much as 250 million miles apart when the two planets were on the opposite sides of the sun. The Moon was the talcum face in the night sky. Mars was the out-of-reach gem. Mars would take at least half a year to reach, using optimal energy cost. It would take a year and a half for the planets to realign,* and then it would take another six months to return. Elon said he thought such a mission sounded entirely doable. Earlier, in May, Elon had attended a Mars Society event with Jim Cameron, who was working on a six-episode TV miniseries on the Red Planet. Over breakfast the next morning with Mars Society cofounder Robert Zubrin, Elon had pledged $100,000 to the cause.

Peter, who had been in constant pitching mode for what felt like an eternity, tried to sit back and listen, but he kept finding himself back in the mode of selling. Elon and Adeo asked whether a manned mission to Mars was possible for less than $10 billion. *Now the budget is edging closer to reality,* Peter thought. Peter said enticingly, "I've got a way for you to do it for a tenth of the cost, for one billion." Everyone leaned in. "You could build a one-way mission with existing Russian hardware. You send a few people with the goal of their living on Mars for five years until a resupply or rescue mission gets there. They will be the world's first Martians." Adeo and Elon loved the idea and spent the next hour in a fast-paced discussion, going over details and obstacles. Peter then told them that whatever they did in space, they needed to first prove themselves, "step by step."

Even though Peter had figured out quickly that Elon and Adeo had no

*A craft departing Earth naturally has all of Earth's orbital velocity plus the rocket velocity toward Mars. You time the journey to intercept Mars, taking into account that it is also orbiting. Going back to Earth, the craft has Mars's orbital velocity plus the rocket velocity toward Earth.

interest in Blastoff, he admired how these guys were willing to gamble on space. And the evening offered a pleasant surprise: Elon loved the XPRIZE idea. "I could be a supporter of that," he said. Elon thought that unlike Blastoff, the XPRIZE could jump-start an industry and rekindle public interest in space. Elon and Adeo appeared keenly interested as Peter talked about the *Spirit of St. Louis,* Charles Lindbergh, and the teams that had signed up.

"I'd love to meet some of the teams," Elon said. Adeo offered to join the XPRIZE board.

Well after midnight, the men finally walked out of the restaurant and were greeted by the white-clad valet. They had plans to meet the next morning to continue the discussion at the Idealab office.

As Peter drove back to his apartment in Santa Monica, he turned on his recorder and began to reflect. It was strange, he said, how until tonight, he had felt like the kid in the room. Now, at forty, he felt more like the elder statesman. Both Adeo and Elon were weeks shy of turning thirty. They were just starting on space; he had never been anywhere else.

"They only have ten million to fifteen million to spend," he said as he drove. "It's not going to be on Blastoff. They could be backers of the XPRIZE. Regardless, I get the feeling we'll be friends. I really liked this guy Elon. He was quieter, but sounded serious about space. I think I planted some ideas, some seeds, maybe some direction, tonight."

He continued, "It amazes me when I hear there is apathy about going to the Moon. Not from these guys tonight so much as from people who are not drawn to space at all. They have this kind of been-there-done-that mentality. People are fickle. What gets them excited?

"I think about sports teams and to me—BFD, so what? Why would people spend so much on sports? I guess they want to relate to that person. They want to be out on the court. Here I am, talking about sending robots to the Moon, and it's an apathy-inducing agent? Space has to be developed for more than science. It's about daring to do great things. That alone is valuable."

Peter paused as he pulled into his garage, the sound of his car blinker magnified by the silence, a kind of metronome to his thoughts. "Blastoff is dead," he said. "But my heartfelt passion of space is not."

Another few moments passed. "Dammit, I can't let this go," he said. "My mission in life is to make all of this happen. I've learned a ton of lessons. Whatever I do from here on out is going to be my show."

20

Burt and Paul's Big Adventure

"Let's go fly some models," said an excited Burt Rutan, rounding up a handful of engineers at lunch hour on a cool spring day in Mojave. The crew at Scaled Composites grabbed small foam airplanes and headed outside, walking down the flight line to the Mojave Tower. There, on the seventh-floor deck, eighty-seven feet up, Burt and the others began chucking model space planes into the air. The models, with variable shapes, were designed to have extreme drag and stability. They had models with fishing line, and models with streamers to measure the angle of attack. Some dropped straight down; others rode the air like gentle waves.

Burt watched, smiled, frowned, and studied. The foam models had different drag mechanisms, or ways to slow the plane: crude sorts of elevators on the trailing edge of the wing or tail for nose-up, high-drag descent; rudimentary ailerons on the ends of the model plane's wings to roll to the left or the right. One model had a large top flap intended to provide high drag and to raise the nose to an extreme angle. The strings helped to visualize the angle of the airflow during descent. Burt, who was right-handed in everything except controlling flight line models, suggested different angles for the planes to be thrown from. He and the crew were assessing whether the models were stable, tumbled, or descended at bizarre angles. If a plane looped, did it recover or flip over backward? Today's field trip was the latest exercise in Burt's quest to design a first-of-a-kind, affordable, and reusable spaceship. Burt knew that these crude models did not give an accurate answer, because a real atmospheric spaceship entry would be done at supersonic speeds, where the aerodynamics are very different. Burt loved the idea of the XPRIZE—and even more the idea of winning it—though he still found it disconcerting that Peter had announced the race without funding for the prize. Peter now had $5 million—for a $10 million prize. Burt was aware there were a number of teams in contention, with ideas ranging from rocket and balloon combinations to vertical-takeoff, vertical-landing missiles. Peter, who called Burt regularly to fish for information on what he was doing, kept assuring him that full funding of the prize was close at hand. And no one else had launched anything—yet.

At this point, Burt no longer considered one of his latest creations, the Proteus, to be a viable launch platform for the space competition, though its design was still very much influencing his thinking. Burt had toyed with the notion that the craft could be used to launch a single-seat rocket from high altitude to the edge of space, in the same way the B-52 had shot the winged X-15 skyward from 45,000 feet. But Burt had concluded that the Proteus would not work carrying a three-passenger cabin, an XPRIZE requirement. Since then, he had talked with a few engineers about other ideas he had for suborbital spaceships. From time to time, he shared sketches: some looked derivative, others futuristic, others uniquely Burt Rutan. He sketched on paper and by computer, and drew in pen and pencil on restaurant napkins. There was a delta wing (isosceles triangle) space plane with a bubble-shaped cabin up front to hold a half dozen space tourists, including the pilot. There was a Redstone-inspired rocket with a *Friendship 7*–like capsule on top. Burt's version of the rocket and capsule would detach after boost, return to Earth by parachute, and then be set down in front of Scaled by helicopter. He'd also drawn an oblong capsule tapered on one end, with two small feathered wings jutting out from the sides like arms.

The visit to the tower with the foam models would allow Burt to do basic hand calculations to determine the efficacy of his latest, more advanced designs. The calculations on lift and drag were not unlike those he'd done on planes as a kid, when he would build whatever he imagined using balsa wood pieces bought for sixteen cents apiece. Now, of course, he could turn to the computer and work with his aerodynamicist to look at hundreds of different calculations involving how air flowed around a design.*

The foam models were concepts, but they gave clues. Burt would see how they would do subsonically first. Like his doodles on napkins, the models were a part of Burt's process. With every new type of plane, Burt plotted and planned and worked out hundreds of details in his mind before testing anything in a computer. There was never an epiphany, a single "aha" moment; only iteration after iteration, layer after layer. Aesthetics were a part of performance. If a wing reached its performance goal, it was beautiful. If he put a sweep on a wingtip that looked like a shark fin, it was to improve performance. The fact that it looked cool was an unexpected benefit. He worked on every project until he reached the point where something in his gut told him he was right. Once he had that gut feeling, he was in a hurry.

Already, the spaceship project had Burt tapping into his likes and dislikes. In his mind, the most impressive manned launch system ever was the Northrop Grumman Lunar Module, designed less than four years

*By 2000, computational fluid dynamics had advanced pretty far. The first computer programs for numerical simulation of Navier-Stokes–derived equations appeared in the sixties. By the 1980s there were a number of them running quite effectively. Every aerospace company had its own. By 2000 these computer programs had become quite sophisticated and were approaching the hard problems: dealing with viscosity, and dealing with vortices. If you include both, you have the Navier-Stokes equation. If you include only vortices, you have the Euler equation. If you exclude both, you have the so-called full potential equation.

after the world's first manned spaceflight and flight-tested three years later. It descended from orbit, landed on the Moon, took off, and then ascended to lunar orbit. The most impressive plane was Lockheed's SR-71 Blackbird, a Mach 3-plus reconnaissance craft, designed just a decade and a half after the world moved away from slow piston-powered aircraft. Burt's philosophy was that small improvements to designs were not interesting. What mattered to him was "reaching way out."

Burt didn't like parachutes for recovery and couldn't imagine space tourists wanting to be inside a capsule that parachuted back to Earth and landed in the ocean. And despite his early drawing of a rocket and a capsule like the one that lobbed Alan Shepard to suborbital spaceflight forty years earlier on a fifteen-minute ride, Burt wanted his pilots to fly and land the craft. Amazingly, there had been only four manned suborbital spaceflights in U.S. history where the astronaut or pilot reached higher than sixty-two miles: Mercury-Redstone with Alan Shepard and Gus Grissom, both in 1961; and Joe Walker in the X-15, who in 1963 reached 354,200 feet, or sixty-seven miles. The gunmetal black X-15, built to test manned suborbital spaceflight, had smashed world speed and altitude records. Carried aloft under the wing of a modified B-52, the X-15 rocket plane was drop-launched for a rocket ascent followed by an unpowered glide back to Earth. Burt was at Edwards in the 1960s when the X-15 was flown, and was now consulting on his space plane with former X-15 test engineer Bob Hoey.

The most challenging part of the space plane equation was figuring out how to slow the descent after the craft coasted through apogee. Everything changed in space. There was no air, and therefore no aerodynamic forces. Burt needed to come up with a configuration that would give the ship maximum drag and natural stability as it returned through the upper atmosphere, and could be built quickly and cheaply and be safer than anything done before. He had begun looking at ways to have the craft returned in the thin atmosphere belly first. That would put the vehicle at close to the theoretical maximum drag: zero lift from the wings, maximum cross section in the direction of motion. If he didn't reorient the craft, it would plunge nose first at dangerously high speeds. This situation resulted in high aerodynamic loads (high stresses on the structure) and heating of the airframe. He wanted something that could approach the atmosphere at any attitude or flight path, then naturally reorient itself to the right angle without relying on the pilot or computer to do it. The X-15 had relied on a special nickel alloy with a space-age name, Inconel-X, for heat protection up to 1,200 degrees Fahrenheit and speeds up to Mach 6.7. The space shuttle used thousands of individually made tiles filled with high-grade sand. Anything Burt made would be built of composite materials, and have far lower heat tolerance than metals or custom-made tiles.* Additionally, his lightweight, high-drag ship would decelerate much higher up in the atmosphere than the X-15, where the loads and heating are considerably less.

Now, after years of thinking about what a homebuilt spaceship would involve, Burt was convinced that he was closing in on designing a craft that would work. To cut fuel costs and add safety, his rocket would be carried aloft like the X-15 by a mother ship similar to the Proteus. The rocket-powered plane would drop-launch from the mother ship's belly like the X-15. This would mean Burt would have to build his own B-52 equivalent *and* his own rocket. And here's where things got interesting. His idea for reentry—after sketches, models, debates, dreams, and analysis—was to bend the tail booms of the spacecraft upward while in space, where there is no aerodynamic load on the structure. The design was tricky. The feathering configuration would be used only for the flight phase where the ship is decelerating as it enters the atmosphere. After deceleration, it could descend to any altitude desired because it is flying slowly. However,

*The space shuttle tiles were six by six inches. There were about twenty thousand unique tiles. The shuttle was thus sometimes referred to as the "flying brickyard." The tiles, including the borosilicate coating, weigh less than a comparable piece of Styrofoam. They were 90 percent air. The heating of the tiles was due to convection, not friction. At Mach 15-plus the shock wave prevented air from slamming the shuttle underbelly. Heat was transmitted from the shock front to the surface via convection of a superhot layer of plasma. The tiles insulate so well that you can hold one by the bottom corners, even when the rest is red hot.

it must defeather to be able to make a proper gliding approach to land on the runway. The only design remotely like it was something that Burt had made as a teenager and brought to the Academy of Model Aeronautics national competition in Dallas. He was seventeen years old in 1960 when he built and flew a Nordic towline glider that used a mechanism called a dethermalizer to flip the horizontal stabilizer of his model glider upward midflight. He had soaked cotton in saltpeter and lit it before takeoff. The ignited cotton burned through a rubber band slowly, flipping the tail up.

Burt asked X-15 engineer Bob Hoey, who lived nearby, to help by looking at the hazards of the program. Hoey took one look at his friend's design for bending the tails and booms and thought it was as strange as anything he'd seen before. But he said the idea just might work. Hoey had also been a modeler as a youth, and had used dethermalizers. The concept was practical, but the application in space was entirely unknown. It had never been done before. Having lived through the high-risk, high-reward X-15 program, where every week brought agitation and exhilaration, Hoey told Burt that he and his team at Scaled were about to fly into the unknown. He said they should prepare to live with the likelihood of catastrophe. But Burt knew he could do this; he now had that feeling in his gut.

After a few more launches of the foam models from the Mojave Tower, twenty-six-year-old engineer Matt Stinemetze looked around. Burt was long gone. Matt chuckled. Burt made you feel like anything was possible, that the out-of-the-ordinary was ordinary. This was Scaled, where it wasn't unusual for someone to show up in the lobby with a suitcase of cash and insist he had the next great idea for aviation. Pilots flew in to show off their tricked-out, homebuilt planes, including one number that resembled a walrus. The running joke about living here was "How many people live in Mojave? . . . About half of them."

Matt had moved to Mojave shortly after graduating and a week after his honeymoon. He understood why Burt settled here: cheap rents, space to build and fly, and little oversight. The area was classified as "unpopulated" by the Federal Aviation Administration, perfect for testing and flying experimental and homebuilt aircraft. As Matt saw it, it was also a place that brilliantly weeded out those who were not serious about the job. Mojave's main drag was nothing more than truck stops, motels, gas stations, and fast food. Trash blew in and got caught on cyclone fences. The desert was not for the meek. One of the longest-living plants on Earth—the creosote bush—prospered in Mojave, along with the lethally venomous Mojave green pit viper. But this was also home to the *Voyager,* and to dozens of original planes made from unexpected materials. Twenty-three miles down Highway 58 was Edwards Air Force Base. Mojave was made by the frontier spirit. It remained a place where outsized ideas were not to be tamed.

Stinemetze, in T-shirt, jeans, and with hair halfway down his back, had been at Scaled Composites for more than two years, and was still intimidated by the man he'd dreamed of working for since he was a kid in Kansas. This was the reason Matt had come to Scaled. As he put it, wonderfully weird shit happened here all the time. He'd known about Burt as the face on the cover of aviation magazines his dad read. He thought Burt's backward-looking planes were the coolest things he'd ever seen outside the Kansas Cosmonaut Space Center.

Matt was having a great time flying the models off the tower; it was a hobby that had filled his childhood. He'd never had the money to spend much time in or around real planes, but he always made and flew models, the stranger the better. He was seven years old when the space shuttle flew for the first time in 1981, and he watched as his dad, a city sanitation worker, snapped pictures of it on TV. His parents let him use sharp tools with big blades when he was just a kid, and he nearly cut off a finger or two. He got stitched up and never made the same mistake again. He worked on old cars and even built his own all-composite sailboat modeled after the

Koch brothers' America's Cup catamaran.* In college at Wichita State, where he earned his aerospace degree, he led a team that built a small jet. When Matt interviewed at Scaled, Burt seemed most interested in what he was building for fun. Scaled was filled with guys like him who were not Ivy League, but who were aerospace graduates or had simply been born with a need to make stuff, including their own planes.

When Matt arrived at Scaled, business was slow and the company took on projects to keep everyone employed. Matt was given the job of building a twenty-one-foot-long carbon fiber spine of a cow for a prominent artist in Washington, D.C. He then worked in the bids and proposals department with his mentor Cory Bird, an artist, perfectionist, and self-taught aerodynamicist who arrived for work on his first day at Scaled with his own toolbox. Matt was then involved in a cruise missile project, followed by the building of a prototype plane for the carmaker Toyota. It was in a meeting in 2000 when Matt began hearing Burt talk in detail about doing a spaceship. Burt spent an entire lengthy meeting one day talking about how he'd make the cabin door of the ship. Matt had never sought out projects directly run by Burt, finding him unnerving. Matt was wiry and watchful; Burt was big and had blue eyes that cut right through you. Matt had intentionally kept his head down while learning the ropes.

Now, flying a foam model straight off the tower deck, Matt thought about the idea of sending a privately built ship to space from Mojave. He felt he was ready for something new, maybe even working with the man himself.

In a matter of weeks, in fall 2000, Burt was seated across the table from Microsoft cofounder Paul Allen in Allen's office in Seattle. Soup and salad were served, but Burt didn't touch the food. He knew his time was

*Burt and the team at Scaled designed the first hard sail ever used in an America's Cup race. The wing sail was for Dennis Conner's catamaran *Stars & Stripes*, raced in 1988.

limited. Burt got to the point. Since their last discussions about using the Proteus not just for broadband communications but possibly for single-person spaceflight, Burt had looked at the problem of reentry with the goal of finding a generic solution to get people to and from space. When he told Allen in earlier meetings that he didn't have the solution yet, Allen said to come and find him when he did. Now, Burt told him that he'd figured out the technology. He was confident that he had come up with a breakthrough design that would provide a "carefree" reentry.

Allen listened and said little, occasionally looking out the window. The view to the west was of the city's waterfront and the Olympic Mountains. To the southwest was the Seattle Seahawks' new football stadium, under construction. Allen had bought the NFL team in 1996. To the south was Mount Rainier and the Cascade Mountains. Allen had followed Burt's career from afar, thought he had a brilliant body of work, and had studied his safety record, noting that none of his designs ever crashed during testing. He and Burt had talked several times about a shared desire to get America back to space. They also shared a passion for coming up with new and unconventional solutions to old problems. Allen wanted to bring back the sense of daring, imagination, and technological skill that gave NASA its successes, but get there with private industry. Both men believed that the "little guy" needed to restart manned spaceflight, or it wouldn't be done at all. Allen was intrigued by the XPRIZE, but had been told that the financing for the award was shaky—they were four years in and $5 million short.

During the lunch in Allen's office, Burt talked about the fatal crash of the X-15 involving his friend Mike Adams, and about his plans for slowing his vehicle. He used the term "feather" to describe how the wings would bend up to create drag and turn the ship belly first for its trip back to Earth. Allen, who loved playing badminton as a youth, likened the feather concept to a badminton birdie, which, launched from a racket, oriented itself nose first in the direction of flight. *This is a genius approach,* Allen thought. *It can go subsonic, supersonic, go to space, and come home as a benign glider*. Allen had great respect for NASA, but believed its systems had become too expensive and that innovation had gotten lost. It was time for the torch to be passed to entrepreneurs.

Burt said that based on his decades of experience building airplanes quickly, using composite materials and techniques similar to the building of boats or surfboards, he could do a space project "real quick." He would keep costs down by building it basically with cable and pushrod flight controls. He didn't want a computer as the pilot. He loved the idea that the last time anyone went supersonic in a mechanically controlled plane was Chuck Yeager in 1947.

"Have you done a rocket before?" Allen asked.

"No," Burt replied. "But I believe in this so strongly, that I would fund it myself if I had the money."

At that, Allen stood up from his chair, reached his hand out, and said, "Let's do it."

It took almost a year from the handshake for Burt and Paul Allen to sign an operating agreement. There were provisions in the contract that Burt was not about to accept. The biggest point of contention was the stipulation that if Allen for some reason ended the program at his discretion, Burt would not be allowed to use any of the ideas he had developed. Allen would own the intellectual property—ideas, concepts, drawings, and models developed even before the two started working together. When Burt protested, Allen's lawyers told him that they'd had people before who would "bail and go compete with Mr. Allen with someone else." The lawyers reassured him that they were protecting Mr. Allen's liabilities. "This is normal, don't worry," they said. Burt wasn't worried; he was *haunted* by the idea that he could be the only guy in America who could not do a space plane. Burt told his employees at Scaled, "I'm going to stick with it until we get something that doesn't look so bad."

Slowly, the two sides found compromises and set up a company called Mojave Aerospace Ventures. Allen's point person on the venture was Dave Moore, who handled a number of Allen's investments and had been an early Microsoft employee. Moore thought the deal should be simple: At the very beginning, Burt and his company would own the majority share in the joint venture. As Allen funded it, that share—and their ownership—would be diluted.

The tricky part for Moore, a self-described "hard-ass" when it came to negotiations, was how to put a valuation on an idea that was still unproven but came from someone who was a "preeminent star," and was considerably smarter than just about everyone in the field. Moore always had the advantage of negotiating from a position of strength; his boss had the billions.

Moore had started at Microsoft as a software designer in 1981, when the company had fewer than seventy employees and personal computers were just getting started. When Moore went for his interview at Microsoft, he was asked math-related questions and given a technical test. When he passed the test, he was moved up to talk to Paul Allen. Allen asked him what he'd been working on in his previous job, and Moore said he was creating an electronic exchange to design parts on a computer and then send the parts to manufacturing to run the machines. When he mentioned he was using "spline," a mathematical method to determine shapes, Allen said excitedly, "You know about *splines*?" Allen got up and went next door to Bill Gates's office, came back and said, "Bill would like to talk to you." Moore then gave Bill Gates a ninety-minute tutorial on splines.* Moore was hired to work on Microsoft's first business graphics. He left Microsoft in 1997, when the

---

*Splines are complex curves formed by gluing together simple curves. Simple equations can represent only simple forms. They are "simple" precisely because they are easy to describe; you need only a few numbers—or parameters. But a simple curve described by few parameters will be "rigid"—you can't force it to pass through very many points in given directions. A spline, though, can pass through as many points as you like in whatever direction you like, because it's stitched together from little pieces, each of which is simple and passes through a few of the points. The next piece passes through the next set of points, et cetera. The trick is how you stitch the curves together. Naturally you want the ends to meet up. That's a given. But you want more. You want the ends to meet up nicely, for some definition of "nicely." The most common splines are "cubic splines," polynomials of degree 3.

company had 25,000 employees, and began doing technical due diligence for venture capital firms. He was soon pulled back into the Microsoft fold when Allen asked him to look into some of his potential investments. Moore had worked closely over the years with a number of innovative and "unique" thinkers, so he welcomed the chance to work with Burt Rutan.

Before the handshake between Burt and his boss, Moore had met with a number of people who wanted Allen to fund their space companies, even talking at one point with Buzz Aldrin. Aldrin was on the board of directors of a green technology company, and when the subject turned to spaceflight and advanced avionics, Burt's name came up. Moore asked Aldrin whether he thought Burt could build a manned suborbital spaceship, and Aldrin said no; he thought Burt "wouldn't be successful." Moore had also met with Peter Diamandis, who pitched him on Blastoff and made an argument for Allen to come in as the title sponsor of the XPRIZE. Moore listened to Peter's pitch, but never let on that his boss was in talks with Burt. Moore met with Gary Hudson, who'd been at the Montrose, Colorado, weekend event when the XPRIZE idea was hatched. Hudson had a company called Rotary Rocket, funded in large part by none other than Walt Anderson. Moore had flown to Mojave in his own plane, a Socata TBM 700, and sat in the Rotary's simulator. Even though Moore was an experienced pilot, and was in the process of getting his helicopter rating, he couldn't come close to landing the simulator. Moore thought the propulsion system on the craft, called Roton—with a spinning wheel of engines on its base to lift the rocket and another set of spinning rockets on rotor blades to recover the vehicle like a helicopter—sounded "crazy." He returned to Seattle and told Allen, "You shouldn't invest." Allen didn't. In 2000, Rotary Rocket closed, after having burned through about $30 million. The Roton, standing 63 feet high—the same height as the tail of a 747—was now on display at the entrance to the Mojave Airport.

When contract talks started, Burt gave Moore a bid based on twenty-one tasks to be completed during the spaceship program, from building a flight simulator and performing glide tests to flying "three people twice above 100 km in less than 2 weeks with same spaceship . . . achieve the goal defined by the XPRIZE organization." The twenty-first task was to fly the spaceship every Tuesday for five months to show its reliability and direct operating costs. Burt thought it was a neat task, but got cold feet as he couldn't put a price on the costs before development was under way. Task twenty-one was nixed. The jointly owned Mojave Aerospace Ventures would hold the intellectual property developed during the program. Moore would be CEO of the new venture. During the drawn-out negotiations, Burt didn't get a nickel from Allen and wondered whether he ever would. But he still charged ahead with the design of the rocket and mother ship and began studying types of propulsion.

After months of back-and-forth, Burt finally got a contract he could accept. Allen would be the majority owner of Mojave Aerospace Ventures, and Burt would have a strong minority interest. In March 2001, the two sides signed a deal. They agreed on the need to do the program covertly. They didn't want anyone—including NASA—to find out what they were doing.

In April, Burt and Moore met to go over insurance, launch licenses, the timing of permits, patents on new technology, and when powered tests might begin. When the meeting ended, the two walked down the hall, pausing before heading outside. The price tag of the whole program was put at about $18 million. Turning to Burt, Moore said, "Well, how much do you need to get started?" Burt went over figures in his head: *Thirty thousand? Eighty thousand? One hundred thousand?* He was calculating what would come first when Moore said, "Why don't I just put $3 million in a bank account now and let me know when you need more."

Burt tried not to smile too big. Three million to start. That would work.

By early June 2001, Burt had completed his preliminary design of the launch aircraft, the rocket motor, and the basic spaceship. Burt had given the rocket the in-house name of *SpaceShipOne*, which wasn't popular. Dave Moore came up with names close to his boss's heart, including "Faye"—for Paul's mother—as well as some Native American names for various birds. Burt defended his name, saying that kids fantasized about flying to space and always used the term "spaceship." He wanted to emphasize the informal, fantasy nature of the first nongovernmental manned space program. Fabrication of the mother ship was set to begin. He sent employees armed with wrenches to the local junkyard to pry handles off cars to be used temporarily for the simulator and spaceship. He sent one of his builders, who happened to look like a biker, to another junkyard to buy J-85 engines on the cheap. He found his hybrid rocket propulsion expert in Huntsville, Alabama. The guy hadn't flown any rockets to space, but he had built a rocket-fueled bike that he rode at impressive speeds along the back roads of Alabama.

In a cordoned-off area of Scaled's hangar, the mother ship—which would carry *SpaceShipOne* to altitude for a drop launch—was given a name: the *White Knight*. It was named by Cory Bird, who referred to the windows as slits in a knight's helmet and the sharp slender boom fronts as jousting spears. In another curtained-off area, the rocket-powered *SpaceShipOne* program appeared stalled. A few things were being built, but it didn't have the momentum of the *White Knight*. To get the rocket part of the equation moving, a new project engineer was named: Matt Stinemetze.

Stinemetze, thinking about the pressure of running the spaceship program, realized the goal of a manned space program was so preposterous that it fit with Scaled. Burt didn't give a shit about doing stuff that was impossible or nuts. Why should anyone else give it a second thought?

21

A Lifeline for the XPRIZE

When the invitation came in to join a group of friends on a ski trip in early January 2001, Erik Lindbergh hesitated. He was doing better, thanks to the breakthrough arthritis drug Enbrel, but he hadn't skied in more than seven years. After returning from the Mayo Clinic, Erik had taken stock of his past and future. He looked over his treasured Atomic telemarks, Dynamic VR27s, and Völkl Snow Rangers, and said, "Who am I kidding? I'm walking with a cane." He sold nine pairs of skis, each representing a different kind of joy: skis for powder, skis for downhill, skis for slalom racing.

But now here he stood, on his oldest and sturdiest pair of cross-country skis, at the top of a gentle sloping hill, situated about twenty miles east of Stevens Pass in Washington. He was wearing the one pair he hadn't sold, his Black Diamond Vector cross-country skis, and his leather telemark boots. He figured that if he could walk, he could still cross-country ski. His wife, Mara, was with him, and they had a new addition to the Lindbergh clan: their six-month-old son, Gus, bundled up in a carrier on Erik's back. The skies were blue, the sun shined, and four inches of fresh powder beckoned.

Starting out that morning from their friends' cabin in Scottish Lakes High Camp, Erik had planned to just tool around and do the best he could. He had two relatively new knees, one fused foot, and the wear and tear from years of living with rheumatoid arthritis. He didn't know how he would hold up. But after he and Mara had cross-country skied about a quarter of a mile, Erik felt great. The only thing feeling old and stiff was his gear. Still, there was a difference between walking and running—between cross-country and downhill.

"I have to try it," he told Mara after they reached the top of the hill. It was a gentle rolling slope, a far cry from the slalom courses and cornices of his youth. But the gentle slope looked like a blast. He took nothing for granted anymore, not walking, sitting comfortably, or sleeping through the night. This was all a gift: like the gnarled lumber of his sculptures, he had been handed a second chance. And now he was a dad who could carry his son on his back.

He pushed off with his skis and had that familiar feeling of gravity's tug and momentum's pull, and of air—cold, fresh air. It was effortless.

"I can ski again!" Erik yelled out.

A moment later, he added with a laugh, "And I need better gear again!"

Erik was back home working in his woodshop and putting the finishing touches on one of his sculptures when he stopped what he was doing to call Gregg Maryniak. He had been toying with an idea that was both rebellious and reverential. It was something friends and strangers had asked him about for years, something he would dismiss before the person could finish the sentence. The idea had been hatched months earlier when he was holding wood pieces up to the sunlight, and the plan had taken shape in his mind as he finished the *Spirit of St. Louis* sculpture. The idea was affirmed by his days on the ski slopes and by the mornings when he woke up feeling great.

Erik caught Gregg at work. Gregg was the only salaried employee left at the XPRIZE, and he was working part time and hustling for outside sources of income. He kept the office open in the St. Louis Science Center in case someone called about a possible sponsorship. Everyone else, Peter included, was off salary. Erik was on the XPRIZE board and attended meeting after meeting where desperation had set in around fund-raising. Peter, back to the XPRIZE after the letdown of Blastoff, was calling on friends and family to keep operations afloat. He was putting in his own money to sustain his dream.

Erik told Gregg that he had landed on an idea that could save the XPRIZE, get a ton of attention, and be a huge personal challenge. "I want to recreate grandfather's flight," Erik said.

There was silence. Finally, Gregg said he thought it was a very bad idea. "It's way too risky," Gregg said. "Erik, you have a nice wife, a young child. *Why?* It's thirty-six hundred miles over wet stuff. A helicopter can rescue you the first fifty miles or the last fifty miles."

Even in a modern plane, nonstop solo transatlantic flying was risky, Gregg noted. Erik was still living with the residual damage from rheumatoid arthritis, and the flight would have him seated, alone, in a cockpit, for close to twenty hours straight. Modern technology could help to a degree. Gregg likened it to the difference between climbing Mount Everest when Edmund Hillary did it in 1953 and climbing Everest today; the modern climber had better equipment, communications technology, and training, but the potential perils remained: unpredictable weather, equipment failure, human error, and exhaustion.

Erik was undeterred. He explained he'd come up with the idea while making the *Spirit of St. Louis* wood sculpture, and it became more plausible when he found himself gaining strength after finding the new treatment for his rheumatoid arthritis. He told Gregg he had even started skiing again. His reasons for wanting to do the flight were threefold: to better understand what his grandfather had experienced; to show people—especially kids—that if they were suffering, they could get their lives back; and to spotlight how far aviation had come and, with the XPRIZE, how far it could go.

"I think we can raise a lot of money and attention for the XPRIZE," Erik said, noting that the flight would be done in spring 2002, on the seventy-fifth anniversary of his grandfather's flight.

As Gregg listened to the reasons, he started warming to the idea, with certain caveats. They needed to do a feasibility study of what would be required to make the flight. He suggested they meet with Peter, Byron Lichtenberg, Marc Arnold, and a fellow named Joe Dobronski, a well-regarded chief test pilot and engineer for McDonnell Douglas who could advise them on the type of plane. Gregg was emphatic that even though transatlantic flights by small craft were common, Erik would have to undergo ditch and survival training. If he crashed at sea, Erik would need to know how to survive until a rescue operation could arrive.

The other part of the feasibility study was possibly the riskiest: Erik needed to broach the subject with his reticent and reclusive Lindbergh family. The idea of Erik's flight was taboo, like adding a new layer of paint to the *Mona Lisa*.

One of the first people Erik talked to was his aunt Reeve, Charles and Anne Morrow Lindbergh's youngest daughter. Her response was supportive, but she was the only family member who was publicly a Lindbergh. And this assertiveness had taken time. When she married her first husband, Richard Brown, she was happy to be Mrs. Brown. There were no expectations. But over time, she became Reeve Lindbergh Brown and then Reeve Lindbergh. Her way of dealing with the Lindbergh legacy was to write about it; her newest book was a memoir detailing her mother's final seventeen months living with her on her farm in Vermont.

Reeve told Erik that the flight sounded daring. It seemed to her only yesterday that Erik could barely walk because of his rheumatoid arthritis. But she cautioned him that others in the family would be less gracious in their responses. They would probably say he was doing a "big commercial publicity stunt." For decades, a family lawyer named James W. Lloyd had been charged with enforcing Charles and Anne Lindbergh's instructions of "no use of name, likeness or signature [of Charles or Anne] for commercial purposes." This left the nonprofit Lindbergh Foundation dancing around the delicate issue of fund-raising while trading on the legacy. And in the end, it prevented the foundation from securing a sustainable endowment. Family members referred to Lloyd somewhat lovingly as "Dr. No." Although, when Erik brought fake "official Charles Lindbergh merchandise" to his attention, nothing was done. Reeve advised Erik that if he was to do the flight, he should emphasize *his* name, and reference who his grandparents were.

Erik's mother, Barbara, was troubled by the idea of the flight. She feared for her son's physical safety during and after. While long divorced from Erik's father, Jon Lindbergh, she still felt constricted by the tacit rule that you don't publicize who you are. If you do, bad things happen. When Erik and his siblings were growing up, she told them they could use her maiden name, Robbins, as their last name if they wanted. A few did, but most—like Reeve—eventually returned to Lindbergh.

News of Erik's plans spread fast among family members, and Barbara was soon getting calls from her other children. The men in the family were opposed to the idea, some more vociferously than others. The women in the family were concerned about Erik's safety. Erik's brother Morgan was a supporter of the XPRIZE and could see how the seventy-fifth anniversary flight would generate money and attention. But what if Erik's flight failed? How would that affect Grandfather's legacy? And why did Erik have the right to take this piece of family history and make it his own?

After weeks of back-and-forth with family members, Erik had another conversation with Reeve. She concluded that the women in the family feared he wouldn't make it, and the men feared he would.

But Erik had made up his mind. He was going to do this. Erik was done being crippled by this disease and hobbled by his legacy. As his grandfather once said in response to critics and skeptics, "Why shouldn't I fly from New York to Paris?"

Erik met Gregg and Peter in St. Louis to go over plans and budgets for the flight. Dobronski, Arnold, and Lichtenberg joined them. Peter had first heard of Erik's idea through Gregg. Peter didn't share Gregg's initial worries. He thought the flight would be life changing for Erik—affirming both his legacy and his restored health—and told Gregg, "That's awesome. I'm all in." He found it profound that Erik just might save a competition inspired by his grandfather.

A pilot for four decades, Gregg agreed that the XPRIZE financing effort had reached a point where it had run out of "altitude, ideas, and airspeed." They were scraping by and needed a cash infusion to keep operations going long enough so they could find a sponsor. Doug King, director of the St. Louis Science Center, had already rescued them several times.

Erik began sketching out a budget and goals, and talk turned to the type of plane that Erik would use for the transatlantic journey. Someone suggested they buy a Beechcraft Bonanza, a single-engine, six-seat general aviation plane, and then sell it after the flight. Dobronski said, "No, why don't you talk to Lancair, in Oregon? They make the best state-of-the-art airplane" for this type of flight. Erik and Gregg—now project manager for Erik's mission—agreed to go and take a look at what Lancair had to offer. They set a fund-raising goal of $1 million for the XPRIZE, and additional sums for the Lindbergh Foundation and the Arthritis Foundation.

Erik left the meeting thinking: *I'm going to recreate Grandfather's landmark journey in my own single-engine light monoplane and fly the same path over the Atlantic without stopping.* Peter left the meeting thinking, *This is oxygen for the XPRIZE.*

22

A Display of Hardware

Steve Bennett stood on the wet thick sand of Morecambe Bay on the northwest coast of England, preparing for the launch of his four-story-high rocket, Nova 1. The sand flats were beautiful and brutal, a place where pink-footed geese and delicate hairstreak butterflies coexisted with shifting channels, deadly quicksand, and a tide that came in fast and quiet like an advancing army.

It was early morning on November 22, 2001, and Bennett was here for the test flight of Nova 1, a two-stage rocket that he'd been developing in the five years since hearing of the XPRIZE. When he first started telling people he wanted to build rockets to go to space, he got looks like there was something seriously wrong with him. But when he began telling people he wanted to win an international competition involving *ten million dollars,* it suddenly became about beating the Yanks at their own game. Ordinary people were taking interest. Though the prize was only half funded, that had done little to dampen the enthusiasm of Bennett and other space entrepreneurs around the world: they had forged ahead full steam and now had hardware to show for their efforts, even if some were more viable than others.

Nova 1, one of the most serious entries in the field, would be the first rocket actually flown as a part of the XPRIZE. Soon after hearing about the XPRIZE, Bennett had left his job at the toothpaste maker Colgate—which at first supported his rocket pursuits by offering six months of paid leave—and founded Starchaser Industries. He spent the next three years living off his credit cards. A part-time teaching job at the University of Salford, just outside of Manchester, had kept what he described as "the wolf from the door." The job gave him access to a small laboratory, an office and telephone, and students who were interested in space technology and wanted to help him. He landed a huge break when one of his rockets featured on a Discovery Channel segment attracted support from a deep-pocketed, space-minded benefactor.

If this morning's launch went according to plan, Nova 1 would be the largest privately built rocket ever flown from British soil. The gleaming white rocket with the tapered throat—all Bennett's design—had the XPRIZE logo splashed across its fins. Nova 1 was big enough to carry one person, but Bennett would need a capsule fit for three to win the XPRIZE. Although no one was inside the Nova capsule, Bennett had every intention of one day flying to space in his own ship, and this flight would be an important stepping-stone toward that goal.

Today's flight would test the hardware, including the mobile launch tower, the Nova airframe, booster system, capsule parachute descent, and avionics. Bennett's rocket, the stuff he'd dreamed about both as child and adult, was big: more than 37 feet high, 4 feet in diameter, and weighing 1,643 pounds on liftoff. The estimated speed would be 500 miles per hour.

Bennett surveyed the scene, with its earthly beauty and untested machine. The shellfish grounds and shrimping channels of Morecambe Bay had been plied for generations, first by using horse and cart and later with tractor and nets. When Bennett first started coming here to test out smaller rockets, he met with the local cocklers, fishermen who eked out a hard living collecting the small, heart-shaped mollusks called cockles.

The cocklers knew the lay of the land in Morecambe Bay: the fast-moving tides, unpredictable channels, and invisible areas of quicksand. Today, launch day, the cocklers were being paid to shuttle media and VIPs on the back of their tractor beds from the main road to the rocket staging area.

The biggest question on Bennett's mind was whether the solid rocket motor propulsion system would work. He had nineteen solid motors that needed to ignite at precisely the same time. If they didn't ignite at once, the rocket would probably cartwheel over. The igniters had been tested three times in the shop, and there were four igniters in each of the nineteen motors—they had quadruple redundancy to make sure everything worked. Another concern was the nuclear power station five miles away. It would not be a good day if the rocket crashed into a nuclear power plant.

Bennett had arrived that morning before sunrise, when ground and air were blurred by heavy moisture, reminding him of a J.M.W. Turner landscape. The team relied on floodlights to set up. The cockle fishermen were happy to earn a little extra money and to have a break from the tedium of their days. Bennett had limited time before the tide returned. At around nine A.M., carrying duct tape and wearing a white construction helmet and a jacket with the Starchaser logo, Bennett was up in a cherry picker at the top of Nova 1. Three helicopters flew nearby. After years of dreaming, scrimping, building, and testing, the moment of truth was here. This was what he'd worked for—the answer to his fears that he would go through life as a "conventional person," that he'd be lying on his deathbed, bills paid, but unfulfilled.

Back down on the ground, Bennett positioned himself with a vantage point to take in the guests, media, and Mission Control, which in this case was a cabin resembling a carnival ticket booth set down on the back of a tractor bed. He had a full-time team of twenty, a legion of volunteers, reporters, and television crews, his benefactor, Paul Young, who had made his money in cell phone technology, and a group of cocklers, who had never seen anything like this. The Civil Aviation Authority, England's version of the Federal Aviation Administration, said Bennett needed to keep Nova 1 from flying above 10,000 feet. He was fine with that; just under two miles high was sufficient to test all of the systems.

At 10:30 A.M., a hush fell over the crowd. The countdown began: 10, 9, 8, 7—Bennett took a deep breath—5, 4, 3, 2, 1 . . . ignition!

The rocket lifted off with an earsplitting screech. A fiery plume became a thick white line, straight at first, then jagged, then billowing. The nineteen motors lit perfectly. At around 10,000 feet, the capsule separated from the rocket, and both began to float back to Earth by parachute. The cocklers cheered; Paul Young had a tear in his eye. Bennett tracked the trajectory of the parts against the cool blue sky. The wind was blowing at around 15 miles per hour, enough to cause concern. Both sections landed on target, but the capsule was picked up by the wind and dragged for some distance before it could be stopped.

The boy whose mom wouldn't let him stay up late to watch the Apollo 11 landing on TV had staged his own show. It wasn't the Moon, but he had succeeded without any help from the government. He'd built and flown the largest nongovernmental rocket ever launched from the United Kingdom mainland. His next challenge would be even bigger, even better: Nova 2, with room for three.

Bennett remained at Cape Morecambe Bay, as he'd taken to calling it, after everyone but the fishermen had gone home. The tide would soon wash in, erasing any hint of what had happened that day.

For their new rocket company, video game legend John Carmack and his wife, Katherine Anna Kang, came up with the name Armadillo Aerospace—a nod to the nocturnal animals that were well represented across Texas, and that often scurried around the Carmacks' property. A small armadillo in a flight suit was their team logo.

After his "larval stage" of research into rocketing, and after giving small amounts of cash to a few aerospace companies going after the CATS Prize, which no one won, Carmack had called the Dallas Area Rocket Society and asked whether anyone wanted to help him build experimental, high-powered rockets. Carmack hinted over the phone that he wanted to work on "a little bit of a special project, something extreme."

Neil Milburn, a member of the Dallas rocket group, was one of a handful of guys who responded to Carmack's invitation, drawn in by the promise of "something extreme." He and the other rocket enthusiasts went to meet with Carmack after hours at the id Software office. Milburn watched a guy with long hair, John Lennon glasses, T-shirt, and shorts amble down the stairs. Realizing it was Carmack, he thought, *What the heck am I getting myself involved with here?* But after an hour of talking, it was obvious that Carmack was exceedingly sharp and had done his homework; soon, a core group of nine people was in place, including Carmack and his wife.

Beyond the XPRIZE competition, Carmack's long-term goal was to create private suborbital manned spaceflight. He wanted to be credible in what he was doing and told the volunteer group, "I want this to be a rich man's hobby, not a poor man's aerospace company." That meant he would fund the company out of his own pocket, and the team would focus on building, testing, and flying. They would not have to create simulations of what they would do if only they had the money. Their goal, Carmack said, was to operate more like a software company, to be open source, post their successes and failures, and "celebrate the positive but don't get torn up about the negative."

And so the development process began on their rocket, Black Armadillo. Carmack bought one hundred acres of land east of Dallas to use for high-energy tests and helicopter drop tests of the capsule. Four months after team members had a space to work in, they managed to get a small craft to hover. Then the building of Black Armadillo proceeded. The rocket would be cylindrical, the nose cone providing the space for the occupant, similar in design and flight mission to the DC-X Delta Clipper with vertical takeoff, vertical landing. Russ Blink, an entrepreneur considered the electronics whiz of the group, was also the daredevil, doing freefall parachuting for fun. He would be Black Armadillo's pilot. Carmack had found a used Russian spacesuit on eBay for Blink to wear.

The all-volunteer group met twice a week, for four hours on Tuesday nights and eight to ten hours on Saturdays. Carmack had committed $500,000 a year to cover overhead, buy parts, and pay for launch costs. Everyone volunteered out of a shared goal of getting to space quickly and cheaply. The key members of the team were Blink, Milburn, and Phil Eaton.

There were many successes, and even more failures. Carmack was surprised that rocket building was more difficult than he'd expected. And while team members faced an array of technical challenges, they also found the bureaucratic side equally painstaking, reminding them of Wernher von Braun's quote: "We can lick gravity, but sometimes the paperwork is overwhelming." There were times when there were more people at the FAA's Office of Commercial Space Transportation working on their launch license than Armadillo had building the rocket.

Their propellant of choice was rocket-grade hydrogen peroxide, over 90 percent concentration compared with the household 3 percent variety. But team members soon found they would have a hard time buying large quantities of rocket-grade peroxide in the United States because the company they'd first purchased it from was concerned about lawsuits if Armadillo had a fatality or a major accident. As an alternative, Carmack had been enticed by the idea of concocting a mixed monopropellant of their own, that is, mixing a fuel with an oxidizer, in this case 50 percent strength hydrogen peroxide. This still had the simplicity of just hydrogen peroxide, and the purchasing requirements were straightforward. So they ended up using a mixture of 50 percent peroxide and alcohol, a relatively safe combination, but one that was harder to get working.

Black Armadillo would stand thirty feet tall and be up to six feet in diameter. It would do a DC-X-style landing, freefalling back through the atmosphere until it reached 15,000 feet above ground, when the engines—two banks of four engines—were relit. It would then continue falling tail-first before the engines would slow it to a safe landing, Buck Rogers style. The crew cabin was beneath the fuel tank, directly above the engines. Each engine would have about five thousand pounds of thrust. By far the most dangerous part of the mission was the return to Earth: if the engines didn't relight the way they were supposed to, there would be little chance of survival.

By early 2002, team members were getting ready to do more drop tests of their rocket. They hoped the rocket would fly in 2003. As Carmack and the team raced ahead, Carmack's wife, Katherine Kang, served a different role: she became the adult in the room.

Since their marriage in 2000, Katherine, a self-described "type A personality," managed the business side of things. She supported the idea of her husband's starting a rocket company, but she wanted it run as a business, in an orderly way, with the goal of one day making money. She expected results beyond flawless engine firings and a rocket that shot up and returned safely.

When she and John first started dating, she learned that he had a significant amount of money parked in a zero-interest checking account. He didn't know what to do with his millions, and didn't have time to think about investing. She told him that he needed to at least think about moving his money from a zero-interest account to a money market. She laughed years later remembering it. When he saw he was making money on his money, he thought it was "neat."

They shared a similar background of growing up without much parental support and having to pay their own way at a relatively young age. At around the time she and John began talking about a rocket company—he was still working at id Software—they separated funds into His, Hers, and Family. Katherine had begun looking into costs of insurance, launch licenses, lawyers, and environmental impact studies, and told John, "We need some reasonable ceiling. Any other hobby is manageable by comparison." The $500,000 a year was an underestimate. She could see that rockets burned through money faster than fuel.

As time passed, Katherine knew that some on the Armadillo team thought of her as the bad guy. She was monitoring the money. Still, the cash kept flowing out at Mach speed. She realized she needed something to illustrate to her husband just how much they were spending. She needed something tangible, something that would get his attention. After pondering different ideas, she sat him down one night and told him that for every dollar he spent on rockets, she was going to spend, too. On diamonds.

At the time, she wasn't particularly interested in jewelry, though that changed as she started collecting. If John wrote a check for $50,000 for insurance for a launch license, she would have $50,000 to spend shopping for diamonds. As her diamonds got bigger and bigger, her plan worked. John took notice. One day, seeing Katherine's haul, he said, "Whoa! *How much am I spending?*"

But she could see that John was happier than he'd been in years, working with a small team on a hugely challenging project. He told her he felt like he was back in his apartment in Wisconsin, bringing 3-D *Wolfenstein* to life. But instead of a virtual game where a World War II spy goes after Nazis, John was now writing software and building hardware to fly to space. They had engines to fire and launches to realize. And she had more diamonds to buy.

Across the globe, teams were at work on a range of rocket concepts. A group called TGV Rockets (Two Guys in a Vehicle) in Bethesda, Maryland, had a design for a rocket called Michelle B., described as a "suborbital bus service" that would take people sixty-two miles up and back. The group was heavy with aerospace veterans and military test pilots, but light on cash. Nonetheless, they were determined, fixed on the goal of offering no-frills suborbital flight priced by the pound.

One of two Canadian XPRIZE teams, Canadian Arrow, run by Geoff Sheerin, had constructed a full-scale engineering mockup based on the World War II–era V-2 rocket. Sheerin shopped the model around on the back of a flatbed truck in hopes of getting funding, hauling it to New York to display it in Rockefeller Center—quite a sight in the months following 9/11—and going on the *Today* show to talk about the rocket, the XPRIZE, and his space dreams. The team also built a V-2-style engine, which was made of steel, had brass propellant injectors, and would use liquid oxygen and alcohol as propellants to burn for fifty-five seconds.

The other Canadian team, da Vinci Project, was led by Brian Feeney. Feeney had been living in Hong Kong when he first read about the XPRIZE. His vision was of a rocket-powered spacecraft called Wild Fire, which would be air-launched from the world's largest reusable helium balloon at about 65,000 feet. Feeney was also building hardware as he searched for backers of his manned "rockoon."

In Hitchcock, Texas, Jim Akkerman, who spent his youth making and racing turbocharged go-karts and spent thirty-six years working as a NASA engineer, was toiling away in semiretirement (though he preferred to say he "graduated" from NASA) on his XPRIZE rocket, the Mayflower II. His plan was unlike any of the others, at least in terms of where it would start. He planned to launch the massive vehicle, weighing fifteen thousand pounds, from about thirty miles offshore in the Gulf of Mexico. According to his plan, the titanium rocket would bob upright like a buoy, and would have two ten-thousand-gallon fuel tanks: one for liquid natural gas, one for liquid oxygen. A cockpit would be on top of the rocket, and passengers would ride below. It would be powered by eight TRW engines, producing forty thousand pounds of thrust.

Akkerman had a pilot signed up. He named his company Advent Launch Services to represent the start of a new and private era of space. For now, its base of operations was a rice field near his house. He estimated that the Mayflower II would cost $10 million, which he was trying to raise. In the meantime, he was funding what he couldn't get donated out of his retirement savings. A devout Baptist, he wanted to win the XPRIZE, but he was really out to make the world a better place by giving more people access to God's great universe.

And down in a town south of Buenos Aires, Argentina, Pablo de León was making progress on his Gauchito "Little Cowboy" rocket. He had recently performed the first drop tests of a fully instrumented, reduced-scale model of the capsule out of a C-130 Hercules aircraft at 54,000 feet. Another drop test of the tomato-red Gauchito capsule was done from 90,000 feet, which set an altitude record for XPRIZE tests at the time. With both tests, de León was able to track the capsule and record GPS data and video from the descent, parachute opening, and recovery. At the same time, de León also did stratospheric glider testing and thermal testing of the spacesuit he had designed and constructed to use for eventual Gauchito launches.

De León had an impressive résumé, having managed the project that sent the first Argentine-made payloads to space on space shuttle *Endeavor* in 2001. His house was filled with rocket parts, including a small satellite that came to rest on his kitchen counter. He had made pressure suits for underwater use and spacesuits for NASA. But despite public enthusiasm for his project in Argentina, he was pulling in only around $50,000 a year in sponsorships, nowhere near what was needed. He and five engineers worked full time on the project, and he had more than thirty volunteers, most from local universities. Through scrounging and relying largely on donated time and goods, they built Gauchito's full-scale capsule out of wood and fiberglass, and also built a full simulator. The capsule was a concept demonstrator with running simulation software. The next goal was to build a 50-percent-scale rocket.

De León, who had recently met Peter Diamandis at the ISU summer session he attended on scholarship, felt a sense of camaraderie with the other rocket makers in the XPRIZE competition. Like everyone else, he searched for any news he could find on Burt Rutan. The last thing he'd heard was that Rutan was building a rocket to be drop-launched.

De León was the only competitor from Latin America. All he needed to do next—he told his team with a laugh—was come up with six unmanned flights before he could certify Gauchito for occupancy. But even without the funding he needed, De León was happy. He was certain he was taking part in something important; it felt like access to space was about to crack wide open.

Dumitru Popescu was in Bucharest, Romania, at the same Internet café where he'd first come across a story about the XPRIZE when he heard the news that Steve Bennett had successfully launched Nova 1 in England. Popescu got on the phone and called his wife. "We need to move faster," he told her.

Popescu had by now dropped out of the university where he was studying aerospace and worked around the clock in his father-in-law's backyard in Dragasane, a town about 100 miles west of Bucharest. When Popescu wasn't building the rocket, he was reading about building rockets. His parents told him he was wasting his time, and friends weighed in to say he was nuts. But his father-in-law, Constantin Turta, a skillful mechanical technician who prepared molds at the largest shoe factory in the area, was happy to share all he knew. Popescu's wife, Elena Simona Popescu, a French major, supported him by learning about rocketry, and was soon casting composites herself. The building of their rocket and engine had begun. Like Nova 1, their two-stage vehicle would separate in the sky and return to sea by parachute.

Earlier in 2001, Popescu had landed a meeting with Romania's first and only cosmonaut, Dumitru Prunariu, in the hope that Prunariu would support his team. Prunariu had flown aboard the Soyuz 40 in 1981 and been given a hero's welcome by both Soviet leader Leonid Brezhnev and Romanian president Nicolae Ceauşescu. The cosmonaut had coauthored several books on space and space technology and was now president of the Romanian Space Agency, ROSA. Popescu arrived at the Ministry of Research building in Bucharest and was welcomed into Prunariu's office.

Popescu found the cosmonaut predictably impressive and outgoing. Prunariu talked about his 1981 spaceflight, telling stories about the difficulty of sleeping in space and the challenge of walking when he first set foot back on Earth. Popescu told the cosmonaut about his rocket and engine designs, propellant ideas, and hopes of winning an international competition started in America called the XPRIZE. He shared photos of his work as well as drawings and simulations. Prunariu listened and smiled, but said the space agency would not be able to help. Before the meeting ended, though, Prunariu told him that the agency was holding a competition to generate new ideas for aerospace projects. He would be happy to include Popescu's ideas as a possible way to get him funding.

When the winners of the state-sponsored competition were announced, Popescu and his team had not been considered. When Popescu asked Prunariu why his team, ARCA—Aeronautics and Cosmonautics Romanian Association—was not included in the competition, the cosmonaut became considerably less friendly. Not long after, when Popescu's team began attracting attention from the local media, Prunariu was quoted as saying that ARCA was a group of "amateurs" with no idea what they were doing and no chance of winning the XPRIZE. Relations worsened from there.

In an e-mail exchange between Popescu and Prunariu, the cosmonaut said that in light of the terrorist attacks of September 11, 2001, Popescu and the work of his team "could be used for terrorist activities." Prunariu said that Popescu did not have the "clearance to manufacture missile guidance systems," and such work could only be done under government control. Popescu was worried by the implication that his work would be of potential interest to terrorists. It felt like the head of his country's space agency was linking him with terrorism. He learned that some in Parliament had begun asking Prunariu how it was that a group of students with no obvious funding source was managing to build a rocket when the government-funded ROSA appeared to be doing nothing.

Popescu and his volunteer team did their best to ignore Prunariu's increasingly public criticism and forge ahead with their work. Popescu wanted to build something that flew *and* looked great. He set out to create rockets the same way Apple created its products—the company had just released its first iPod—with attention to shape, color, and symmetry. Working outside year-round presented challenges, though. The weather was beautiful in the fall and spring, but freezing in the winter and stifling in the summer. One summer day, when Popescu and three others were building their orange launchpad, the temperature was 104 degrees Fahrenheit. They tried to protect their welder from the sun by having two people hold a blanket over him and another fan him with paper.

Popescu kept thinking about the launch of Nova 1 and figured other teams were soon to follow. Despite being labeled amateurs in public and linked to terrorists in private, ARCA picked up momentum, even finding a sponsor who donated hydrogen peroxide and other combustibles. The building of the fuel tank, the feed lines, and the launchpads was under way. A few neighbors who knew of the project occasionally wandered over, donating cash, tools, or old machine parts. A few stayed on to volunteer, and others set up chairs to watch.

After months of work on a rocket motor and the completion of the test stand, operations were moved to an open field at the far end of Popescu's father-in-law's property. Popescu had dug a deep trench one shovel of dirt at a time, until a bunker was made about 100 yards from the test stand. Construction helmets and ski goggles were donned as Popescu, his wife, father-in-law, and two volunteers readied for the test. The propellant was hydrogen peroxide and ethyl alcohol. They were confident. It was their first big test of whether the engine would be a success.

The moment of ignition came—and went.

Nothing. They looked at one another. Seconds later came the loudest explosion anyone had heard in peacetime Romania. The Popescu team peered out from the bunker. Everything was lost: test stand, fuel tank, rocket engine. Blown to smithereens. A perfect bomb; a very bad rocket engine.

Within minutes, police had swarmed the property. Windows were apparently shattered for a two-mile radius.

Popescu, still rattled, did his best to appear calm and play the whole thing down, saying they were just university students testing a rocket for a school project. He hoped the police had not read stories about them. He didn't want the explosion to end up in the press and be connected to the XPRIZE. He feared he could be disqualified.

"Come with us," the police said, taking him to the station. That afternoon, waiting to be questioned, Popescu considered what had gone wrong. He was pretty sure they had mixed the ingredients prematurely *and* had a delay in the ignition, allowing too much fuel to accumulate in the chamber. After some time, Popescu was released with the promise that he would be more careful next time.

He arrived back at his father-in-law's house and found the team quiet and sullen. He told them they'd lost expensive parts but learned invaluable lessons. He said he'd had an idea while riding to the station in the back of the police car. They needed to have a build-a-rocket party and invite the town.

23

Another Lindbergh Takes Flight

Erik Lindbergh had been home on the morning of September 11, 2001, when he heard over the radio that a jetliner had smashed into the South Tower of the World Trade Center in New York and that another plane had hit the North Tower. He heard that a third plane crashed into the Pentagon in Virginia, and a fourth plane, originally headed for San Francisco and identified as United Airlines Flight 93, crashed into a Pennsylvania field. Erik and his wife, Mara, held their son, Gus, who was one, especially close. Erik felt shock, anger, sadness, and a changed reality.

In the days that followed the attacks, the entire airspace of the United States and Canada was closed to civilians. Under a little known and never before used national security plan, only military and medical planes were allowed to fly. Most small private planes wouldn't fly for weeks.

Erik was scheduled to make his transatlantic flight in six months. In committing to this flight, he had overcome not only debilitating health issues, but also the considerable doubts of his family and friends. But in the wake of the attacks, he had wondered whether his flight was irrelevant. Erik had posed the question to Peter. As it turned out, Peter was being urged by some XPRIZE board members to give up on the space travel competition; because of 9/11, it simply wasn't going to happen. Peter was told, "No one will fund you now."

Peter refused to give in and told Erik he believed that the terrorist attacks made it even more imperative that Erik embark on his flight. The country needed something positive now more than ever. It needed everyday heroes. Others urged Erik to fly as a reminder of one of the most inspiring chapters in American history. Gregg also pushed his friend to go ahead with the flight. He wanted it for Erik, and for the XPRIZE.

Erik didn't know about being anybody's hero, but he did know that he wanted to help the XPRIZE raise much-needed operating income. The XPRIZE had lit him up when everything else in his life was dark. It made him think about how to solve problems in a different way, whether it was something grand like getting out of Earth's gravity well into space, or running when he should probably be walking. Once reduced to a life of sitting, of barely walking, he now wanted to soar.

But first, he had to live through his survival training.

The fuselage of the airplane began to fill with water, first to Erik's feet, then to his thighs, next to his waist. He would soon be submerged, strapped into his seat in the cockpit. As the water rose to his chin, he told himself: deep breaths, prepare to close your eyes. He raced through his mental checklist. The water will be saltwater. Maybe filled with oil, debris, gasoline. Don't unfasten early. Water pressing against the window and door makes it impossible to get out. Wait for the pressure to equalize, inside and out. Final deep breaths. He was under.

Eyes closed, Erik fumbled to free himself from his four-point harness. Feeling for the exit handle, he pushed open the door, kept a hand on it, and established his path out. He pushed himself up to the surface and opened his eyes. His heart was racing. He did it. There was his friend

Gregg, also in a soaked flight suit and helmet, having made it out the other side of the cockpit. Their trainer, close by in the deep pool in Groton, Connecticut, gave them a thumbs-up. It was time to do it again.

Erik's transatlantic mission was now three months away, in May 2002. Within minutes, Erik and Gregg were back in the simulator, which resembled the body of a Huey helicopter. It was raised eight feet above the pool. Their trainer gave a hand signal from behind his back to the simulator operator to dunk them. Erik's mind hurried to review what they'd gone over in their morning classes: Lean forward before impact to lock in the shoulder harness. Move thumbs to the front of the yoke. The impact of a crash landing could break your thumbs. Know the primary and secondary exits. Know the location of the emergency equipment.

The fuselage hit the water; they were going under. Faster this time. And they were being inverted, turned upside down. Erik told himself to breathe sooner this time. The water scrambled him, swirled bubbles all around. He was strapped into his seat, upside down. He closed his eyes. Same steps as right side up. Find egress, unfasten harness, feel for the path out.

He did this eleven more times that afternoon, in every crash landing simulation imaginable: fuselage right side up, upside down, jutted sideways, spinning, lights on in the facility, all lights off, dark as night. Hours later, waterlogged yet adrenalized, Erik and Gregg climbed out of the pool. Erik did additional training that day: using underwater emergency breathing devices after his sinuses had already filled with water, practicing how to brace for impact, and escaping from a cockpit filled with smoke. He even did a simulated, plucked-from-the-water helicopter rescue.

Survival Systems, the company providing their training, was just across a parking lot from the Long Island Sound. The company had been one of the first to sign on as sponsors of Erik's flight and was providing training at no cost. For Erik and Gregg, their second day of lessons would be even tougher: they were going to be tossed into the frigid waters of the Atlantic Ocean.

The next day started early. Gregg and Erik boarded a small boat with their trainers and headed out onto Long Island Sound. The wind whipped at around thirty knots, turning the gray sea into menacing swells. The air temperature hovered at around zero degrees Celsius (32°F), and the water temperature was around thirty-five degrees. Erik and Gregg stepped into their orange immersion suits, which left only their faces exposed. The suits were recommended for pilots flying over water below fifty-five degrees, and provided a thermal barrier and up to fifty pounds of buoyancy.

Erik and Gregg had practiced climbing into life rafts in the pool, but not in roiling water like this. The two took Dramamine in hopes of staving off seasickness.

Erik pulled the operating cord—the painter's line—to activate the CO_2 and inflate the raft. The raft was in the water, and Erik and Gregg took the plunge. In the water, Erik was breathing fast in shallow interrupted breaths as three-foot waves slapped his face. Without an immersion suit, he would be incapacitated in no more than twenty minutes, as blood moved away from his extremities toward his core. Fatal hypothermia would set in quickly. Heavier people would survive longer, given that they had more of an insulating layer. Erik was tall and lean like his grandfather, who was six feet two and weighed 170 pounds when he made his 1927 flight.

The raft separated from the Survival Systems boat, which needed to head toward shore to get out of the waves splashing over the sides and avoid hitting the raft. Erik and Gregg were on their own. They had already learned that climbing into a raft wasn't as easy as it looked, even in a pool. And now they had to do it at sea. Erik imagined how hard this would be if he were injured. Using a boarding ladder, they got themselves up and into the raft. Erik had turned green. Seasickness was stronger than Dramamine. From inside the raft, only waves and ocean were visible. They knew the critical steps they would need to take from here—get the canopy up if it didn't pop up on its own, rid the raft of any water, ensure all sharp objects were secured, and find first aid, flares, flags, and signal lights. Erik was having a hard time just sitting up.

When the Survival Systems crew returned, Erik and Gregg made their way slowly, unsteadily, from raft to boat. Still in their immersion suits, they slumped down on the deck, backs against the side of the boat. They had been in the raft for a little over an hour, and neither one felt like moving or uttering a word.

Finally, after a year of planning, training, and drama, the morning of the flight arrived—May 1, 2002. Erik walked toward his Lancair 300, ready to begin the trip to Paris. Like his grandfather, Erik had actually begun his journey at the San Diego International Airport—now known as Lindbergh Field. Erik flew from San Diego to St. Louis and days later flew to New York, where the most famous and difficult part of the flight was set to begin.

Charles Lindbergh's plane, the *Spirit of St. Louis,* was built for $10,580 by the scrappy Ryan Airlines company in San Diego. Its engine was a 220-horsepower (at sea level) Wright Whirlwind J-5C. It weighed 5,250 pounds fully loaded, including 450 gallons of fuel. It had a maximum airspeed at the start of the flight of 120 miles per hour, and 124 at the end, when the plane was lighter. Its range was put at 4,110 miles with zero wind and ideal economical airspeeds of 97 miles per hour at the start. Made of wood, metal, and treated fabric, it was painted silver and given the registration number N-X-211 (X was for "experimental"), which—along with *Spirit of St. Louis*—was painted in black. Charles was twenty-five when he set out on his journey. His pilot's chair was wicker, intentionally uncomfortable to keep him awake. He brought water and five ham sandwiches with him. For navigation, he had a compass, the stars, a wristwatch, and paper maps—with edges trimmed off to save on weight.

Erik's plane was a modified Lancair Columbia 300, sleek, single engine, and built for just under $300,000 by the Lancair Company in Redmond, Oregon. Christened the *New Spirit of St. Louis*, the plane had a 310-horsepower Continental engine driving a three-bladed prop. Erik's expected cruising speed was more than 180 miles per hour, and he would fly at an altitude of between 7,000 and 17,000 feet, depending on the weather. The airframe was all composite, painted white, and had the registration number N142LC. It weighed 4,260 pounds loaded. Erik was thirty-seven years old and would spend the flight in a leather seat, one of four in the plane. Erik had five ham sandwiches, his smeared with mustard. It was Erik's first trip to Europe, as it had been for his grandfather.

At this point, Erik had already flown three hundred hours in the Lancair 300 and knew its quirks and handling qualities. It was a fast, fixed-gear aircraft with controls that were nimble at lower speeds and stiffer at higher speeds. He had flown cross-country, dodging storms, icing, and turbulence. A high-tech, state-of-the-art Mission Control was set up in the St. Louis Science Center with Gregg at the helm. He would be there from takeoff to landing. Erik's plane was equipped with a global positioning system—which told him where he was, his speed, and the expected duration of his flight—and an Iridium phone. The Lancair's position would be updated every five minutes.

Erik would be taking off from Republic Field in Farmingdale, New York. Roosevelt Field, where his grandfather flew from—and where Peter once launched rockets—was now a shopping center.

On the tarmac of Republic Field, Erik stopped to hug Peter, who was going to hop on a commercial plane with Erik's luggage and beat him to Paris. Peter could not have imagined when he first read *The Spirit of St. Louis* that Erik Lindbergh would become so important to realizing his dreams. Erik was rescuing the XPRIZE during otherwise bleak times. Peter was also struck by the picture of Erik now versus when they'd first met in the restaurant in Kirkland. Erik looked healthy and strong—something miraculous to behold—and ready to repeat his grandfather's historic and physically demanding feat. Peter hugged Erik one more time and told him, "See you in Paris."

With his orange immersion suit unzipped at the waist, Erik climbed into the Lancair cockpit. Unlike his grandfather, who had barely slept the night before his flight—awakened at 2:15 A.M. by the very man he had stationed at his door so as not to be bothered—Erik slept perfectly, a solid seven hours. And unlike his grandfather, who took off on a cold and misty morning—the tires of the *Spirit of St. Louis* were rubbed with grease to keep the mud of the runway from sticking—Erik would begin his flight on a picture-perfect day.

In a matter of minutes, Erik taxied off. Soon, the *New Spirit of St. Louis* was in the air. Just as his grandfather had done, Erik tilted the wings to the left and the right to wave good-bye, delighting the hundreds of spectators below.

Erik's communication with Mission Control began immediately. Gregg was running the show, pacing, talking, and checking data. Meteorologists studied weather patterns. A team from Lancair monitored the plane. Byron Lichtenberg, Marc Arnold, and Joe Dobronski were the three "Capcoms," capsule communicators, authorized to talk with Erik during the flight. Gregg had search and rescue groups at the ready. Erik would do a systems check every thirty minutes, going over oil pressure, fuel flows, cylinder head temperatures, and more.

About an hour into the flight, Erik was finally able to settle in. His grandfather's words, recounted in his book, were with him: "My cockpit is small, and its walls are thin; but inside this cocoon I feel secure, despite the speculations in my mind. . . . Here, I am conscious of all elements of weather, immersed in them, dependent on them. Here, the earth spreads out beyond my window, its expanse and beauty offered at the cost of a glance. Here are no unnecessary extras, only the barest essentials of life and flight. A cabin that flies through the air, that's what I live in. Through months of planning, I've equipped it with utmost care. Now, I can relax in its solitary vantage point, and let the sun shine, and the west wind blow, and the blizzard come with the night."

In between the talking and the checklists, the satellite communications and Iridium phone, Erik came to relish something unexpected—solitude. It wasn't quiet, as the sound of the engine was constant. But just being in the air on his own was a relief. Like the exhausted but exultant feeling at a marathon's end.

Erik was flying roughly the same course that his grandfather flew. To figure out the shortest path from New York to Paris, Charles had placed a string on a globe. He found that in the higher latitudes, the shortest route from New York to Liverpool was not an east-to-west parallel, but—because of the Earth's curvature—an arc across New England and Canada, west of Nova Scotia, and through Newfoundland.*

Erik was happy to leave the coast of North America behind and begin his journey across the ocean. In the hours that followed, he saw storm clouds and clear skies. He saw Mother Nature's finest monochrome paintings in changing palettes of blue and gray. When night came, there was tedium, but also moments of unexpected humor. He got himself on the same air-to-air frequency of commercial pilots who were flying 20,000 feet or so above him and listened to their chatter. Soon they were asking Erik about his plane and flight. They were enthralled to hear he was the grandson of Charles Lindbergh. At one point, Erik asked, "Hey, can you check to see if you have a guy named Peter Diamandis on board?" He was flying from New York, JFK, to Charles de Gaulle in Paris. Erik heard the Delta pilots say they would check. Then Air France chimed in

*The great circle path between two points on a globe is determined by stretching a rubber band between the two points. The rubber band will naturally find the configuration with the least tension, and this will be the shortest path.

to say they would check. Finally, someone from American Airlines came back and said, "We've got him. What do you want me to do?" Erik shot back, "Wake him up!" A few minutes later, the pilot said, "We went to wake him up but it was someone else." Erik laughed. Peter had apparently been bumped to business class. Not long after, they found him, woke him up, and told him they'd been sharing the sky with his friend Erik Lindbergh, who would see him in Paris.

As Erik flew over the open ocean at night, spotting icebergs below, he was again reminded of his grandfather's observations of "white pyramids below—an iceberg, lustrous white against the water." Grandfather wrote: "I've never seen anything so white before. Like an apparition, it draws my eyes from the instruments and makes me conscious of a strange new sea." Erik saw clouds as textured and dramatic as the Grand Canyon, and—like his grandfather—found in the Moon a "forgotten ally."

Erik knew that sleep haunted his grandfather, whose eyes were dry from the air. Charles didn't have communication satellites. He didn't have anyone to talk to. He had himself, and the reminder of his good partners in St. Louis to keep him awake. Grandfather wrote, "How could I ever face my partners and say that I failed to reach Paris because I was sleepy?" He continued, "[I've] lost command of my eyelids. When they start to close, I can't restrain them. They shut, and I shake myself, and lift them with my fingers." His grandfather had resorted to diving down and pulling the plane up sharply to try to jolt himself awake. He got to the point where he wanted sleep more than anything in the world. During the seemingly endless stretch between dusk and sunrise, his grandfather expressed relief that they didn't make his plane more stable.* Erik grew weary, too—the early morning, predawn hours were the toughest—but he had constant chatter, systems checks, and questions from his friends at Mission Control.

---

*Charles Lindbergh had designed the plane without dihedral, which is less stable. He did this because if he started to fall asleep, the plane would bank and the noise created from this would wake him up. Erik joked that this was a primitive form of autopilot.

The communications went down just once during Erik's flight. Erik hit a stormy patch, and contact was lost. While Mission Control feared something terrible had happened, Erik found himself relishing the seclusion. In those fleeting moments of quiet, he felt cradled in the same sky as his grandfather. He marveled at the quality of the air—invisible but with weight and substance during flight—and the power of wings. He imagined going far higher and getting to see the splendor of Earth from space. He was in a modern plane, safe, dry, and warm and with technology that surely would not be mute for long. He realized that his grandfather, barely insulated from the elements, experienced a poetry that he could only glimpse.

"The *Spirit of St. Louis* is a wonderful plane," his grandfather wrote. "It's like a living creature, gliding along smoothly, happily.... *We* have made this flight across the ocean, not I or it."

When Erik saw the coast of France, the joy was a shared joy. His grandfather had been amazed to find he was on course. Erik was on course. "Ireland, England, France, Paris!" Grandfather wrote. "The night at Paris! This night at Paris! Yesterday I walked on Roosevelt Field, today I'll walk on Le Bourget. What limitless possibilities aviation holds when planes can fly nonstop between New York and Paris!"

Erik was now in his sixteenth straight hour of flight, and he was over the coast of France. Grandfather said it came to him "like an outstretched hand to meet me." Erik had maybe an hour to go. He was sleep-deprived and rummy. AIIRRRIK. That's what his childhood friends called him because he was always flying—skiing, jumping, climbing high. That was before, when all that mattered to him was physical. When he won competitions and tournaments, when he was the only kid in his high school who could climb the rope to the gym ceiling, using only his hands and not his legs. Then came Mount Rainier. There was life before the summit and after the summit. His body became foreign. All he knew went away, like a beloved house shredded by a tornado. Hope was lost in that squalid hotel room in Minnesota. Knees replaced, titanium in his foot. An old man afraid of his shadow. His blue eyes welled with tears. Life—this plane, this day, this body, this moment—was a gift. He was AIIRRRIK again. But so much better.

Erik would land at Le Bourget in the morning. His grandfather came in at night. Charles had never landed the *Spirit of St. Louis* at night. The final lines of Grandfather's book were with Erik now. Brilliant. Emotional. Pure. The sod came up to meet him. Grandfather faced nothing but night. No lights on the plane, or the field. He questioned whether he should climb back up for another try at landing, then—the wheels touched gently, the nicest greeting he'd ever had. He kept contact with the ground. There was still only blackness. The plane swung around and soon stopped rolling, resting on "the solidness of earth, in the center of Le Bourget." As Grandfather started to taxi back toward where there were now floodlights, he suddenly noticed it: "The entire field ahead is covered with running figures!" More than 100,000 people were there to greet him.

Erik prepared for the final descent, going through his own checklist. In the clouds, flying on instruments, he was vectored around Paris's Charles de Gaulle airport.

I have to grease this landing, Erik told himself. *I can't come down and bounce it.* He made some small turns and limbered himself up. He was stiff, and the plane felt stiff. He came out of the clouds and there was Le Bourget. He was ready. *Grease this.* Seconds later, he was down. *Smooth as butter!*

"Are you down?" Gregg asked anxiously from St. Louis.

"I AM DOWN!" Erik replied to whoops and hollers in Mission Control. Grandfather had made the flight in thirty-three hours, thirty minutes. Erik's flight took seventeen hours and seven minutes.

Erik climbed out of the plane, taking a moment to look at the sky. The golden light on the horizon had been even better than he'd imagined. To applause and cheers, Erik walked a few feet on the tarmac—amazed he wasn't terribly stiff—got down on his knees, and kissed the ground. Nearby were Peter and Erik's mother, Barbara, both reveling in this unlikely moment of history. Soon he would get a call from President George W. Bush, thanking him for "inspiring the country" after the attacks of September 11. Erik had raised more than $1 million—keeping the XPRIZE alive at a critical time, and supporting the Arthritis Foundation and the Lindbergh Foundation. Stories about his flight had reached half a billion people. The History Channel filmed the flight and planned to air the documentary later in May. Erik took it all in. The flying had felt easy compared with the obstacles he'd faced on land. Having rheumatoid arthritis had leveled him, and being a Lindbergh had limited him. He was leaving all of that behind.

The XPRIZE had thrown Erik a lifeline when he needed it most. Now Erik had returned the favor.

24

A Hole in One

Since Erik's dramatic transatlantic flight, the XPRIZE had gotten a new lease on life, receiving more public attention than ever. The renewed spotlight brought in significant cash infusions, keeping the operation afloat and motivating the teams to race ahead building rockets and engines. But the XPRIZE still needed a title sponsor to seal the deal, something that Peter had chased for six long years. Peter was still $5 million short of what he needed, and the First USA funds would disappear after December 17, 2003—now less than a year and a half away.

Bob Weiss was sitting with Peter and Gregg in his office at Paramount Pictures in Los Angeles when he presented an "out-there" concept to get the XPRIZE fully funded. The idea, which indeed sounded crazy—at least at first—was to get a group of professional gamblers to wage a $10 million bet *against* anyone winning the XPRIZE.

Peter and Gregg laughed. But Weiss was serious. He told them how an insurance company had agreed to pay out $1 million if a randomly chosen fan at a professional basketball game could make a three-quarters-court shot on his first try. The insurance company was of course betting that the fan would miss the extremely difficult shot. In the case of the XPRIZE, the insurance company would have to analyze the probability that one of the contenders—Burt Rutan, Dumitru Popescu, Jim Akkerman, Brian Feeney, or someone else barely on the radar—could make a lob shot to space, not once but twice within two weeks.

Peter welcomed any and all ideas. He'd recently pitched the XPRIZE concept to filmmaker George Lucas, and to the heads of Intel, Sony, Discovery, Fox, Rolex, Emerson, and Ford. Plans A, B, and C had failed, as had D, E, and F. Now he was moving on to G, H, I, and J.

Weiss said that when he was going to school in Illinois, the archdiocese of Chicago had offered a $1 million prize to someone who could catch a fish tagged with a religious medallion in Lake Michigan. Weiss remembered asking himself how the Catholic Church could afford to host a $1 million charity fishing contest. Was the archdiocese going to write to the Holy Father and request the money? Was the church going to cut the check? What he learned, he told Peter and Gregg, was that the archdiocese had secured something called hole-in-one insurance. Local church officials paid a premium to purchase a policy that the insurers believed would never be paid out. In this case, the insurance company bet right. No one caught the fish. But in the case of the random fan winning $1 million by making a shot from the foul line on the other end of the court, American Hole 'n One insurance bet wrong. The fan nailed the long shot on one try, winning a promotional gimmick in front of more than seventeen thousand cheering fans at Chicago Stadium in April 1993.*

"Here's the deal," Weiss said. "We're going to get someone to bet against us and then we're going to pull the rug out from under them because we are going to pull it off. Then we'll have our prize money."

Peter considered Weiss's proposal; he'd had people betting against him from the start. He found something poetic in the idea that he might

---

*The insurance company ended up refusing to pay the fan, saying he lied about not having any pro or college basketball experience. His $1 million still got paid, but by the Bulls, Coca-Cola, and the Lettuce Entertain You restaurant.

actually raise money from those who were sure he would fail. Peter reached out to two friends, Bruce Kraselsky and Jean-Michel Eid, of Aon, a risk management company, to see if they could take brokerage experience for insuring satellite launches and spin it into hole-in-one space insurance.

The man tasked by Aon to take on the important first step in weighing the probability of anyone winning the XPRIZE was aerospace engineer Jim French, who had worked for NASA for years and was now an independent consultant viewed as both extremely competent and unbiased. He had helped test and develop rocket engines for the Apollo and Saturn launch vehicles, and worked at the Jet Propulsion Laboratory on the Mariner, Viking, and *Voyager* missions. Peter asked him whether he would be interested in interviewing as many of the teams as possible, assessing credibility, and writing a report that could be given to Aon, which was trying to broker the hole-in-one deal for the XPRIZE with XL Specialty Insurance Company.

French was open about his longtime interest in seeing commercial space become a reality, an affliction he attributed to reading way too much Heinlein as a kid. He felt that humanity's long-term survival depended on an expansion into space, though he didn't think private industry was going to get there anytime soon. He was happy to do the analysis for the XPRIZE and began contacting the contenders, meeting with some in person and interviewing others in far-flung places by phone. He didn't have time to interview everyone, but he was willing to look at any team whose idea "didn't violate the laws of physics." He asked detailed questions about hardware, funding, and credentials of team members.

French started with Burt Rutan. The two had worked together years before on the DC-X Delta Clipper, where French was in the program office that funded it. They had talked at the time about Burt's interest in space and the different designs he had in mind for a spaceship. French also remembered in the eighties when Burt and his brother Dick first talked about a plane that could fly around the world nonstop without refueling. French felt at the time that it couldn't be done. He'd come to see that Burt was exceptionally good at setting his sights on a specific goal and building a plane to meet that goal.

French visited Burt at Scaled Composites in Mojave, listened to his plan for a spaceship, and asked: "Can you do this?" Burt knew French as an experienced space guy. He pondered the question. It wouldn't be honest to say his chance was 1 percent, and it wouldn't be honest to say with certainty he could do it. He told him the truth: he was taking big risks, including "building a Mach 3.5 plane with no wind tunnel testing." French asked whether he could do it by December 17, 2003, the target set by the insurance company and the date established by First USA. Burt said he didn't think so. Then French asked about his funding. Burt again replied honestly. "We are adequately funded," he said. No one knew just how adequately funded he was, or who was backing him.

French went through the list of contenders. A few teams were composed of a handful of smart guys with bright ideas and no money. Several were not credible. But others had a chance of doing something *if* funding came in. He really liked what Pablo de León was doing in Argentina and felt if funding came in he had a serious shot. He talked to Dumitru Popescu and thought the Romanian team could surprise everyone, but needed money like a rocket needs fuel. He thought the V-2 rocket and engine of Canadian Arrow was a reasonable approach—though expensive—and he was impressed that the self-taught Steve Bennett had successfully flown a serious rocket from the United Kingdom. He looked at Kelly Space & Technology in San Bernardino, California—a company formerly focused on satellite launches that was now building a rocket-powered delta-wing aircraft intended for suborbital flights. The technical smarts were there, but the funding was not.

In the end, French wrote a report saying he thought Burt had the best chance of winning. But French believed that Burt wouldn't be able to accomplish the feat by the Wright brothers' anniversary and possibly even by the end of 2004—if at all. For as long as French could remember, private space travel was always just a few years away. He was skeptical that someone could pull off what was needed to win the XPRIZE. But at the same time, he saw how teams were intoxicated by it. Inventiveness had been sparked. Pablo de León told him he'd grabbed the leg off a chair to use as a last-minute lever in the testing of a small rocket engine. Armadillo Aerospace in Texas used a trunk latch mechanism for a variety of things, including parachute release. Brian Feeney of the da Vinci Project in Canada worked out of the back of a scuba shop and bought parts at Home Depot. Jim Akkerman had recruited retirees to work with him out in a field in Texas on a rocket that he dreamed of water launching. Scaled Composites was trying out used paintball gun tanks to hold air for the spaceship's reentry control thrusters. Many of the teams regularly went shopping for odd parts in a McMaster-Carr catalog.

Once French had completed his report, the insurance company had the job of taking French's findings and doing its own analysis of probabilities. The challenge of the XPRIZE was that it was a "unique event"—something where no prior data existed. A nonunique event, by comparison, like the roll of a die, could be repeated ad nauseam, results tabulated and outcomes predicted with high precision based on the main assumption that things would continue to work the same way in the future as in the past. Making a wager on a unique event was more textured. How would you determine the probability of someone flying to space without the government's help when it hadn't been done before? Probability, in this case, had to rely on a *belief* about the outcome. It had to rely on something called Bayesian inference, the idea of taking any a priori belief and then systematically accumulating evidence that conditioned the belief. Any and all evidence under Bayes's law would be considered.

A unique event usually fell into a set of categories that the insurance company would assess, collect evidence on, and then come to a conclusion largely based on the disparate evidence. The conditions for the XPRIZE would be historical data regarding Rutan's success rate, tech prizes in general, failure rates of all launch vehicles, failure rates of known private launch vehicles, related historical evidence, and the informed opinions of people like Jim French.

The actuaries at XL Specialty Insurance began calculating the odds. Their hard work could yield surprisingly good results, but in the end, it still came down to systematic guesswork. It came down to beliefs.

Peter, Gregg, and Bob Weiss had the uncomfortable feeling of talking out of both sides of their mouth. When they talked to potential sponsors, the XPRIZE was possible and near term. When they were dealing with the insurance company, the prize was daunting and somewhere in the unknowable future. In the end, both sentiments rang true.

On August 18, 2002, Peter received a call from Bruce Kraselsky, who delivered the news: he had his deal with a group of doubters. XL Specialty Insurance wanted to move ahead with a policy for the XPRIZE, inferring that it was the easiest money they would ever make. The policy had two target dates. First, XL agreed to insure a $5 million payout if a team could win the contest on or before the one hundredth anniversary of the Wright brothers' famous flights at Kitty Hawk in their first powered aircraft. First USA would kick in the other $5 million. The second part of the XL contract dealt with what would happen if no one won by the 2003 date. Peter got XL to extend the contract by a year and double it to a $10 million payout.

But the insurance company wasn't the only one making a multimillion-dollar wager. In exchange for the promised prize money, Peter would have to come up with $50,000-a-month premium payments for more than sixteen months, as well as make a one-time balloon payment of $1.3 million. The XPRIZE was operating with a handful of people—Peter, Gregg, Bob, and Diane Murphy—who all relied on outside jobs or other sources of income. If at any point the XPRIZE missed a payment, XL would keep all the payments made to date and the hole-in-one policy would be null and void. Peter was at a loss for how he would cover the monthly premiums or the $1.3 million balloon payment, until, that is, he got the phone call that he'd been dreaming of for six years.

The good news came from Anousheh Ansari; her husband, Hamid Ansari; and her brother-in-law, Amir Ansari, who had recently sold their telecommunications software company, Telecom Technologies, to Sonus Networks in a deal worth $1.2 billion.* Peter wouldn't be getting the title sponsor he had sought, but the family pledged $1.75 million up front—an amount that would cover the insurance balloon payment and some of the related expenses. The Ansaris also said they would help Peter fund-raise. If they raised an additional $4.5 million or greater, they would get their $1.75 million back.

Months earlier, in the spring of 2002, Peter and Byron Lichtenberg had flown to Dallas to meet the Ansaris for the first time. Peter had come across the Ansari name as he flipped through *Fortune* magazine's list of the forty wealthiest self-made people under forty in the United States. Anousheh, who was thirty-five years old, was number thirty-three on the list. But it wasn't her money that caught his eye. It was a single word she used in the interview: *suborbital.* He read the paragraph three times:

> *Anousheh Ansari wants to see stars. But not the Hollywood kind. Sitting in the lobby of Manhattan's Peninsula Hotel, the 35-year-old Sonus Networks VP—one of two women on this year's rich list—is*

*The value of the sale of the Ansaris' company was between $400 million and $1.2 billion based on the stock price and whether certain targets were met. All targets were met, so the payout was $1.2 billion.

talking about her desire to board a civilian-carrying, suborbital shuttle. "It would be nice," she said, "to get outside the planet and see the universe for what it really is."

Both Anousheh and her brother-in-law, Amir, had listened intently to Peter's pitch during the meeting in their Dallas office. Amir felt guilty for having given up on his space dream so easily compared with Peter. He and Anousheh both grew up watching *Star Trek* in Iran and dreaming of interplanetary travel. But no one from Iran had ever flown in space. NASA was not taking paying customers up there. There had been only two tourists—Dennis Tito and South African entrepreneur Mark Shuttleworth—and they'd forked over tens of millions of dollars for rides into orbit aboard Russian launchers.

Anousheh listened to Peter in that first meeting and tried not to smile. She had never met anyone with Peter's passion and commitment. *With this guy, if you close one door, he's going to open another,* she thought. Amir thought a space prize was an odd approach, but just cool enough that it may work. Peter also talked about his efforts with Byron to get the Federal Aviation Administration to issue permits so they could offer civilians zero-gravity flights in their modified Boeing 727, something they'd been pursuing for eight years. Peter said their ZERO-G Corporation would offer the world's first parabolic flights to the public and could be used to train NASA astronauts, scientists, and engineers. After a labyrinthine legal battle, they believed they were close to winning the necessary permits.

Anousheh's husband, Hamid, watched Peter make his pitch at what seemed like 100 miles per hour. "If someone wins the XPRIZE, we can all go to space," Peter said. Hamid didn't have the "space DNA" of his brother and wife. Their interest in space was not a wish or a game; it was who they were, like the color of their eyes. Hamid was drawn to the idea of the international competition, and the daredevilry required to win it.

He wasn't at all afraid of the obvious risk of something going wrong along the way. Risk had been good to the Ansaris. His family had fled Tehran at the start of the revolution under Ayatollah Khomeini, arriving in America with little money and speaking no English. Anousheh and her family had left Mashad a few years later, when conditions in Iran had worsened. Hamid and Anousheh had met in the United States in the mid-eighties, when Anousheh, a computer and electrical engineering major at George Mason University, applied for a summer internship at MCI, where Hamid worked. Anousheh, Hamid, and Amir all worked together at MCI and eventually took $50,000 in savings from their jobs to start Telecom Technologies.

As Peter showed pictures of XPRIZE events and talked about the contenders and what was being built, Anousheh, Hamid, and Amir exchanged looks. Anousheh did her best to keep a poker face while listening to Peter. She had dreamed of space for as long as she could remember. Her favorite book as a child was *The Little Prince,* which she'd read in Farsi. As a child, she loved sleeping outside on her grandmother's balcony. She would stare up at the stars and fall asleep to the same fervent prayer: that aliens would come and take her away. She was incapable of being outside without looking up to see the stars. Anousheh asked Peter a few questions, including how these private teams could afford to build a rocket. Then she asked why commercial space hadn't happened yet. "Who is the enemy?" she wanted to know. "NASA?"

Peter had been living with this question for decades. "The opponents are multiple," he said. "The opponent is the government by being the regulatory body that makes it difficult. The opponent is NASA for establishing the mind-set in the sixties, seventies, and eighties that only governments do this. The opponent is our risk-averse society that is resistant to any of this. The opponent is the laws of physics. The opponent is the capital markets that don't value these risks and want the cheap and fast return of an Internet play. Those are the opponents."

When the meeting concluded, the Ansaris said they needed some time to think about it. Peter then heard the familiar refrain: *We'll get back to you.*

As the group headed to the door, Peter walked with Anousheh. He turned to look her in the eyes and said, "I promise you I will do everything in my power to get you to space."

She believed him. And now, Peter had a fully funded hole-in-one policy, and a $10 million XPRIZE, *there to be won.*

At home in his Santa Monica apartment, Peter sifted through paperwork and found the John Galt charter written eight years earlier in Montrose, Colorado, when the XPRIZE came to life. He reread his impassioned words and realized how naïve he was when it came to funding. But he still believed every word in the charter. He cut out a paragraph and posted it above his desk:

> *There is a strong technology available which helps humans in achieving difficult, sometimes seemingly impossible feats, this technology is a forcing function which helps to focus the whole of human ingenuity at the same well articulated goal.... This concept, the forcing function, this technology, is the competitive "Prize." Not prizes for spelling bees or prizes for a lifetime achievement, but prizes which lay out impossible goals and tempt man to take great strides forward. Prizes such as those which were set out to the aeronautical world for speed, distance, endurance. Prizes which brought forward adventurers, dreamers, and doers. Prizes such as the $25,000 Orteig Prize. Where no government filled the need and no immediate profit could fill the bill, the Orteig Prize stimulated multiple different attempts. Where $25,000 was offered, nearly $400,000 was spent to win the prize—because it was there to be won.*

Six years after the announcement of the prize under the arch in St. Louis, Peter had his promise of $10 million. He wondered whether this was how Charles Lindbergh felt after securing his financial backing in St. Louis. Lindbergh had the money, but he still needed to build the right plane, get to the starting line, and make it to the finish alive. Peter still had to come up with $50,000 a month in premium payments, something he'd taken to calling his "fifty-thousand-dollar Fridays." He and others would start dialing for dollars on Monday, needing the fifty grand by week's end. Peter had moved one big step closer, but he still had his ocean to cross. And now the clock was ticking: If no one succeeded by 12:01 A.M. on January 1, 2005, the $10 million would vanish.

Part Three

A RACE TO REMEMBER

25

A Fire to Be Ignited

On November 21, 2002, after delays, false starts, and missed deadlines, Burt Rutan's Scaled Composites was ready to do a first hot fire test of its hybrid rocket motor. At this point, the still-secret *SpaceShipOne*—Burt's hope of getting beyond the atmosphere—remained a work in progress, and time had become an issue. But a successful motor test would be a turning point in what Burt considered the most important project in his company's history.

Getting to this day had been difficult. Scaled Composites had wanted to build everything related to the outside of the motor, including the propellant tank, case, and nozzle. But Scaled was an airplane company, with no experience making rocket engines. Fabrication problems quickly surfaced with the case, throat, and nozzle. There was uncertainty that Scaled's nitrous tank design could handle the required pressures. Alabaman Tim Pickens, hired early on by Burt for his work in propulsion, design, and fabrication—and because his side projects included rocket-powered bicycles, rocket-powered backpacks, and rocket-powered pickup trucks—agreed with Scaled's call to outsource the nitrous oxide tank, which would transport the nitrous. Pickens found a guy in Texas who owned a scrap yard and said he could help build a nitrous trailer. The Texan already had a tank with a generator for refrigeration that could hold ten thousand pounds of nitrous. A deal was made, and not long after, the Texan pulled up to Scaled with nitrous trailer in tow. An old beat-up truck was thrown into the deal to haul the tank and generator around.

Burt was confident he could build the solid motor case, but didn't feel he had the expertise to build the parts that see the highest temperatures—the ablative throat and nozzle. Scaled engaged a specialty company, AAE, to supply these components, knowing they had supplied ablative nozzles for all the big companies that made rocket motors. But Scaled also needed to find some source for many other components it did not have expertise building: injectors, igniters, valves, controllers—the critical metal components on both sides of the big tank.

In order to maintain secrecy, Scaled sent out RFPs (requests for proposals) to all the big rocket companies, using a cover story about building a hybrid rocket motor for an unmanned sounding rocket whose mission was to measure the top of the atmosphere for NASA's earth-sciences programs. But they got only two types of responses: no bid at all—apparently revealing an attitude that the project was hopeless—or bids to build custom-designed components at a cost that exceeded the entire *SpaceShipOne* budget. Quickly switching to another plan, Scaled set up visits to the community of small operators, including Gary Hudson, eAc (Environmental Aeroscience Corporation), and SpaceDev. Two of them immediately self-unselected by staging failed tests during their visits.

Burt decided to fund the two most impressive small shops with the promise that the best components identified during the tests would be used on a historic new space program. Scaled set up a fixed-price competition between SpaceDev, based on the West Coast, and eAc from Florida to see whose components would fly on a manned ship out of the atmosphere.

The hybrid motor had been sketched on a napkin in Huntsville, Alabama, by Burt and Pickens. Instead of using the common rocket fuel of liquid oxygen and liquid hydrogen, their motor would use laughing gas and rubber (nitrous oxide and hydroxyl-terminated polybutadiene, or HTPB). The rubbery part was pliable and could be touched without gloves. Some Scaled employees had coffee coasters made from the stuff.

The promise of the first hot fire test energized the crew. A quote by Plutarch was scribbled on a whiteboard: "The mind is not a vessel to be filled but a fire to be ignited." Ignition was the word of the day. This motor test would be by SpaceDev, and eAc would get its chance six weeks later in early January. Burt went over safety procedures and talked about how far the company had come in its design of all the components. He talked about the safety of hybrids and said there wasn't much that could go wrong with the day's main event.

On a bright and cool Mojave day, with light winds out of the east, the SpaceDev motor, about the size of a van, was mounted on a stand in the desert. The transfer of the nitrous began. A small control room, about the size of a horse trailer and called the SCUM truck—for Scaled Composites Utility Mobile—was situated two hundred feet away, protected by steel shipping containers. Among those inside the truck was Scaled pilot and engineer Brian Binnie, who watched as both a hopeful future pilot of the spaceship and as the person managing propulsion development.

Standing about three hundred feet from the test site were Burt and Tim Pickens. Burt could feel the excitement in the air. And Pickens was more than ready. He had, after all, grown up in Huntsville, aka "Rocket City," where the booms of Saturn V motor tests were a part of life, like the honking of taxis in New York.

Joining Burt and Pickens at their vantage point were Dave Moore, who was running the spaceship project for Paul Allen, and Jeff Johnson, whom Moore had brought in to try to get Scaled on a better production schedule. Paul Allen had not been pleased by the delays, saying in one meeting: "I know you had a slow start, but you mean to tell me that after three months of work you're three months behind?" At the next meeting,

Allen said, "You mean you've slipped another three months on top of the other three months you told me about last time?"

Just before test time, Moore said to Pickens half jokingly that he was going to stick with him just in case something went wrong. But Pickens was thinking about nitrous—nitrogen and oxygen combined—as an energetic substance on its own, even without the rubber. Nitrous usually had to be hauled around at close to zero degrees, but when team members filled the rocket tank, they'd need it at around sixty-three degrees. This nitrous would perform two thirds as well on its own as opposed to when it was combined with rubber fuel. As soon as the command was sent from the control room, a valve in the tank of nitrous would open, and nitrous would flow in a controlled way. That was the idea at least.

The countdown began. When it reached "zero," there was white smoke, a small flame, and then—a violent explosion. Binnie jumped inside the SCUM truck, thinking, *This is what we're going to fly on top of?* The motor was supposed to fire for fifteen seconds and then shut off. Fifteen seconds came and went. Hands were raised for high-fives. The hot fire was successful! The program had reached a milestone. A few seconds later, though, congratulations turned to watchful silence. All eyes were focused on a small potential problem: a flame continued to flicker out the nozzle, like a snake's quick tongue. Jeff Johnson was the first to say, "It shouldn't be doing that."

Dave Moore turned to ask Pickens what he thought, only to find that Pickens was gone. He looked around again and spotted their propulsion expert—50 feet back, crouched behind a truck. Moore darted toward him to get more information, and Pickens said, "This could be bad. Real bad." Pickens said the valves had shut, but the seal was blown. The system was full of nitrous oxide, and the plumbing could only support a short test. He feared what would happen next, as extreme heat soaked back into the nitrous tank. He told Moore that the whole motor could blow up, sending huge metal chunks flying. There were still people on both sides of the tank; the one fire engine on the scene was not moving.

Moore, crouched behind the truck with Pickens, watched the persistent flame. This was not the start they were looking for. What if the thing exploded? It didn't help that Pickens was wondering aloud whether the nitrous inside the tank was turning to gas. An American flag flapped in the wind, looking vulnerable next to the semidormant giant. A full five minutes later—watching and waiting for Armageddon—the fire truck finally moved close to the motor and began spraying foam.

Burt, exasperated, pointed out that the fire truck was spraying *the wrong end* of the motor. It needed to spray the *nozzle end,* where the fire was coming out. About fifteen minutes later—an eternity when an explosion feels imminent—the flickering flames were extinguished. They were lucky; the tank hadn't blown up. One of the problems, Pickens believed, was that Burt had asked for three igniters instead of two, the way SpaceDev had originally designed the motor. Burt wanted the additional ignition energy. Before the test, Pickens had told Burt and a few others that he didn't have a good feeling about how the day would be run. SpaceDev engineers had said they would start flowing the nitrous *before* they hit ignition. Pickens had said, "This is a really bad idea." Burt responded, "Well, this is a competition. We have to let them learn."

Dave Moore and Jeff Johnson headed back to Scaled for a debriefing to review the video and telemetry from the test. One of their biggest concerns was that the tank and the test stand were damaged, which would create more setbacks and delays. Moore was already thinking about the report he would need to send to Paul Allen detailing the day's events.

Moore had brought Johnson in to get a better sense of what was going on inside Scaled. Johnson had a knack for ingratiating himself with the right employees, the ones who dealt in reality over fiction. Moore had learned that although Scaled had talented and resourceful builders and engineers, the company lacked some basic project management. The top-down structure—Burt ruled—was not working for this project. Moore needed real schedules, not a guessing game or wishful thinking. Moore had twenty years' experience at Microsoft, and Johnson had ten. The program management that Moore was trying to bring to Scaled was something he'd worked on with Bill Gates to ensure that software projects remained on track.

At one point, Moore told Burt, "You need to walk around and ask people when they think a certain part is going to be done. They really have to believe it." Burt took this to mean he had to *convince* people. Moore said, "No, it's the other way around. They have to believe intrinsically that this date is the real date." Moore also said he'd rather see them pick conservative dates and stick to them than string them along with "guesses." They also needed people assigned to specific jobs. At one point, Burt had said he wasn't going to assign engineers to the program. He was going to give them different tasks at different times. Burt had said, "The engineers are like mothers-in-law. Once you assign an engineer, they move in, take over, and never leave."

As Moore and Johnson saw it, Burt was the solution—he was the genius. None of this would be happening without him. But he could be elusive when it came to scheduling. There were points in the program when Moore looked at the Scaled crew and thought, "They're a bunch of motorcycle mechanics in the desert building a spaceship!" He said this with admiration or exasperation, depending on the day.

Before the less-than-ideal rocket motor test, the Scaled team had already succeeded in flying the *White Knight*—the mother ship. Resembling the Proteus, but bigger and more beautiful, the *White Knight* would take *SpaceShipOne* to around fifty thousand feet up before releasing the rocket for its final ascent to space. The *White Knight*'s first flight had been memorable: it lasted two and a half minutes. Pilot Doug Shane reported after the flight that "everything was good"—except for the minor issue of the J-85 engines producing "a small fire." There was also the matter of the spoilers flapping and banging, prompting Burt to order them bolted down. The first flight had earned the *White Knight* a new nickname, "White Knuckle Knight." Fortunately, the flying had been fire-free ever since.

Things were not going so smoothly with the production of the spaceship—it was, after all, a spaceship. Scaled was put on what came to be known as the "blood schedule." Johnson and Moore showed up regularly to play good cop, bad cop. The Scaled team worked nights and weekends. The main area of concern was the fabrication of the spaceship. Matt Stinemetze, who spent his first two years at Scaled avoiding Burt out of intimidation, was now the one who had to needle Burt to stay on schedule. They had set a target of spring 2003 for the public unveiling of the whole program. The XPRIZE was now fully funded, and there was a chance someone could beat them to the starting line.

By early 2003, the spaceship was in the build phase; everything was started but nothing was finished. The craft was in a walled-off hangar next to the *White Knight* and looked more like a picked-over carcass than a supersonic rocket. Its dark gray shell had unfinished portholes and wires dangling out. The Scaled team was in the midst of what felt like a million "cure cycles," where the uncured fiber resin was shaped and "cooked" into desired parts. The team needed fairings—smooth composite panels—for everything from landing gear to wingtips. The plumbing and fittings weren't done. The gear was being assembled. The reaction control system and the feather—both pneumatic, driven by high-pressure air fired through thrusters—all had to be built, from bottle to tubing. They had a chemist mixing up different recipes of thermal protection for the rocket. The difficulty of getting to space wasn't so much distance as it was the required speed, which in turn generated heat. The team had looked into what NASA used for thermal protection and learned it was too expensive and sold only in large quantities. So Scaled had its own scientist cooking up mixtures for the day its ship would go supersonic. One of the more recent mixtures, tested at high temperatures, had begun to sizzle, like the sparklers Stinemetze had lit as a kid. The chemist was returned to his recipe cards.

On the positive side, the *White Knight* was now flying beautifully, and was photographed by locals every time it was taken out of the shed. Stinemetze and others started bringing in clips of the latest blogs speculating on their space program. They tacked them up to the inside wall of the hangar.

The Scaled team began to gain momentum on the rocket, employing the disciplined style of Moore and Johnson while preserving the best of Scaled's creative culture. The myriad components—from landing gear to nose cone—were finally coming together in this one-of-a-kind puzzle. Two new hot fire tests of the hybrid motor—by SpaceDev and eAc—went off without a problem. No unwanted flames. No explosions. And the gray carcass was starting to look like the spaceship of Burt Rutan's dreams.

On Saturday morning, February 1, 2003, Stinemetze was at home with his wife, Kathlene "Kit" Bowman, who had joined Scaled as a process engineer. She came into the bedroom with a worried expression. "*Columbia* came apart," she said slowly. He didn't understand at first, as he hadn't been watching the news. But then he learned: space shuttle *Columbia*, STS-107, was returning to Earth after sixteen days of research in space when it broke apart. It had been traveling at eighteen times the speed of sound at 200,000 feet above Earth when communication was lost. The seven astronauts were twelve minutes away from making their landing. Instead, their one-hundred-ton spaceship disintegrated in the blue sky.

Later that day, a stony-faced President George W. Bush appeared on television to make a statement: "The *Columbia* is lost. At nine A.M. in Mission Control in Houston, we lost contact with space shuttle *Columbia*. Debris fell from the skies over Texas. There are no survivors." He went on to reference "the difficulties of navigating the outer atmospheres of the Earth." Stinemetze could not pull himself away from the news, thinking of the astronauts, and thinking, inevitably, of Scaled's own space program. What was required to reach orbit and stay there was very different from what was required to get to the start of space. The space shuttle flew at Mach 25. The Scaled pilots would need to fly at Mach 3-plus. Still, they were trying to reach space with a team of only a few dozen people. NASA had been at this for decades and was spending more than a billion dollars for every shuttle flight. Stinemetze asked his wife: "What are we doing?"

On Monday morning, Stinemetze pulled into the Scaled lot, turned off the engine, and sat in his car for a few minutes. It was raining and gloomy, and the wind was whipping the flags on the tops of buildings. He was still thinking about *Columbia*. The space program he grew up believing in felt bereft. Early reports on the cause pointed to damage to the left wing by foam insulation that had come loose from the orange external tank on liftoff.* As Stinemetze got out of his car, he realized his job was more like a mission. There could be no motor misfirings, no loose or faulty pieces. They had to get everything right.

The rollout of *SpaceShipOne* was now scheduled for the morning of April 18, 2003—Good Friday. Burt was working around the clock, drinking copious amounts of coffee. In earlier days, he liked to rib Mike Melvill for wasting time cycling and staying fit. Burt would say his best exercise was at home on his couch, raising his spoon from tub of ice cream to mouth. Nowadays, he didn't even have time for the ice cream.†

Finally, the morning of the unveiling arrived. Guests were driving

*Bits of foam insulation from the external tank came loose and hit the shuttle several times previously. When the program restarted after *Columbia*, NASA instituted mandatory external inspections of the shuttles once in orbit. Also, they launched only when a backup vehicle was ready to go in case a rescue was needed. For *Columbia*, the *Atlantis* shuttle was queued up for STS114, but it was not ready to fly.

†Burt had been on a healthier program in recent years, after suffering serious heart problems in the late 1980s.

and flying in from near and far. Hundreds of people were expected, with a few notable no-shows. Paul Allen wasn't going to make it. He didn't think it was a good idea to announce his sponsorship so soon after the *Columbia* tragedy. While rumors had swirled that Allen was Burt's backer—Burt's customer, really—Burt dispelled such questions with comments including, "I hadn't heard that one." Allen was also being hit with more than his share of bad press. A harsh biography, *The Accidental Zillionaire,* had just been released, portraying Allen as a great party thrower, a lucky Microsoft stockholder, and a bad investor. A *Newsweek* story in February had described Allen as having a "reverse Midas touch."

Burt scanned the crowd as guests began seating themselves in the Scaled hangar. He was thrilled when he spotted Maxime "Max" Faget. Faget had designed the shape of the Mercury capsule, been involved in the designs of Gemini and Apollo, and was a lead designer of the space shuttle. Burt had worked with him in 1992, when he, Faget, Antonio Elias, and Caldwell Johnson met in Houston to discuss the preliminary design of a carrier aircraft that would have a range of capabilities, including air launch to orbit.* Burt had called Faget a few weeks earlier to invite him to the *SpaceShipOne* event. "Max, come and tell me if my 'feather' reentry idea will work," he said. Faget declined, saying, "I'm in my eighties and I don't travel much anymore." After a pause, Burt said, "Max, what do you plan to do with the rest of your life?" The conversation ended with Burt assuming he wouldn't come. A day later, Faget's daughter called and said, "I am bringing him to your rollout."

Burt had never invited the public to see any of his planes before they flew, but he was making an exception for his rocket. Peter was there, having driven over at four that morning from his apartment in Santa Monica. Erik Lindbergh was nearby, along with the Ansari family, and Pete Worden, now brigadier general in charge of the U.S. Air Force's center for space transfor-

*This preliminary design would eventually become Stratolaunch, a mobile launch aircraft designed by Burt and funded by Paul Allen that will be the largest plane ever built, with a wingspan of 385 feet.

mation. A few seats away in the crowd was millionaire adventurer Steve Fossett, whose *GlobalFlyer* was being built by Burt to try to set a speed record for an around-the-world solo flight. Also present was space tourist Dennis Tito, and Kevin Petersen, head of NASA's Dryden Flight Research Center. Buzz Aldrin was in the front row. Burt was introduced by Academy Award–winning actor and good friend Cliff Robertson.

Suffering from a terrible cold and a hoarse voice, Burt began, "This is not just the development of another research aircraft. This is a complete manned space program." Flashing his inimitable isn't-this-cool smile, he added, "We are not seeking funding and are not selling anything. We are in the middle of an important research program to see if manned space access can be done by other than the expensive government programs. Nothing you will see today is a mockup. I didn't want to start the program until we knew this could happen."

The star of the show, cordoned off behind a blue curtain dotted with yellow stars, was about to take center stage. At Burt's command, the curtain was pulled away and the spaceship was revealed. Guests in the back stood up to get a better look. Cameras were pointed at the small, strange-looking rocket—white, pristine, blue stars painted on its belly, a nozzle out the back. The name *SpaceShipOne* was on the side, with the FAA registration number N328KF—for 328 kilofeet (about 100 kilometers), the designated start of space and chosen finish line of the XPRIZE.* Its body reminded people of a bullet, a bird, even a squid. Aldrin sat forward in his chair studying the design. Burt, two months shy of his sixtieth birthday, was smiling ear to ear. He was a kid again, back at his model airplane shows, wowing the pros and confounding the traditionalists.

After the commotion died down, Burt talked about Scaled's history, saying proudly that they'd never had a significant accident or pilot injury during flight test activity. Looking toward the spaceship and trying again

---

*The registry number N100KM was already taken.

to speak up despite his nonexistent voice, he said, "This program, if successful, will result in the first nongovernment manned spaceflight above one hundred kilometers altitude. If I'm able to do this with this little company here, there'll be a lot of other people who will say, 'Yeah, I can do it, too.'"

He noted that suborbital manned spaceflight had been achieved before by Mercury-Redstone in 1961 and by the B-52/X-15 in 1963. Burt marveled, "Even though the experience—as described by Alan Shepard, Gus Grissom, and Joe Walker—was *awe-inspiring,* suborbital spaceflights were ignored for the next forty years. Our goal is to demonstrate that nongovernment manned spaceflight operations are not only feasible, but can be done at very low costs."

At this, Burt asked his crew chief Steve Losey, who had grown up in Mojave and was now seated in the cockpit of the spaceship, to activate and extend the "feather"—his solution to giving the craft drag and slowing it for reentry. It took thirteen seconds to raise the wing to full extension of sixty-five degrees. Burt then asked Losey to fire up the thrusters for the pneumatic jets. The crowd cheered.

The plan, Burt said, was to attach the three-person spaceship to the turbojet *White Knight,* which would climb for an hour to reach 50,000 feet. The spaceship would then drop from the *White Knight,* the pilot would light the motor, and the rocket would "turn the corner" and make a vertical climb at speeds of 2,500 miles per hour. After the engine shut off, the ship would coast to its target altitude of 100 kilometers (62 miles) before falling back into the atmosphere. During that time, the pilot would have weightlessness for three to four minutes, and the ship would come back into the atmosphere "carefree," thanks to the feather—which had only two positions, up or down. Then, the ship—feather back down—would become a conventional glider, allowing a "leisurely" seventeen-minute glide down to the runway in front of Scaled. Max Faget couldn't take his eyes off the spaceship. He thought the feather was clever and unique, and had a hunch it would work.

"The program is a lot like the X-15," Burt quipped, "but we had this minor annoyance: we had to build our own B-52." The crowd laughed. Burt was referring, of course, to the *White Knight*, which was not only a launch platform, but also an in-flight systems test bed. The cockpits of *SpaceShipOne* and the *White Knight* were functionally identical, so the *White Knight* could be used as a training simulator for the rocket.

Burt also took a moment to single out Peter, saying that the "rules of the XPRIZE had stood the test of time" and were as valid today as when they were announced under the arch in St. Louis seven years earlier.

After the unveiling, the crowd made its way outside to see the *White Knight*, which by now had flown fifteen times, including more than twenty hours at altitudes of about 50,000 feet. On the tarmac, Mike Melvill and Doug Shane climbed into the cockpit of the *White Knight* and taxied out. Minutes later, the *White Knight* came roaring back, dove down in front of the crowd, pitched its nose up, and tore eighty degrees skyward. At around 10,000 feet, it rolled over and then did a few more high-octane air show–style maneuvers. The crowd loved it.

Burt smiled and exclaimed, "Dammit! Those guys are having too much fun. I'm gonna have to stop paying them." Then he said, "You ain't seen nothing yet!"

That night, Peter and a small group of friends met up at an Asian fusion restaurant called Typhoon, which overlooked the runway of the Santa Monica Airport. He was joined on the second-floor terrace by some XPRIZE board members, including Adeo Ressi, technologist Barry Thompson, Anousheh Ansari, and Erik Lindbergh. The sake flowed and wine was poured, but the fuel at this table was optimism. Both Ressi and Thompson had saved the XPRIZE on different fifty-thousand-dollar Fridays, stepping in with last-minute cash infusions to cover the hole-in-one insurance premiums. Peter had also paid the premiums, as had his parents.

Ressi, who had joined the board in 2001 and had been at the morning's unveiling in Mojave, congratulated Peter for driving innovation. After the dot-com crash, Ressi feared that innovation had died, but the XPRIZE was bucking that trend and finding a way to resurrect inspiration. What he had seen in Mojave—and as he traveled to meet some of the XPRIZE contenders—was creativity unleashed. And he saw real determination, a willingness to get down in the dirt and dig a ditch for a homemade rocket bunker; sacrifice a steady job for a crazy dream; or spend retirement savings to make a giant spaceship.

Peter relished the excitement that the *SpaceShipOne* unveiling had brought. As he watched planes take off and land on Santa Monica's 4,973-foot runway 21, he was reminded that *SpaceShipOne* was as small as a private plane. He pictured it being pulled out of a hangar, rolled out onto the runway, and flying off to space. That was his dream—spaceships for personal use.

As if on cue, in walked Elon Musk. Since meeting Peter shortly after the demise of Blastoff, Musk had set out on his own quest for space, motivated by the question: if one was to make a rocket, what would be the best choices to make it cost effective? Ressi and Musk had gone to Russia in 2001 to try to buy rockets, only to find a sort of criminal-filled Wild West, where missiles of any sort could be had for the right amount of cash. The Russians got them drunk on vodka, and by their next visit, the price of the rockets had tripled. Ressi had kept one of the vodka bottles that their Russian hosts had made for them, complete with a logo featuring Ressi, Musk, and a palm tree on Mars.

Ressi and others had tried to talk Musk out of starting a rocket company. Ressi reminded him that there had been "a long line of rich guys who had lost fortunes on space." He and Peter and a handful of others had shown him clip after clip of rockets blowing up. For Peter, the reason was simple. He told him, "Building rockets is hard. Most people fail. A better mechanism is to fund the XPRIZE." Ressi told Musk, "Dude, don't do it. Don't do it. Don't do it." Musk responded, "I'm going to do it."

In June 2002, around his thirty-first birthday, Musk started his own company called SpaceX. His dealings with the Russians had convinced him that he should build his own engines, rocket structures, and capsules. His first launch vehicle would be the Falcon, a two-stage, liquid oxygen and kerosene–powered rocket, named after the *Star Wars Millennium Falcon*. He was hoping to launch by the end of the year.

Musk was impressed by what the XPRIZE was trying to do and saw how it was getting the general public excited about space again. He liked Burt's idea for the feather, saying he thought it was a good solution for suborbital flight. "It's something that only works for suborbital," he said. "It won't work for orbital."

Ressi, looking at Musk, playfully raised the possibility of certain well-funded "secret teams" entering the XPRIZE race at the last minute. Talk turned to Amazon founder Jeff Bezos, who was also getting into the space field and had a new, under-the-radar company called Blue Origin, headquartered in Seattle in a one-story warehouse—its windows covered in blue paper.

Soon, the discussion at the dinner table turned to the *Columbia* shuttle disaster and the national outpouring of grief. Elon remarked that it had shown him just how much people still cared about space and admired their astronauts. There was a feeling among the group that night that they had arrived at an unexpected moment in history. NASA had grounded manned flights, but because of the XPRIZE—and because of entrepreneurs including Elon—the prospects for getting humans off planet had never looked better.

Peter listened to the animated chatter. He looked at the daring group around him and thought of the morning in Mojave with Burt. There were teams building rockets in backyards, rice fields, and deserts. They were willing to risk everything, from ridicule and debt to their personal safety. They were modern-day explorers who shunned federal sponsors. Peter knew there would be errors of illusion and assumption. There would be imperfect starts and certain failure. But this moment felt as real to him as anything before. A changing of the guard. A new beginning. Hypergolic—that's how it felt. Parts coming together and igniting.

26

The Test of a Lifetime

Inside the shop of Scaled Composites, Matt Stinemetze frowned as he studied the nose of *SpaceShipOne*. Instead of looking sleek, forceful, and aerodynamic like the tip of an arrow, the front of the rocket had the look of a dry lake bed. There were cracks and superficial pieces missing, apparently lost during the morning's latest glide test.

The Scaled crew was searching for the perfect homebrew of thermal protection. Types and ratios of fillers and resins had been changed, quantities added and subtracted, in pursuit of the formula that could protect *SpaceShipOne* as it cut through the atmosphere on its return to Earth. But so far, nothing was working.

The rocket plane had already gone on seven successful glide tests and was fast approaching its first powered flight, set for December 17, 2003, the one hundredth anniversary of the Wright brothers' milestone in Kitty Hawk. Even if all went well today, Scaled was still months away from a trip to space. Burt's deadline now was the end of 2004, when the XPRIZE's hole-in-one insurance was set to expire.

Before *SpaceShipOne* could fly to the start of space, it needed to prove that it could safely go supersonic. The December 17 powered flight would be telling, revealing the integrity of the motor, the craft's sturdiness, and the pilot's ability to handle the spaceship. But the concoctions for thermal protection were failing time and time again.

Burt kept saying they needed something light and easy. Stinemetze examined the cracking coat on *SpaceShipOne* and stated the obvious: "This is not going to work when we go to high altitude." Burt took one look and said, "Take this stuff off. Clean up the lumps and bumps. Put some body putty on it." At one point, his in-house chemist had tried to talk Burt into putting forty-two pounds of thermal protection on the ship. Burt said "bullshit" and talked him down to fourteen pounds, which eventually became four pounds. In the end, though, Burt felt the mixture was overdeveloped. *Body putty should do the trick.*

"Body putty . . ." Stinemetze said, raising his eyebrows.

"Body putty," Burt said, walking away.

Stinemetze exchanged looks with a couple of the guys in the shop. "Okay," he said. "Body putty it is."

Before he could paint the putty on the parts of the ship that would be exposed to the greatest heat, Stinemetze would need to test the substance to around eight hundred degrees, as he'd done with the other pixie dust concoctions that ended up sizzling, sparkling, cracking, or falling off. As he worked, he kept thinking, *body putty?* It was standard stuff in aviation. Like Bondo, only epoxy based, highly versatile, and used to smooth over dents and nicks. It was basically glue and filler. A smooth layer of the stuff would dry like hard candy.

Stinemetze tested a layer of putty. The results astounded him. *It worked.* The space shuttle had used fancy, high-grade synthetic sand for its thermal protection. Could Scaled really go to space with body putty? Stinemetze tested it again. And again. Finally, he headed off to find Burt.

"This stuff is awesome!" Stinemetze said. "Body putty. It works!"

Burt got the same excited look in his eyes as when those rare clouds over Mojave beckoned to be flown through, or the guys in the shop gave him the idea to put a saddle on one of his planes. "You need herbs and spices!" Burt said. "You can't just use *body putty* for thermal protection *on a spaceship.* Thermal protection has to be high tech! It has to be a proprietary recipe. Get some herbs and spices."

Stinemetze smiled. He understood. He had heard Burt tell stories about his days at Bede Aircraft. People would ask about a certain high-tech goo that was apparently uniquely effective. Burt loved to tell people that he couldn't reveal the top secret proprietary mixture. He would cite ITAR—international regulations limiting the disclosure of certain information—before confiding conspiratorially, "Okay, I'll tell you. It's made from the eyelashes of Nicaraguan racing spiders." In reality, Bede's secret "goo" was a new fiber invented by DuPont. It was Kevlar. In those days, Bede was testing it for possible use in aviation.

Stinemetze and one of the shop guys, Leon Warner, headed to the grocery store for herbs and spices. They bought red dye, oregano, and cinnamon. Back at Scaled, Stinemetze mixed the herbs and spices with the putty, turning their homebrew potion a lovely pinkish-red. The color would complete the patriotic triumvirate—with the white paint and scattered blue stars. The top-secret thermal protection was painted under the nose, belly, and wings. If someone looked closely enough, they might even see a sprinkling of oregano.

Scaled Composites had three pilots doing the flight testing of *SpaceShipOne*: Mike Melvill, sixty-three, Burt's longtime go-to guy to bring the plane home; Pete Siebold, thirty-two, a whiz kid pilot and a Cal Poly engineer who irritated people because he was so skilled; and Brian Binnie, fifty, a straitlaced, Ivy League–educated engineer and U.S. Navy–trained pilot. All three were talented and smart and could handle different planes the way a professional jockey can ride any horse. But their courage—that ineffable quality that couldn't be taught—was about to be tested like never before.

Binnie was first up, chosen to pilot the spaceship's maiden rocket-powered flight test. Scaled had never built a plane that broke the sound barrier; in fact no private company had ever independently built a piloted craft to fly faster than the speed of sound. But that's what would be required before Burt's spaceship could reach space: it would need to show that it could push from subsonic to transonic—which started at about .7 Mach—and then to supersonic.

Binnie knew the *SpaceShipOne* motor better than most. He was managing the development of all of its parts, from nozzle to propellant. Plus he was the only pilot at Scaled who had experience flying supersonic jets, courtesy of twenty years in the U.S. Navy, where he flew dozens of combat missions over Iraq. Binnie had also survived the harrowing test flights of Rotary Rocket's Roton, which was Gary Hudson and Bevin McKinney's answer to a reusable rocket. Binnie had been sure that if the test flights had continued on the Roton, the day would come when he climbed into the cockpit of the cone-shaped contraption and didn't climb back out. It was just a matter of time.

Burt had seen past the short-sleeve, button-down Navy guy with the perfectly combed hair, creased pants, and polished shoes. He saw a man who turned down a job pushing paper in a comfortable office in the Pentagon to fly experimental rockets in the Mojave Desert. There was nothing buttoned down about that.

Born in 1953 in West Lafayette, Indiana, Brian was that kid who loved anything that flew or otherwise left the ground. His father was a physics professor at Purdue University, and the family lived in university housing behind a golf course. Brian spent hours running around the golf course with his two sisters, flying a rubber-band-powered airplane that launched like a rocket and slowed with wings that popped out for a glide back to the ground. When the family lived in Scotland—his mother and father were Scottish born—Brian's school was close enough that he would walk home for lunch. At around age seven, in 1960, his mother asked what he wanted to be when he grew up. Brian contemplated the question, thinking "football players are famous, so that would be nice. Or maybe a policeman or a fireman." His mother shook her head. "No, if I was a wee laddie, I'd want to be an astronaut," she said, talking in a dreamy way about rockets, stars, and planets. She was the one in the family with the adventuresome spirit. When the Binnies moved from Scotland back to the United States, Brian had an experience he wouldn't soon forget. It was a hot summer day when Brian walked off the plane at Logan Airport in Boston. The smell of asphalt and plane fumes overwhelmed him; he loved it. It became both inspiration and anodyne. He went on to graduate from Brown University with a bachelor's degree in aerospace engineering and a master's degree in thermodynamics and earned a second master's degree in aeronautical engineering from Princeton.

While at Princeton, Brian ran into some Navy pilots who urged him to consider flying with them. He'd been dreaming of planes, studying planes, and working on planes, but never had the money to actually fly planes. As it turned out, his lack of flying experience proved to be a good thing, as he was open to the Navy's way of teaching. Brian took to it readily, passing one test after another. He spent three years training in Florida and eventually earned his Navy wings—as cherished as his wedding band. The training continued with flight refueling, simulated air-to-air combat, close flight formation, night flying, and low-level night flying. Every day presented an opportunity to fail. The scrutiny was intense. But Brian loved what he was doing and thrived in the single-seat aircraft, where he was alone in the cockpit with his instructor flying in his own plane alongside. It was one man, one machine, make-or-break flying. Before he got into a plane, Brian flew every formation in his mind. He drew flight patterns out on the carpet in his room. He dreamed about airspeed and maneuvers.

His next challenge combined the precision of a high-wire act with the power of a drag race: landing an F/A-18 Hornet on an aircraft carrier at sea. It was a moving target where any mistake could be his last. The runways were no more than 500 feet long, and landing was done at speeds of up to 150 miles per hour. He had to come in and catch with his tail hook one of the four "arresting" cables strung in parallel lines across the deck of the ship. Catching the third wire was the goal. A counterintuitive challenge was to touch down with engines full throttle. If the plane didn't catch a wire, Brian needed enough speed to take off for another pass. All of this was to be done on a heaving and swaying ship.

Brian would spend twelve to thirteen hours flying the same pattern, all under the unforgiving watch of the meanest guy on deck, the air boss. The air boss was impossible to please, barking orders and running the show. Brian spent much of his training on the USS *Lexington,* which had only three trip wires. Every pass was graded A through F. If a pilot got an F on a flight, he'd be talking to the admiral and his career was in danger. Brian also trained on the USS *Enterprise,* which had four wires. One day while training off the USS *Independence,* a typhoon was sweeping in through the North Philippine Sea, and all the pilots were called back to the aircraft carrier. Brian was on his landing approach and gave his call signal—with his type of plane and amount of fuel—so the cables could be set at the right tension. As he neared the back of the ship, the bow of the *Independence* dipped down and the stern came way up. Brian stared at the propellers whirring out of the water. The landing signal officer—charged with keeping his pilots alive—said calmly, "Deck's up. Looking good. Keep it comin.'" The bond that developed between pilot and landing officer was intense. *I gotta trust in something,* Brian thought. "Okay, I'm comin' in." Sure enough, the deck reversed pitch and he saddled into the wire, pulling off a respectable landing in roiling seas.

Night training was even more demanding. "Okay" was the language of the Navy. Not "great" or even "good." "Okay" was what every pilot wanted to hear, making him feel like King Kong for the next fifteen minutes—or at least until he had to do the run all over again. Brian flew in Operations Desert Shield, Desert Storm, and Southern Watch. He flew thirty-three combat missions over Iraq, primarily in the F/A-18. He was a Navy commander when he retired in 1998 at the age of forty-five.

Not interested in being relegated to a cubicle in the Pentagon, he started looking around for other jobs that would keep him in the air. When he saw a job posting to flight-test an experimental rocket in the Mojave Desert, he was intrigued, applied, and got the job at Rotary Rocket. He, his wife, and three kids had been living at Point Mugu in southern California, finding it paradise compared with many of the bases they'd called home. It had mountains, the Pacific Ocean, golf courses, and strawberry fields. They left that for the dry and flat town of Rosamond, thirteen miles south of Mojave and just west of Edwards Air Force Base. Brian had long considered his wife, Valerie, who went by "Bub," an angel—now he had proof.

Just as his family wasn't prepared for desert dwelling, Brian struggled with civilian life. Going from the military to Mojave was like entering the Twilight Zone. At Rotary, it felt like there were no rules. Everything was "fluid." Schedules were amorphous. Some of the employees looked and sounded more like agitators than engineers. Brian took one look at a guy with green hair and nose rings and asked, "Who is this guy? What does he do?" He was told, "You better get over it, pal. This guy is designing your flight controls." When he overheard some employees talking about recreational drug use, he made it an issue, saying he wasn't going to fly a rocket built by guys on drugs. Again he was scoffed at. He didn't believe a word of Rotary's "schedule" for approvals, licenses, and flight tests. He had seen how slow the government moved. Still, there was money coming in to Rotary, and the promise of more.

Brian started at Rotary in 1998 as one of two pilots—the other was Marti Sarigul-Klijn, another Navy-trained engineer and pilot. Brian also served as program manager. He knew of Burt Rutan by reputation and had seen him around, but the two had never met. Scaled had been hired by Gary Hudson to build the shell for the Roton. When it came time for a design review of the project, Burt and a handful of others were invited over. Brian had the last presentation of the day, from 1:30 to 2:30 P.M. At the end of his presentation, Brian asked whether anyone had any questions. Brian saw Burt's hand shoot up. His heart sank. He felt like he was in the crosshairs of the lion. "Yes, Mr. Rutan."

"If we quit right now," Burt said cheerfully, "we can still get in nine holes. What do you think?"

Brian thought it was a great idea. The two became golfing buddies, and when Rotary closed its doors in 2000, Burt brought Brian to Scaled.

On Wednesday, December 17, 2003, one hundred years to the day after two bicycle mechanics named Wright managed a powered and sustained flight in a flimsy aircraft that no one thought would fly, a bunch of "motorcycle mechanics in the desert" were ready for the first powered flight of *SpaceShipOne*. The *White Knight* and *SpaceShipOne* were rolled from the hangar of Scaled into the early morning light. The plan for the flight was straightforward: achieve a fifteen-second burn of the rocket motor and supersonic flight. Brian would assess how the motor lit and performed, how it felt going through transonics, and how the feather did at a relatively high altitude. Of these things, the sound barrier posed the greatest challenge. The little craft with the pointed nose sprinkled with oregano had never flown so fast or so high.

Brian was in the cockpit of *SpaceShipOne,* to be carried aloft by the *White Knight.* He focused on the instruments before him. The rocket controls were simple. Two switches: one to arm, one to fire. The avionics suite had a dedicated propulsion display that showed important motor parameters that could be monitored by the pilot and Mission Control. The pilot couldn't throttle the rocket. At around 48,000 feet—Burt liked to say the first 50,000 feet were free—*SpaceShipOne* would be dropped like a bomb from the hooks of the *White Knight*.

With relatively little fanfare—a few invited guests and media—the *White Knight* took off into the hazy blue sky. The ground below was a palette of beige and sand. Six miles north of Mojave, Brian did a systems check. They were close to their desired altitude. Brian got the green light; he and *SpaceShipOne* were ready to be dropped from the *White Knight*. The countdown began: 5-4-3-2-1-mark!

"Good release," came the call from Cory Bird, piloting the *White Knight*.

"Nose wants to drop," Brian reported. Then, "Control checks are good."

"Status—go for light."

"Armed," Brian said. "Green light."

Brian, in the rocket and free from the mother ship, stared at the FIRE switch. He had seen, heard, and felt the rocket motor hot fires. It was the Brahman bull angered, and today was the day the bull left the chute. With his head back into the seat and his finger on the FIRE switch, Brian girded himself. He flipped it. Both his hands clutched the stick like the horn of a saddle. He was jolted up and back until he could barely keep hold of the stick.

He had landed on carriers and flown in combat, but this was something else. Chuck Yeager had brought his own leather football helmet, knowing the X-1 would knock him around. Brian was jostled. "A lot of oscillation," he told Mission Control euphemistically. He tried to focus on the velocity vector, which told him where he'd be in the next few seconds. His trajectory was nearly perpendicular to the ground. He feared he would go over backward, but the flight path maker told him otherwise. Brian was now in the transonic corridor, where he was hit by even more buffeting. A sad-sack cowboy being flung around like a rag doll.

Just then, a giant boom was heard across the desert floor, signaling that *SpaceShipOne* had flown faster than the speed of sound. Brian was at Mach 1.2, wondering how much longer he could hold on. Then, as fast and as violently as it all began, the rocket motor shut down—marking the end of the longest fifteen seconds of his life.

Brian still had to get back to Mojave. He activated the feather that bent the wings in half, locking them in place at sixty-five degrees. After his fifteen-second foray into the supersonic, he was returning through the unfriendly transonic region with more pitching and swerving. He almost laughed when he got the call from Mike Melvill, in a chase plane: "Brian, you look great!"

Then, at around 60,000 feet, there was a gift, better than the blue expanse or a pillow of clouds. It was calm and quiet. No bucking, jostling, jerking. No noise. One man, one machine.

Brian exhaled. "That was quite a wild ride from Mr. Rutan," he said, prompting laughter from Mission Control. At 50,000 feet, he got the nod to "defeather" and move the wings back into glide position.

As he made the final descent into Mojave, everything finally felt familiar. Mike was off to his left in a chase plane, monitoring and encouraging him and ready to call out wheel height. This was good. Brian could sense impending triumph. He had fired the motor, gone supersonic, and operated the feather. It all worked. The runway was just ahead. The years of hard work on *SpaceShipOne* were paying off. If the Scaled crew could go supersonic, they could go to the start of space. The celebrating would begin with touchdown. On the ground, to mark this important next step on the way to space, were the few invited guests and media, and their benefactor, Paul Allen.

Brian lined up with the runway and dropped the landing gear. He was almost home. Then suddenly, the nose started to slip to the right. He tried trimming it, but that made things worse. The wings wobbled, refusing to stay level. Mike called wheel height as Brian wondered whether he would land upright. He relaxed pressure on the stick, but this just dropped him closer to the runway. *Was the airspeed indicator off?* He could *not* crash-land, not after all this effort. Brian pulled the nose up and flared the plane before touchdown, but that didn't help, either.

He was going down.

SpaceShipOne hit the runway hard, its left landing gear crunching and collapsing. Brian skidded down the runway and veered off it into the sand. He was enveloped in a cloud of dust.

Brian threw off his helmet and used every expletive in the book. Burt was among the first to reach him and dust him off. Paul Allen, who was using the day to officially announce that he was backing Burt's space program, was conspicuously absent from the crash site. His handlers didn't think it was such a good idea for him to be pictured with his newly crashed spaceship.

Burt told Brian that he'd done a great job: he'd gone supersonic and he'd feathered. The hard landing was caused by last-minute modifications they'd made to make the plane more stable, Burt said. He told him they'd have the plane flying again in weeks.

But Brian was inconsolable. He believed that he had been working up to this flight all his life. He was a war veteran, a skilled Navy fighter pilot, and a family man. He had done everything right, but in this moment, all of those accomplishments seemed for naught. He felt like an Olympic athlete who had won every heat, jumped every hurdle, and then stumbled right before the finish line, when everything mattered most. That's how he would be remembered—he was sure of that. It even seemed irrelevant to him that he had made history that day, piloting the first private plane to reach supersonic speeds.

Brian's colleagues at Scaled only reinforced his doubts about whether he had the right stuff as a pilot. In the wake of the test flight, some were saying the Navy aviator had landed the spaceship as if he were landing an F/A-18 on a flight deck—hard and fast. Even Mike, who'd watched from the chase plane, said Brian had flown the craft into the ground. The name *Brian Binnie* was suddenly synonymous with *crash landing*. But all the negative talk couldn't make Brian feel worse than he already did. As repairs were made to the spaceship, Brian feared he'd be relegated to writing up the test cards for missions he'd never fly.

That Friday afternoon, only two days later, Burt, Brian, and Scaled colleague Kevin Mickey took off early for a round of golf in Bakersfield. Mickey had grown up in Mojave and spent most of his youth peering through the chain-link fence to see what this guy named Burt Rutan was building. He had started as a floor sweep at $6 an hour and was now a vice president.

Out on the course, Burt took out his laminated spreadsheets detailing different clubs and holes and projections for how far the ball would carry and roll on the fairway, greens, and in the rough. Brian had worked his way down to a five handicap, and Burt was a solid ten. Mickey could hold his own. Brian's strength was off the tee, while Burt had a strong short game. Once Burt got inside a hundred yards, he could be formidable. He had built his own putter, which was similar to a belly putter, with a "T" as the head. He called it the Titanic because he trusted it to sink any putt. While relaxing on the links was the primary goal, it wasn't unusual for Burt and company to get into discussions around the aerodynamic properties and ballistic coefficient of different clubs, club heads, and even golf balls—talking dimple depth and shape, including hexagons. Would a bigger driver head create more drag, or give the golfer more confidence in hitting the ball, thus a better chance at striking the ball with more power? How did the grooves in a golf ball reduce drag?

Burt could see that Brian was still depressed about the crash landing of *SpaceShipOne*. As they were finishing up their round in Bakersfield, Burt asked Brian whether he knew the story of Doug Sanders, the American golfer who had been one shot away on the eighteenth hole from winning the British Open at St. Andrews in 1970. Brian said no.

"He missed a three-foot putt against Jack Nicklaus to win the Open," Burt said. He won twenty PGA tour events, but that one thing changed the course of his life. Burt had heard Sanders asked years later how often he still thought about that missed putt. Sanders said that "maybe nine or ten minutes" went by when he didn't think about it.

The story stayed with Brian, distracting him as he drove home. Brian thought, *I didn't come this far to have* that landing *define me. I'll be damned if that's the way it's going to be for me.*

27

Flirting with Calamity

It was the middle of the night on June 21, 2004, and a warm Mojave wind whipped up the sand and tipped over chairs, tents, and anything else not tied down. Even the Porta Potties had to be pushed back up. Not far from Scaled Composites, a makeshift RV park was now full. A long line of cars snaked from California Highway 58 into the newly renamed Mojave Air and Spaceport. Big crowds were amassing here to witness the test flight of all test flights—*SpaceShipOne*'s first journey to space—and, if successful, a preview of its first XPRIZE attempt.

Burt Rutan arrived at Scaled before three A.M. The sky was dotted with stars, and the Moon was a waxing silvery crescent. Pilot Mike Melvill had spent the night in an RV in the hangar because gridlock was predicted. A preflight meeting was scheduled for 4:45 A.M. Inside the hangar, *SpaceShipOne* and the *White Knight* were gleaming and ready. The rocket plane had flown two powered flights since Brian Binnie's supersonic milestone in December. The craft's broken landing gear was replaced and other fixes made. A wider nozzle for the rocket was installed, along with a new fairing that ran from the back of the fuselage over the nozzle to reduce drag. Each test flight had taken the craft incrementally higher and faster—to 211,400 feet and Mach 2.3 the month before. That put Scaled team members at about 64 percent of where they needed to get to today, which was 328,000 feet, the internationally recognized boundary of space.*

Today's flight was the biggest gamble yet. Each test flight—whether glide or rocket-powered—had anomalies. On one occasion, a dangerous tail stall during a glide test had turned into a harrowing loss of control. During a powered flight, the avionics display suddenly blinked off, forcing Mike Melvill to decide in an instant whether to fly blind or abort. As X-15 engineer Bob Hoey had warned, the *SpaceShipOne* team was living with the constant threat of calamity. But team members also got to experience the wonder and validation of what they were getting right. In the weeks leading up to this day, Burt had found it hard to work and hard to sleep. He hadn't even been to space, but he was already drawing up plans for what would come next: orbital vehicles. Only hours earlier, he'd spent the evening hosting Paul Allen and Virgin billionaire Richard Branson for dinner at his pyramid home. Before dinner, he sat them down on his sofa—one billionaire on each side—and showed them a forty-four-slide PowerPoint presentation titled "Manned Space Vision Summit." He talked excitedly about private space stations, zero-gravity hotels, and orbital ships. He shared his renderings for an "upgraded *SpaceShipOne* with room for seven passengers." Between the slides and musings, Burt said to both men, "You're probably asking what we should do this year and next year, right? The only way you can make an intelligent decision on what to do this

---

*The 211,400 feet achieved on the May flight just reached the mesosphere, the third layer in the four layers of our atmosphere. The 328,000 feet sought on June 21, 2004, needed to reach the fourth layer, the thermosphere. The atmosphere consists of four thin onion skin–like layers around the Earth. The troposphere starts at the Earth's surface and extends 8 to 14.5 kilometers high. This part of the atmosphere is the densest. Almost all weather is in this region. The stratosphere starts just above the troposphere and extends to 50 kilometers high. The ozone layer, which absorbs and scatters the solar ultraviolet radiation, is in this layer. The mesosphere starts just above the stratosphere and extends to 85 kilometers high. Meteors burn up in this layer. The thermosphere starts just above the mesosphere and extends to 600 kilometers high. Aurora occurs and satellites orbit in this layer. The mesosphere still contains clouds—although very thin ones—the famous noctilucent clouds. But the atmosphere is still thick enough that meteorites burn up as they cross from the thermosphere into the mesosphere.

year and next is to know the answer to one very important question." He paused. The room went quiet. "The question is, what do you want to see happen before you die?" Allen smiled. He could see that Burt had been bitten by the space bug. Allen was too, but he was satisfied with what they were attempting with manned suborbital flight. Branson was interested in how Burt's ideas and creations could open space to the masses.*

Bouncing around the Scaled offices at 3:45 A.M., Burt realized he was the only one present who wasn't immediately mission critical. He looked at the clock: he had forty-five minutes to do whatever he wanted. So he headed straight down the flight line to find Stuart Witt, a no-nonsense former top gun Navy pilot who was now the airport director. Witt had recently pushed the Federal Aviation Administration to authorize the transformation of the airport into a spaceport. The plan had been Burt's idea, presented with his typical urgency. Burt had marched into Witt's office—he always came through the side door by the flight line—and said, as if in midconversation: "I need a fourteen-thousand-foot runway and you've only got a ten [thousand]."

"Why do you need it?" Witt asked.

"Oh, I need it as soon as you can build it. Need a throat [a turn in the runway] this big," Burt said, gesturing hugely.

"Who is going to pay for it?"

"Hell, I don't know."

Witt lobbied Congress and got the funds for the longer runway and new turnaround. In the weeks leading up to today's big test flight, Witt had been working around the clock, determined to run the show with military precision. This morning, his busy schedule included orchestrating the uprighting of the Porta Potties, handling security, and fielding last-minute calls, including one from the general over at Edwards Air Force Base, who was asking to land but wasn't on the day's fly-in list.

---

*Paul Allen and Richard Branson had begun discussions about Branson's buying the rights to *SpaceShipOne* in March, three months before this dinner in the desert. But no deal had been made.

Now, here stood Burt in the middle of the night, with another idea, this one more mischievous than ambitious. "I've got about forty-five minutes—now less—and I can't think of anything I'd rather do," Burt said.

Witt couldn't resist. The two hopped in Witt's truck, parked outside the door. It was still dark as the two drove down the runway and past the flight tower. Before Witt came to a full stop, Burt was out of the truck. When he'd first met Burt years before, Burt had told him that the reason he settled in Mojave was simple: he started driving at Brown Field just north of the U.S.-Mexico border and hit every airport heading north until he found one that he could afford and that would give him permission to start his business.

Witt parked and got out of his truck. Burt was heading for an RV parked in a prime viewing spot, suggesting that the vehicle had arrived days before. The Magician of the Mojave wanted to mingle with the rocket fans.

A light was on in the camper, and Burt went ahead and knocked. The door creaked opened. A couple peered out to see a tall man with silvery hair and Elvis sideburns.

"Yes?"

"I'm Burt Rutan, and I want to thank you for coming," Burt said, extending his hand as the two reluctantly opened the screen door. The couple, from St. George, Utah, had two kids sleeping in the back. It slowly dawned on them that the Burt Rutan standing before them was *the* Burt Rutan. The Utah man laughed, and opened the door wide. He explained that he'd come home from work a few days before, listened to his wife talk excitedly about a news story she'd heard promoting the launch of a private spaceship, and he said, "Load up the car, we're going to Mojave!"

Burt continued his meet-and-greet, knocking on a few more campers' doors. He shook hands and thanked people for making the trip to Mojave. He looked around for what to do next. Paul Allen's people hadn't budgeted for public parking. Allen wanted the press and VIPs to attend, but Burt wanted the public to be a part of the day, too. To make it happen, he and his wife, Tonya, decided to foot the hefty parking and police security costs, which were close to $100,000 after all the licenses and red tape. Burt and Tonya paid for about $80,000, and Witt picked up the rest. Burt and Tonya wanted kids to be there so they could later tell their own children that they had attended the world's first nongovernmental manned spaceflight.

In the RV area, Witt could now see that Burt was fixated on some drivers a short distance away who appeared to be struggling to make a ninety-degree turn to park. Burt walked up to a parking attendant, startled him by relieving him of his lighted baton, and said, "Let me help you." He moved into position and began directing traffic, even opening car doors with a "Welcome to Mojave!" Some people recognized him—including a group wearing "GO BURT GO!" T-shirts—and grabbed for their cameras. Others were just happy to have a parking spot.

As a hint of orange light on the horizon announced the dawn, visitors inured to the wind staked out their places in fold-out chairs, ready for showtime. Witt and Burt both needed to be back for preflight meetings. They left the parking area and headed up the flight line. A steady stream of private jets had been arriving. The sunrise would be a gorgeous orange, turning the mountains dark brown. Witt and Burt were silent as they took in the sea of cars, campers, and people. Witt looked at his seatmate. Burt had tears in his eyes.

Inside Scaled, Mike, now in his flight suit, joined the engineers and flight team managers at the conference table, mission notebooks in front of them. As others talked, he was lost in thought. If everything went right, he would become the world's first civilian pilot—not working for the military, not working for NASA—to fly out of the Earth's atmosphere. But Mike knew better than anyone that this was a big "if." A lot could go wrong.

At sixty-three, he was past the mandatory retirement age for an airline pilot. He was also too much of a "cowboy" and didn't take instruction, according to an e-mail from fellow pilot Pete Siebold that got passed on to Burt in the weeks leading up to today's flight. Burt read the e-mail, headed across the hall and dropped the paper on Mike's desk, saying, "See what you're up against?" Mike was the first to admit he wasn't as good in the simulator as Siebold—who built the simulator and wrote much of the software. Siebold was a Cal Poly graduate like Burt; Mike didn't graduate from high school and was a self-taught engineer. Mike had never played video games; Siebold was a master. Mike flew by how the plane *felt*.

Despite Siebold's criticisms, Mike had done the majority—eight out of thirteen—of the manned test flights of *SpaceShipOne*. He was the pilot of *SpaceShipOne* when the avionics went out and he continued to fly blind, something Burt applauded and Siebold criticized. Brian Binnie had done a great job with his flight, until he crash-landed. Siebold pulled off a strong powered flight, but hesitated too long before firing the motor. Once dropped from the *White Knight*, Siebold hadn't trusted the handling qualities of the rocket. He dropped more than a mile as he assessed the problems, until Mission Control told him he had to light the motor. Landing with a full load of fuel would be far too dangerous. Siebold's analytical engineering brain had overridden his test pilot moxie. Still, he had pushed hard to pilot today's historic flight to space and had mission director Doug Shane in his corner. But Burt hadn't been pleased with Siebold's May 13 flight. The cowboy was up.

Burt had never invited the public to watch a critical experimental flight test. There were too many things that could scrub a launch or derail a takeoff. The *Voyager*'s around-the-world trip had drawn global attention, but the ubiquity of the cameras and multimedia was something entirely new. With this flight, there were cameras everywhere, from inside the pilots' locker room to the cockpit of the *White Knight* and the spaceship. If something went wrong, the world would know. The risk was palpable. Burt trusted Mike to get the job done.

During the early morning briefing, the schedule was reviewed and

Shane noted, "We're still trying to taxi out at six-thirty." Crew chief Steve Losey went over the system checks of the vehicle. Looking at his friend Mike, he said, "We're good to go." Mike nodded slightly, but his mind was rehearsing the day's flight. More details were discussed: the weight of the fully loaded spaceship, pilot included, was 6,380 pounds; the target altitude for the *White Knight* to release *SpaceShipOne* was 46,000 feet; and wind limits to takeoff were fifteen knots crosswind.

Shane, who had perfected the art of impassivity, noted with his usual poker face that the "unique nature of the day" presented new challenges, including "lots of people," and the "ears of the world tapped into Mission frequency." They also had what they called a "higher tortoise jeopardy." Under their licensing agreement with the Federal Aviation Administration's Office of Commercial Space Transportation, Scaled was required by the Environmental Protection Agency to do a runway sweep to ensure that runways were free of the endangered tortoise. At an earlier point in the rocket program, an idea was hatched to take one of Mojave's finest turtles to space. But under the licensing agreement, Scaled personnel were not allowed to touch, nudge, or move these reptilian interlopers. Scaled was required to call in a specialist trained in tortoise relocation. Witt, who almost didn't take the Mojave job so that he could catch lots of fish in British Columbia instead, noted drily that they had been doing about three hundred movements across the runway every day—planes, equipment, trucks, aircraft, and helicopters—and had never once been asked to do a runway check for tortoises. "But by God, we're told to do it for *SpaceShipOne*," Witt said, "so we're going to do it."

The meeting ended, and Mike and Brian Binnie headed to the pilots' room to grab their parachute packs. Neither said a word. Brian felt like something of a "bus driver" in the *White Knight*; he was determined to work his way back from the mother ship and into the cockpit of the spaceship again someday. In the hangar, employees signed the nozzle of *SpaceShipOne*. Paul Allen, using a gold Sharpie, signed the nozzle and autographed the inside of the cockpit. Pete Siebold wished Mike a good flight and said with a hint of wistfulness, "This is the flight that says, 'Hey, NASA, we're here.'" Burt and Paul Allen walked out onto the tarmac together and looked in the direction of the wind cones, which hours before were whipping and full of air, but were now quiet. Burt told Paul that the Mojave wind always calmed when it was time to fly, like a child who knows when it's time to behave.

Peter Diamandis was in Mojave with Erik Lindbergh and a group from the XPRIZE, including William Shatner. Virgin's Richard Branson had arrived there, too, as had Buzz Aldrin and aviator Bob Hoover, whom Chuck Yeager had called "the greatest pilot I ever saw." Peter had been hosting parties leading up to the flight and made his way to the tarmac as the sun was coming up. If Scaled succeeded today, the next trip to space would be the first flight for the XPRIZE. Teams were required to give Peter a sixty-day notice of intent to fly. Several teams had told him they were close to launching hardware, and one team hinted that manned flight was not far off. As Peter waited for the test flight to begin, he dreamed of even bigger parties in Mojave—of back-to-back flights to space. He had the $10 million prize money, but under the arrangement of his hole-in-one insurance policy, time was fast running out.

Off to a quiet side of the hangar, away from the clutch of employees and photographers, Mike and Sally stole a minute to themselves. They held each other close and kissed. Mike brushed Sally's hair off her face and said, "I wouldn't trade anything I've done for anybody else's love." Sally looked at Mike and told him, "Come home to me." She pinned their lucky horseshoe on the sleeve of his flight suit. He kissed her again before heading to his plane. Test pilots, like astronauts, had training, simulators, flight plans, flight suits, helmets, talismans, and people praying for them. The wives of test pilots had that last kiss.

It was time. The manning up of the vehicles was under way.

Mike walked to *SpaceShipOne,* helmet under his arm, mind on the mission. Burt had never put him in a plane that wouldn't come home. Mike trusted this man with his life—he had to. Burt had never let him down, and he hadn't let Burt down. He wasn't about to let that change today. Still, Mike was afraid.

Before the cockpit door was closed, Burt leaned in. "This is the big one, Burt," Mike said, his voice full of emotion.

"We got the right guy," Burt said, his voice breaking. He clutched Mike's hand. "Forget about space," Burt said. "It's just an airplane."

The door was soon closed, and Mike was alone in the cockpit. Burt's last message resonated with him: *Just fly it like a plane.*

Standing on the runway, Burt told Paul Allen that Mike's fear would vanish like the winds of Mojave as soon as he was in the air and had a job to do. "Test pilots are like that," he said. Paul was anxious. When he made the deal to sponsor Burt in his quest to build a spaceship, he hadn't been thinking about the pilots, and the pilots' families. Now he was friends with the pilots and their wives and aware of all that was at stake.

Inside the *White Knight*, Brian Binnie was at the controls, with Matt Stinemetze in one of the plane's two passenger seats. Stinemetze's job would be to pull the handle and release the spaceship. Two chase planes were ready on the runway: a high-altitude Alpha jet owned by Paul Allen, and a Beechcraft Starship, one of Burt's early designs. The *White Knight* began its roll down the runway. As the craft taxied, Stinemetze peered out the side windows. He was disappointed to see only a hundred or so spectators. Floodlights were pointed their way, making it difficult to see to the sides. "These are the only people who showed up?" he asked.

But as the *White Knight* passed the tower, the view from the windows changed. It was row after row of people, as far as Stinemetze could see. "Holy crap!" he said. "Will you look at that!" He took in the satellite dishes, cars, and campers. He said it again: "Holy crap!" Brian marveled at the thousands of people who lined the runway fences. "Oh boy," Brian said, "let this be a good day."

From out in the crowds, cheers rang out as the twin-engine *White Knight,* carrying *SpaceShipOne,* approached. A young girl yelled out excitedly, "The *White Knight* is coming! The *White Knight* is coming!" Minutes later, the applause and cheers erupted again: the *White Knight* was taking off toward the southern Sierra Nevada.

It took sixty-three minutes for the mother ship to reach 47,000 feet and the targeted drop point. The countdown began for the release. The spaceship was dropped and the motor armed.

Mike said, "Armed and . . . fire!" The acceleration knocked his head back into his seat. He was flying almost straight up, quickly closing in on the transonic region. The strong wind shear was making it difficult to control the ship, and sending him off course in a way that might sabotage his goal of getting high enough to reach space. Then Mike heard three ominous bangs—one big bang and two smaller bangs. He couldn't see outside the rocket to know what had happened and worried that the ship was damaged or that something had fallen off. The flight had not gotten off to the best of starts.

Seventy-seven seconds into the flight, the motor of *SpaceShipOne* shut down, as it was supposed to do. Mike felt the motor hadn't run well and was still concerned about the earlier banging noises.* He was anxious that the craft might not have the inertia to make it to the Karman line.

In Mission Control, there was clapping when it appeared that the spaceship had made it to 328,000 feet. But the applause quickly stopped. The crew would need to wait on the final altitude report. Burt sat to Doug Shane's right, and Paul Allen and two officials from the FAA's Commercial Space Transportation division stood toward the back.

Down the flight line, thousands of people looked through powerful camera lenses or binoculars. Others held hands to the sky like a universal

*It was later determined that the bangs were a normal noise the motor made as it used up the last of its fuel load in the rocket's motor case.

salute, trying to catch a glimpse of the spaceship, now a white dot with a long white contrail. The spaceship was flying in the direction of the sun. Someone yelled, "Go, Mike, go!" Nearby, a man with a tinfoil hat sold T-shirts reading "SAY HELLO TO MY ALIEN MOTHER" and "TAKE ME TO SATURN." A woman wielded a sign: "WE ARE GOING TO SPACE AND THE GOVERNMENT IS NOT INVITED." Another sign read "I'VE BEEN WAITING 40 YEARS FOR THIS." Sally was on the flight line with her son, looking anguished as she searched the sky.

Mike configured the feather to prepare it for the reentry, which would happen four minutes later. He unlocked and feathered it, and felt the thud as it was forced into its sixty-five-degree angled reentry position. *No problem there,* Mike thought. Studying the display in front of him, Mike noticed that the stabilizers on the feather, used for high-altitude control, were unevenly set. One was at ten degrees and the other at thirty degrees. *This has to be a mistake,* he thought. But it wasn't. *This is bad.* Mike knew it was potentially fatal: a twenty-degree delta in the settings would force the spaceship into a spin from which he couldn't recover. If he didn't find a solution, he wouldn't make it back alive. The only way out was through the nose—an impossibility in the midst of a supersonic spin.

In Mission Control, Shane called for "more stab trim." Aerodynamicist Jim Tighe, sitting one row back from Shane, told Mike to pull the breakers to start the backup motor. Mike had tried that, but the backup motor had failed. Mike audibly exhaled. He took in the beauty with sadness. He was more wistful than panicked. Thousands of people were watching and waiting; he was alone in this experimental space capsule.

Suddenly, Mission Control lost its connection to *SpaceShipOne.* Paul Allen could see a change in the team's body language. Burt and Doug leaned forward in their chairs.

"Ground to *SpaceShipOne,*" Doug said.

Nothing.

"Ground to *SpaceShipOne.*"

No response.

Doug's lips trembled slightly.

"Ground to *White Knight*. Contact *SpaceShipOne*."

Still nothing.

"*WhiteKnight* to *SpaceShipOne*."

A few moments later, Mike, now desperate for a solution, decided to try the trim system again. This time, for a reason he couldn't understand, the left stabilizer moved.* *Oh my God—it worked,* Mike thought.

"*SpaceShipOne* to ground," Mike said, as a collective sigh of relief was heard in Mission Control.

Jim Tighe said quickly, "Back up stab trim, got it. Leave the trim alone."

Shane added, "The trim is adequate for landing."

Mike took in the majestic view outside his windows. He just might live another day. "Wow," Mike said slowly, his heart rate coming down as his eyes followed the curvature of Earth. "You would not believe the view. Holy mackerel."

He was now in his three to four minutes of weightlessness. He unzipped the left pocket of his flight suit to take out an item that he'd purchased at a convenience store en route from home to work: M&Ms. He unfurled a handful of the multicolored candies, which went into the air like a cool spray of water on a hot summer afternoon. The treats floated in the sunlit cockpit, clicking as they bounced off surfaces. Small pieces of a rainbow surrounded him.

"He is twenty miles south of the bull's-eye," Shane said.

"We need to head northwest," Tighe said.

Gravity took hold too soon, and the ethereal M&Ms clattered to the cockpit floor.

"Here come the gs," said Mike, his breathing labored as he came back

*Unbeknownst to Mike and the folks in Mission Control during this flight, the sophisticated motor controlling the stabilizers was programmed to shut itself down to prevent overheating. The motor would work again after a two-minute cooldown period.

through the atmosphere at more than five times the normal force of gravity. Parts of the outside of the plane were heated to one thousand degrees—body putty, cinnamon, and oregano as his shield.

"Coming down, gs past," Shane said when Mike was through the supersonic region. Then, as he looked up at the television monitor, he said, "Spaceship in sight."

"Mike, you doing okay?" Shane asked.

"Doing great," Mike said, though he was still anxious to have one of the chase plane pilots do a visual inspection of the spaceship. In a matter of minutes, the look-over was complete. Mike was told there was only a small buckling around the nozzle, but nothing compromising.

On the desert floor, the crowd caught sight of *SpaceShipOne* and erupted into the biggest cheers of the day. The spaceship looked tiny in the blue expanse, like a kite drifting way up high. Sally clasped her hands to her face and said, "Come home, Michael." People chanted, "Go, baby, go!" A man yelled: "This is what America is all about." No one knew just how difficult the flight had been.

The landing gear was extended. *SpaceShipOne* was coming home, flanked by its mother ship and the chase planes. The birdlike craft cast a dark shadow on the tarmac. Seconds later, it touched down to more cheers. A man drew more applause when he said of Mike: "That guy is a badass!"

In the cockpit, Mike let out a "Yeehaw!" The space cowboy had lived for another day.

In Mission Control, Shane—the man who specialized in showing no emotion—wiped tears from his eyes. Paul Allen patted Burt on the back, and the two headed out to greet Mike. They had seen the data: 326, 327, and some thought it reached 328—the 62-mile point. No one knew for sure, though, whether they'd officially made it to space. They needed to wait for the tracking data from a privately run group and a team out of NASA Dryden at Edwards Air Force Base.

Mike pulled himself out of the cockpit and stood unsteady for a brief moment on the runway. When Burt and Paul reached *SpaceShipOne,* Mike and Burt hugged like best friends kept apart for years.

"If we can do this, we can do anything." Mike beamed.

"How do you feel?" Burt asked.

"I've never felt this good."

"You did a super job," Burt said.

Sally ran to her husband's arms.

Then something happened that had never been seen after any manned spaceflight. With Burt and Paul sitting on the tailgate of a pickup, *SpaceShipOne* was towed right to the crowd line so everyone could see the rocket up close. Burt scanned the crowd. Suddenly, he was gone, running off into the rows of spectators. He returned with a sign he'd plucked from the crowd and handed to Mike. Before long, Burt and Paul were back on the tailgate as *SpaceShipOne* was towed back to the hangar. Mike stood on the top of the spaceship, holding the borrowed sign reading: "SPACESHIPONE GOVERNMENTZERO."*

Scaled soon got confirmation: Mike had made it to space—by a nose: 328,491 feet. He had passed the Karman line by less than 500 feet—one tenth of 1 percent of 62 miles.

In a ceremony that day, sixty-three-year-old Michael Winston Melvill, who didn't become a pilot until he was thirty and was a career machinist before meeting Burt Rutan, was awarded the first ever commercial astronaut wings from Federal Aviation Administration director Marion Blakey and Patti Grace Smith, associate administrator for commercial space transportation. That day, and for the days and weeks that followed,

*Mike said only half jokingly later that standing on top of the spaceship as it was towed was the most dangerous part of the day. The top of *SpaceShipOne* was slick, and he feared he would survive the flight to space only to fall on the Mojave tarmac.

Mike was stopped by strangers and asked to sign T-shirts, coffee mugs, and anything people could find. He was deeply touched, and would say, "I think of myself as an ordinary guy who flies around the Mojave Airport."

For Burt, it was the culmination of a dream that had begun in 1955 when he was twelve years old. He had been transfixed watching Wernher von Braun talk with Walt Disney on the television segment "Tomorrow Land" about his pragmatic vision of space. Von Braun went over the development of pressurized suits for space, John Stapp and his rocket sled, and man's ability to take great acceleration forces. Von Braun said, "I believe a passenger rocket could be built and tested within ten years if we follow a step-by-step research and development program." Later, the German scientist said something else that had always appealed to Burt: "I have learned to use the word 'impossible' with the greatest caution."

By late afternoon, campers were packed up and it looked like the wagons were rolling out of the Wild West. Scaled had opened its hangar to hundreds of invited guests to come by and celebrate the world's first privately manned spaceship. The rocket wasn't cordoned off; the point was to make space accessible. Guests checked out the cockpit and leaned against the ship for photos. They couldn't climb in, but many paused to imagine themselves in the backseat heading to space.

Behind the scenes, there was already talk of the XPRIZE and what had to happen to improve performance. The rocket had been pared down to its minimum weight for the flight and had barely reached the Karman line. To qualify for the XPRIZE flight, Burt's team would need to add four hundred pounds—the equivalent of two backseat passengers—to the spaceship. They would need to do what they'd done today—only better—and twice within two weeks. And they would have to do it soon. The first XPRIZE flight would be on September 29. For Burt, there wasn't a lot of time to bask in the excitement of the day. He needed a rocket with more power, and he had just the idea.

28

Power Struggles

After the June 21 flight, Burt started questioning for the first time whether his homemade craft could pull off the XPRIZE feat. *SpaceShipOne* had barely reached its goal of space, and pilot Mike Melvill had endured the flight of his life.

To the public, it appeared Scaled Composites had hit a home run, hosting an estimated 25,000 people for the first private spaceflight and certifying the world's first commercial astronaut. But privately, Burt was worried that the flight had exposed weaknesses in his spacecraft. During the ascent, Mike had ended up in airspace over the populated city of Palmdale—way off the two-mile-by-two-mile flight path approved by the Federal Aviation Administration.

Burt told his propulsion guys to pull out all the stops and do whatever they could to get more energy out of their hybrid motor in time for the first XPRIZE flight, called X1. Under the XPRIZE rules, the spaceship would have to carry the full six-hundred-pound ballast—the equivalent of a pilot and two passengers. Burt instructed Stinemetze and crew chief Steve Losey to put the spaceship on a diet by finding any way possible to reduce its load. The same was being done for the *White Knight,* in hopes the mother ship could fly higher before dropping *SpaceShipOne.*

But Burt had more on his mind than simply cutting pounds from *SpaceShipOne* and the *White Knight.* He had some shopping to do. For missiles.

If Burt couldn't get the oomph he needed from his hybrid motor, his idea was to strap Sidewinders onto *SpaceShipOne.* The small but mighty air-to-air missile, developed at the Naval Air Weapons Station at China Lake in the Indian Wells Valley, would get them where they needed to go. Burt had an old friend who was confident he could secure the Sidewinders. At the same time, Burt dispatched Brian Binnie to use his military contacts to find alternative suppliers as a backup. Brian, who had been overseeing propulsion development, concluded they would need *nine* Sidewinder missiles. Brian looked at the situation and thought, *This is getting scary.* Flying the spaceship already felt like a ride on a Sidewinder.

Brian then talked with folks at Raytheon, which made a large air-to-air missile called AMRAAM (Advanced Medium Range Air to Air Missile). If Scaled went with those missiles, it would need only two of them. The problem was that Raytheon wasn't sure whether it was such a good idea to pull production away from the armed forces at a time of heavy demand to make missiles for an experimental spaceship program in the desert. Raytheon also feared that it would be sued if something went wrong.

Brian then went to Alliant Techsystems (ATK), which had produced a solid motor used for the Titan IVB rocket. ATK was happy to work with Scaled and sent them blank motor casings so the team could start figuring out the placement of missiles. Paul Allen's guys, Dave Moore and Jeff Johnson, got into discussions with the Scaled team about the logistics, testing, and pricing of missiles.

Burt's idea was to install the solid missiles inside the aft fuselage, alongside the hybrid rocket motor. In the past, missiles had been used to boost gliders, airplanes, and jets, and solid motors had been employed to propel rockets to space. But combining a hybrid motor system with solid missiles for a private space venture had never been done. The basic premise was to use the hybrid motor to get *SpaceShipOne* "around the corner" and pointed up. Then, when past the unpredictable transonic region, the pilot would ignite the solid motors to launch the ship like an arrow to its target. The risks were considerable. If one solid motor lit and another didn't, the asymmetric thrust would mark the end of *SpaceShipOne*. Solids, which carried their own oxidizer, also had a greater chance of exploding.

For Burt's plan to work, the solid missiles would have to be installed at the proper angle, and the exhaust vented so as not to melt parts of the vehicle. All of this required a major reconfiguration of the spaceship less than two months before the big flight. There was the all-important feathering system to think about, the landing gear, and other key parts tucked into the back of the craft. The addition of missiles would require a rewiring from the cockpit to the motors, so the pilot could arm and fire at the right time. Once solid motors were lit, they couldn't be turned off. A pilot could only hang on for the ride. There was also the question of how to test the new system. Would the first "test" be with a human on board for the XPRIZE flight, watched in real time around the globe?

Burt was brilliant much of the time, and right most of the time. But this was a rare instance in which his employees were universally opposed to his idea. For more than a year, the group had been preached to by propulsion experts that solids were dangerous and hybrids were safe. Now, practically overnight, they were talking about bringing the devil to the party.

While Burt and Brian were out shopping for surplus armed forces hardware, the propulsion team was working feverishly to figure out how to get more energy from what they had. Their reputations were on the line. *SpaceShipOne*'s rocket motor, developed almost entirely in-house by

Scaled, was the largest nitrous oxide hybrid rocket motor ever flown and the only one to fly a human outside the atmosphere.*

In the meantime, Stinemetze and Losey scoured the spaceship for what could be removed. They looked at the vehicle piece by piece, even sanding for better aerodynamics. They switched out all of the steel fasteners for lighter titanium fasteners. They removed any extra material from the access doors, literally cutting out fabric between bolts. They removed all extra testing instrumentation typically carried by a prototype and looked for unnecessary wires. Anything overbuilt was cut back. Similar scavenging was done on the *White Knight* in the hangar next door. The mother ship was being wet sanded and fairings were added to reduce drag.

Elsewhere, the hybrid motor component vendor was studying its two options: add more propellant, or make the existing propellant more efficient. They were looking at increasing the motor's performance from its original requirement, set by Burt at 630,000 pounds-seconds—a measure of pounds of force and seconds of duration—to at least 700,000 pounds-seconds and possibly higher. Frank Macklin, the chief propulsion engineer for SpaceDev, had been working closely with Scaled's in-house propulsion expert John Campbell.† The pressure to come up with a solution was intense.

At the same time a solution was sought for the motor, Brian was in pursuit of a second chance. After a long day at work, Brian arrived at his home in Rosamond after six P.M., changed from his work clothes into his running attire, and headed out. It was still close to 100 degrees outside, and the desert town offered little shade. He'd had four knee operations,

*eAc of Miami built the components on the front of the big nitrous tank—valves for fill, dump, and vent system. SpaceDev of Poway, California, built components on the tank's back, the valve, injector, igniters, controller, and the casting of the solid fuel.

†Tim Pickens had left early in the program to return to Alabama and at one point formed a team to compete for the XPRIZE.

all stemming from a single judo injury in high school. Running was painful, but it was now his nightly routine. It was about relieving stress and pressing on despite the pain. Ultimately, it was about getting back inside the cockpit of *SpaceShipOne.*

After his crash landing seven months earlier, on December 17, 2003, Brian had spent Christmas trying to find meaning in what had happened. He would wander back into the hangar on his days off and see the broken spaceship looking like an injured bird.

Though Mike's historic trip to space on June 21 was now in the record books, it remained unclear to Brian and everyone else who would pilot the X1 flight: Mike, Pete Siebold, or Brian. Making the call would be Scaled flight director Doug Shane. Brian thought of him as the oracle of Greek mythology who delivers prophecies where nothing is clear yet nothing is questioned. For clarity on the X1 flight, Brian took to hanging out by the office coffeepot. There was a constant shifting of alliances between the Oracle, the Master (Mike), and the Protégé (Siebold). Brian told his wife half jokingly that the shifting of loyalties made him feel like the reality show *Survivor* had set up inside Scaled Composites. Even Mike wondered about the way Scaled went about its pilot lineup. Shane generally kept the decision to himself until the day before a flight. Mike figured it was Shane's way of getting them all to train hard to be ready, but Mike worried that the approach divided the three pilots instead of bringing them together. Whatever the case, Brian was determined to be ready if the call came his way.

After a quick sequence of stretches in front of his house, Brian headed out on his regular loop from home to a small park with trees offering dollops of shade. He didn't listen to music, as he had his own track to play over and over. It was more of a movie, really. As he set out, the reel began: the four phases of piloting *SpaceShipOne.* The first phase began with him in the cockpit—at 48,000 feet, ready for separation from the *White Knight.* He pictured it: His thumb on the FIRE switch. Ready-set-FIRE! Five seconds of being in combat. Life as you know it erupting, shaking. Control or abort? Picture it. Stay calm, clear.

The next phase—transonics. More shaking, pitching, buffeting. Thunderous sounds. The third phase—the longest segment, maybe forty-five seconds. Nose up to eighty-five degrees. Stabilize. Start burning the oxidizer.

The final phase—endgame. Nitrous at end of liquid phase. Transition. Liquid moves through combustion chamber, followed by gas, followed by slugs of more liquid. Thrust difference is tenfold. Motor bounces between vibration and thrust. Hands shake, head shakes. Wakeup call for final part of flight. Thrust coming out of engine no longer aligns with axis of the vehicle. Body beaten up, shaken. Add fleeting aerodynamic control. Air is thin. Get it right the first time. An error at Mach 3 takes you to Palmdale—or somewhere you won't recover from. Slowing down, coasting to apogee, unlock the feather. Make it to space. Make it home. Not like last time. Nail this landing. Find redemption. Stop on the centerline. Spectators cheer.

Pounding the hot pavement on his nightly run, Brian almost forgot his discomfort—until he came to a stop in front of his house. Every jog was the same: he went through the four phases of the flight four times from start to finish. He never missed a run.

Then the mornings started early. Brian tried to get to Scaled before anyone else so he could grab time in the *SpaceShipOne* flight simulator. At first, he had found the sim "piss poor"—to use a Navy term—but liked the bells and whistles added of late. The sim had the same display as the spaceship, but not the stick force. The views in the sim were not representative of what you would see in the spaceship. Imagination was required, as there was no motion or sound. But a clever feature was added by Jim Tighe to reset each "flight" for different motor characteristics and external forces, such as wind shear. On the rare days Brian couldn't practice in the sim in the morning, he was in it during his lunch break. Sometimes he practiced with Siebold—who also badly wanted to pilot the X1 flight—and the two spent hours trading techniques and strategies to deal with problems that lurked around every corner. Brian and Siebold both worried that they were never going to fly the spaceship again, saying, "The only guy who's going to fly it is the guy who has flown it." That was Mike.

Brian now believed that the hit to his reputation in the wake of the crash landing wasn't justified. Since his December flight, he had discovered that the spaceship's controls had stiffened up in the approach to landing because of the dampers installed to prevent the elevons from fluttering. The dampers—similar in concept to dampers used on a car to smooth a driver's path over potholes—got cold on the December flight and stiffened, making it difficult to control the landing flare. From this flight forward, heaters were installed on the dampers to keep them from binding. Still, Brian knew the shop talk: he was the Navy guy who hit the runway the same way he landed on a carrier—full throttle to catch an arresting cable that didn't exist. That portrayal wasn't accurate, but Brian saw no point in protesting or defending. The only thing he could do was to prove the doubters wrong.

At work in early August, Brian ran into Mike, who had been away from Scaled a great deal since his June flight, doing public talks and media appearances, including the Jay Leno show. Mike's new celebrity status drew old-time celebrities to Scaled. Everyone wanted to take a look at *SpaceShipOne*. One day Harrison Ford popped in, unannounced. Another day, it was Gene Hackman.

These days, Mike was basking in unofficial semiretirement, saying he was never going to top the accomplishment of becoming the world's first commercial astronaut. He didn't announce that he was out of the space travel lineup, and he was still doing flight tests. Yet in his mind—and certainly in the mind of his wife, Sally—the space cowboy had set aside his wings to spend some quality time at the ranch.

On this morning at Scaled, Mike sidled up to Brian, put his arm around his shoulder, and said he had an idea for him. He'd seen how Brian was living in the sim. He knew that Brian wanted back in the spaceship. Mike told him he was going to take him under his wing—in every sense of the word. The high school dropout would mentor the Ivy League–educated Navy pilot.

The two began training in Mike's Long-EZ, the two-seat plane that he and Sally had built by hand and he had flown around the world in formation with Dick Rutan. Mike and Sally considered the craft their own personal taxi, taking it to Alaska, to Death Valley for lunch, to see their kids when the mood struck.

Stinemetze went to work doing the math to produce a cardboard cutout mask for the Long-EZ that mirrored the inside of *SpaceShipOne*, with the small windows on the side and no visibility straight ahead. The idea was to make the cockpit of the Long-EZ feel like the cockpit of *SpaceShipOne*, with a similar window pattern. Burt offered to pay for gas. Mike, who thought Brian was a good pilot who lacked experience in small planes, said he would kick in new tires as needed, since they would be doing plenty of landings.

Mike told Brian that in many of his early test flights, he had to rely on the chase plane pilots to call out wheel height, to tell him how close he was to the runway when visibility was limited. He also said he would go out and make a dozen landings in any plane he could get his hands on. During training flights in the Long-EZ, Mike would station himself in the backseat, simulating wheel height calls on landing.

The Long-EZ was the perfect trainer. It flew toward the runway at the same speed as the spaceship. Mike instructed Brian to come in with an approach speed of 155 knots, then round it out, pull the stick back, and touch down at 115 to 120 knots. After Brian got the approach and speed right, they started honing the landing. After doing a minimum of ten landings in a row every flight, they would work on perfecting the landing. Master and apprentice went around and around, flying over Mojave some days, California City and Tehachapi the other days. At one point, after another round of landings and takeoffs, after the two put the plane away, Brian said, "I really appreciate what you're doing, but it may be all for nothing. I don't think they're going to let me fly."

Mike told Brian that he wouldn't be chosen if he wasn't ready. They kept at it, working every Saturday and Sunday. The Long-EZ went through two new sets of tires. Brian did eighty-four landings. The final ones were perfect.

In August 2004, just a month before the XPRIZE flight, Burt arrived at a meeting of about twenty people at Scaled. Unbeknownst to him, everyone in the room had been anxiously rehearsing how to tell their genius boss that they didn't think his missile idea was going to fly. Burt had always carried the team. It was his vision and his absolute daring that had pulled everyone through; if Burt believed, everyone believed. But now, for the first time, it was Burt who had been having doubts about his homemade spacecraft—and in his team's view, their inventive leader had been overcompensating for his rocket's propulsion problems.

Instead of merely lodging complaints, however, Burt's team members did something remarkable: Taking a page out of Burt's playbook, they'd come up with an ingenious solution of their own. Everyone—from the floor sweep to the crew chief—worked harder and sacrificed more. They made the spaceship and mother ship lighter, smoother, more aerodynamic. Pilots trained pilots. And the propulsion guys came up with what they believed was a missile-free solution.

As the meeting began, Burt's most loyal lieutenants, one by one, voiced their opinion that the missile option wasn't feasible, necessary, or safe. It was unanimous. Burt took the news with a mix of good nature and slight exasperation. After all, Scaled's mantra around safety was "Always

Question, Never Defend." This meant to ask questions up front, don't defend when something goes wrong. After considerable back-and-forth, Burt said, "You are questioning me so much it makes me want to defend!"

Burt was still convinced that his missile idea was workable, and he had made a career of defying skeptics with his revolutionary, and very successful, plane designs. To him, the missile solution would take "simple calculations" to work. Yet he knew that time was running out—the missiles were still not procured—and his team had landed on a different type of a solution. Burt had always promoted moxie and ingenuity, and he welcomed ideas from any corner of the shop.

Talk turned to the non-Sidewinder solution. The answer, according to the propellant guys, could be found in one odd word: ullage. This is the space in a bottle, cask, barrel, or tank not filled with liquid, or there because of loss through evaporation. What *SpaceShipOne* needed, John Campbell explained, was more nitrous, less ullage. The empty space in the tank wasn't needed, because the oxidizer didn't heat up and expand as much as originally thought. The pilot always took off in the cool early morning and climbed to colder altitudes. The nitrous was spent by the time they were heading home, when things heated up. The team would just add more propellant, the same way a hiker adds more water to his pack when he needs to climb higher. While this solution meant adding weight, it was the right kind of weight.

"If we pull out all of the ullage and we fly a perfect trajectory," Frank Macklin said, "we're going to make it."

The trick, the team acknowledged, was that nitrous oxide's liquid density was highly dependent on temperature. It was critical when warming nitrous to allow for sufficient ullage for the liquid nitrous oxide to expand into. Campbell and Macklin dealt with this through pressure-relieving devices. Macklin said, "We'll have no ullage. We'll fill this to the brim and then we'll monitor the heck out of the temperature to ensure we don't overpressurize the tank." Also, the density of nitrous oxide goes up significantly when the temperature is lowered. They would do what they could to load it a little cooler.

Macklin and Campbell were confident that their hybrid motor was up to the task. They knew the temperature profile on the ascent and devised a new process by which to load the nitrous into the vehicle with very little ullage. Campbell figured out how to insulate the tank more efficiently. He then built a small air-conditioning unit to blow cool air on the ends of the oxidizer tank. He and Macklin also devised a memorable—and admittedly primitive—way of mixing the nitrous. They tested their systems out on the Mojave runway at two A.M., the time they would load nitrous before an actual flight. Campbell would climb into the driver's seat of an old tractor and begin pulling the trailer with the nitrous tank. Nitrous stratifies in unwanted ways if left to sit, becoming warm at the top and cold at the bottom, making the pressure higher at the top. Under the new zero-ullage plan, getting the right pressure was more important than ever. The nitrous needed to be mixed before loading. So out on the runway, floodlights trained on them, Campbell said to Macklin, "How about if I just get it going a bit and mash on the brakes?" Campbell's idea was to cruise along in the old tractor as fast as it could go and then suddenly stomp on the brakes to slosh and mix the nitrous around. Macklin would run alongside the trailer checking pressure gauges. Without millions of dollars to spend like NASA, or years to launch a study, the two men had tractor, trailer, and an open Mojave runway. They would stay out until the first signs of dawn. Their shaken and stirred system was working.

Tests showed that zero ullage made all the difference. The propulsion experts had no doubt that if flown correctly, the spaceship would shoot right past the Karman line. They had been able to work fast, thanks to an approach pushed by Burt from the beginning. Normally, motor development begins with subscale testing and moves on to larger testing, and finally there is testing of the real and heavy hardware. Burt, however, had wanted them to do flight hardware testing from the start. He wanted to test what would be flown as fast as possible. The reconfigured zero-ullage hybrid motor was now cranking out between 750,000 pounds-seconds and *1 million* pounds-seconds.

As the Scaled team readied the hardware, Peter Diamandis and the XPRIZE team began transforming Mojave into something that was a cross between a giant fairgrounds and a sporting event. They had three months until their pieced-together $10 million prize vanished, and they needed to see history made not once but twice. Peter now had a dedicated team and an army of space-loving volunteers.

Meanwhile, also in the weeks before the first XPRIZE flight, mission director Doug Shane announced that he was putting in a new system for the pilots: There would be a primary pilot for X1 and a backup pilot. The backup would step in if the primary pilot wasn't ready or got sick or injured. The lineup was announced: Pete Siebold was the pilot for X1, and Brian was the backup. No one was named for X2. Brian was okay with this; he was just happy to be included.

But not long after the lineup was announced, in mid-September, Siebold had a closed-door meeting with Burt. He had serious news. He told Burt that he was taking himself off the *SpaceShipOne* flight program. His wife had just had a baby. He was having lingering stomach pains. And he felt the zero-ullage plan was unsafe and insufficiently tested.

Although the news came at a critical time—just eight days before the X1 flight—Burt wasn't entirely surprised. He had watched Siebold struggle to light the motor after being dropped from the mother ship on his first powered flight in May. Siebold also wasn't comfortable with the handling qualities of the spaceship.

Burt believed that had Siebold been in the cockpit on an earlier flight when the navigation display went black, he would have aborted the mission. Mike had continued to fly by looking out the window, saying afterward that it was a "blessing" because he had gotten to see the sky go from blue to purple to black. He never considered shutting down the motor. Siebold was talented, but he was more careful than Mike. He was an engineer *and* a pilot, and sometimes those things conflicted. He was not as much of a cowboy. In Burt's mind, test pilots had to be cowboys. They had to climb into a cockpit with a leather football helmet on, ignore the fact that they had broken ribs from an accident the night before—as Yeager had done—and fly right through the sound barrier like a bat out of hell.

Mike was in Texas doing a demonstration of the Proteus when he got a call from Doug Shane saying he needed him to hustle back to Mojave.

Arriving at Scaled, Mike dropped his things in his office. He found Burt, Doug, Pete, and Brian in Burt's office. Sally was standing in the hallway. She hadn't been told what was going on, but had figured it out. Doug delivered the news matter-of-factly. Pete was out for X1. Mike was in. Brian sat stunned, thinking, *When did the backup pilot become not the backup pilot?* To avoid having the press get wind of the fact that Pete was out because he wasn't comfortable with the safety level of the program, Scaled would stick to the story that Pete was sick. There was no time to reassure Pete and no time for more tests; they had an appointment in space the very next week.

Little else was said, and the meeting ended. When Mike emerged, Sally looked at him imploringly. Mike nodded. Sally knew he was back in the flight program, despite assurances she'd received from Burt that Mike had done enough already. She and Mike had believed that the dangers of *SpaceShipOne* were in their rearview mirror. Now they were a week out from the XPRIZE flight, and Mike hadn't been seriously training. He'd let down his guard—as had Sally—and wasn't mentally prepared. Mike knew how important this flight was to Burt and the whole team. He also knew that Sally believed he had pushed his luck in the spaceship as far as it would go.

Brian left the office early, afraid of what he might say if he stayed. Arriving home, he stared at his running shoes. For an hour, he did nothing except walk around in a defeated daze. He picked up a book by one of his favorite authors, Dean Koontz. Koontz wrote: “Sometimes life is not about how fast you run the race or with what degree of grace. It’s about perseverance. Finding your feet and slogging forward no matter what.” As the light in his home dimmed, he made a decision. He went and found his running clothes. Brian was heading back out.

29

In Pursuit of a Masterpiece

Dumitru Popescu and his small team were ready for their big event, the first high-altitude launch of a major civilian rocket from Romania. They arrived at the Air Force base at Cape Midia next to the Black Sea on Wednesday, September 1, 2004, and planned to launch within nine days. Their rocket, the Demonstrator 2B, was fifteen feet long and made entirely of composite materials. The rocket fins were emblazoned with the XPRIZE logo. Almost one hundred journalists from four countries would be on hand to cover the flight.

Popescu had launch clearance for two days only, September 8 and 9. Adopting an Apollo 13 imperative, he told his wife, Simona, "Failure is not an option." The launch would be televised live. Popescu and his team had poured all of the money they had into this launch. They were now flat broke and would be unable to try again anytime soon. Popescu had always believed that the XPRIZE was going to make history. He wanted to be part of it, but in his mind, that would happen only if he launched hardware he'd built—and only if he launched it before XPRIZE front-runner Burt Rutan made his first flight, now just weeks away.

After assembling the launch complex, rehearsing procedures and emergency response, testing firing control systems, and giving members of the media an up-close look at the rocket, Popescu and his team were ready for liftoff on September 8. The Demonstrator, gleaming with Popescu's attention to aesthetic detail, had a reusable, all-composite engine. It stood on its platform, smooth white rocket and tall white gantry set against the deep blue waters of the Black Sea. Popescu would watch from closed-circuit television inside an old bunker built during the First World War. Simona Popescu was two miles away, playing host to ninety-six journalists.

Everyone was in position, with the countdown set to begin. Systems checks were done and hopes were high. Suddenly—there was a problem. A hose was leaking propellant. The 70 percent hydrogen peroxide fuel was leaking into the motor and valves. The launch was aborted. Popescu and his wife hustled to reassure the media and VIPs that the launch was being postponed for only a day.

Behind the scenes, though, things unraveled fast. In addition to the hose malfunction, Popescu and his team learned of a problem with the fuel pressure. Next, Popescu was told by Romanian air force officials that a storm was approaching, with winds expected at more than 40 miles per hour and pummeling rain. The military officers were not comfortable with a launch, fearing for the safety of spectators. The nearest major town of Constanta, the oldest continuously inhabited city in Romania, was just thirteen miles away. There was also the matter of all the journalists near the site. Things could go bad quickly.

Popescu and Simona were staying in a small house on the air force base. Neither could sleep that night. "The Romanian Space Agency is going to smell blood," Popescu said, thinking about the unrelenting criticism, innuendo, and undermining he had faced from cosmonaut Dumitru Prunariu. If he didn't launch, Prunariu would win—at least that's how it felt. Late in the night, Popescu ventured outside to study the sky. Trees were being bent by the wind. Then came thunder and lightning. Standing under the dark and hostile sky, he thought, *This is the worst night of my life.* At the first sign of dawn, after about an hour inside tossing and turning and in his futile attempts at sleep, Popescu headed back out for a walk. He needed to think. He reflected on the work of one of his heroes, Hermann Oberth, who took a teaching job in Romania in 1930 and received his patent for a liquid-propellant rocket from the Romanian Patent Agency. Oberth left Romania for Germany, where he became a mentor to a young Wernher von Braun, who would call Oberth "the guiding light of my life." Oberth had to leave Romania to make his mark. Popescu wondered if he would make a mark in the country he had always called home.

As Popescu watched an enormous sun rise behind his rocket, he was approached by the commanding officer of the base. Popescu braced for more bad news—and he got it. He was told that one of his engineers, Andrei Comanceanu, had left the base in the night to go into the small town of Novadari. He had been caught earlier that morning trying to jump back over the wall. Fortunately, Popescu was told, the armed guard who spotted him, recognizing that he was an engineer with team ARCA, didn't fire a warning shot. Popescu apologized and went to find his friend, who had been a year behind him in aerospace engineering school.

When Popescu saw him, Comanceanu fell to his knees and began to cry. He said he was sorry, but he was not able to handle the stress. Popescu, skinny, unshaven, and with dark shadows under his own eyes, propped his friend up and told him they were going to launch and he needed to stand proud. "Two people built the first flying machine," he reminded him of the Wright brothers. "We were brave enough to get this far. We are not giving up." Privately, Popescu berated himself as "probably stupid or something" for refusing to give up.

With little time to spare, he told the now calm Comanceanu that he had come up with an idea for how they could launch without endangering anyone. In the night, when he wasn't out eyeing the angry skies, he had run simulations on what would happen if they used less fuel, given the strong winds that threatened to turn the rocket into a missile. "We reduce the fuel level from one hundred percent all the way to nineteen percent," Popescu said. "The rocket will have a higher velocity off the launchpad, and it will be less affected by the wind." They would not achieve the 28,400 feet they had projected, but they would launch. He told his friend, "When you're in this business of building rockets, you can be a hero or a loser in a matter of seconds. We are going to be heroes today."

Their launch window would close after four P.M. Journalists were reassembled and ready for a noon launch. There were delays. One P.M. passed. Then two P.M. Then three P.M. Whitecaps tumbled on the Black Sea. Just minutes before four P.M., Popescu, back in the bunker, ordered the launch to commence. Everything moved fast from there—go! go! go! were the commands.

Finally, it happened. The Demonstrator shot off the pad. In that moment, time slowed for Popescu. He ran outside. His rocket flew above him, soaring like an arrow shot straight up. Strangely, it felt like it was right over his head. The rocket looked black, not white, and the plume was the color of the storm clouds. This was Popescu's first-ever rocket launch, and it was perfection. It was his. All of the work, struggle, ridicule, and doubts suddenly felt worth it. Flying with its limited fuel, the rocket made it to 4,000 feet before it began to fall back. It crash-landed in the Black Sea. There were cheers all around. The normally reserved military brass hugged Popescu like a brother. It was the purest moment of Popescu's life.

A day later, Popescu learned that Turkish fishermen, fishing illegally in the waters off Romania, had found parts of the Demonstrator rocket. They wanted money in exchange for what was found. The fishermen had watched the news, learned Popescu was involved in a $10 million competition in America, and figured he was rich. Popescu's next source of laughter came when he read several stories in which cosmonaut Prunariu was quoted as saying that the flight was a big deal only to Popescu and his team and "had no real significance." It didn't escape Popescu that these words were coming from a man who had all the government's resources at his disposal—yet had launched nothing.

Popescu was broke, but he had created a company and a brand. He had flown a rocket as a part of the XPRIZE. "We have the energy, the inertia," he told his team. "We will not give up. What we are doing is revolutionary."

John Carmack and his team from Armadillo Aerospace headed out to their one-hundred-acre parcel in Mesquite, Texas, optimistic that it was the perfect day for flying. The skies were blue and the wind was calm. The team planned to fire its forty-eight-inch-diameter rocket, which just days before had hovered in the air for sixteen seconds with no hint of a problem. After arriving at the site and setting up, the Armadillo team loaded the propellant and pressurized the vehicle. Carmack, Russ Blink, Neil Milburn, and a half dozen others holed up a safe distance from the unmanned, cone-shaped rocket, which was ready for a full-throttle test. John's wife, Katherine, was nine months pregnant and had to miss the launch.

Yet there was a hitch: the engine wasn't warming up. Carmack used about 20 percent throttle, and nothing happened. He gave the engine more slugs of propellant until the temperature began to increase the way it should. Until now, the engine had been operated only at hover thrusts. There had been some apprehension about what would happen at full throttle, but with Burt Rutan's first XPRIZE flight on the horizon, it was time to proceed if they had any chance of competing with him.

Finally, happily, they had ignition and liftoff. The rocket shot straight up and was on an impressive trajectory toward 600 feet. The only sound came from grasshoppers in the hot Texas sun. A second or two later, before the rocket reached its targeted height, the ship began to rotate at around fifty degrees per second. This was a problem.

Russ Blink watched, aware that the team—in their crisp white Armadillo team T-shirts—had never crashed anything big before. There'd been a crash of a lander from around forty feet high. But this was very different. The vehicle was falling tail-first. Just before impact, it turned on its side and hit the ground sideways, though with a little tail first. The 450-pound fiberglass tank, pressure still in it, was punted from its impact point about two hundred yards away. Dirt and debris were sprayed across the field of grass and sunflowers.

This was not what team Armadillo had hoped for in the weeks leading up to Rutan's XPRIZE flight. They had hoped to fast-track their test flights.

When the dust settled and the stunned silence ended, Carmack and the team proceeded to the crash site. There was a sizable crater flagged by mangled parts. A few pipe fittings were intact. The fiberglass tank was ruptured. The onboard camera was destroyed, but it appeared the tape had survived surprisingly unscathed. The team spent hours searching and digging for remnants, taking photos of what was recovered. Some parts were splayed open, wires severed and protruding, while other pieces were mangled together from the high-speed, head-on crash. They hauled whatever they could find back to the shop for a thorough postmortem, and used a plasma cutter to open up the engine to figure out what went wrong. They studied the telemetry. They learned that they had blown about two thirds of their propellant out onto the ground during the engine warmup, when Carmack was urging it along with slugs of fuel. Their vehicle had run out of gas. They were using a type of engine that had a prolonged start-up sequence involving the slow heating of the catalyst material. Clearly, this was too difficult to control. They needed a different type of engine for future launches.

Not long after the crash, Carmack wrote in an online post: "We gave up our last glimmer of hope to have a vehicle ready" for the XPRIZE. He added, "The only stacked-miracles path that could have worked was to have perfect test flights with the 48″ diameter vehicle, then build out the 63″ diameter carbon fiber vehicle and have perfect flight tests with it, then get some combination of influential senators and popular support to lean on the AST [office of Commercial Space Transportation] to fast-track our launch license and launch site work. Not a very likely scenario." Despite the loss of the $40,000 vehicle, the team remained in good spirits, calling the collected rocket parts "Armadillo droppings." And Carmack noted of his latest video game, "It's a good thing *Doom 3* is selling very well."

But when Carmack thought about it, he realized that Armadillo had fallen out of the XPRIZE running the day they were forced to give up on using the 90 percent hydrogen peroxide formula that was central to their rocket design. The vendor had refused to sell the propellant to them for fear of liabilities and because Armadillo wanted relatively small quantities. The Armadillo team then had to spend a year developing a mixed monopropellant. Carmack believed that if they'd had an uninterrupted supply of 90 percent hydrogen peroxide, Armadillo would've had a shot at the XPRIZE.

By mid-September, the building of a new and improved engine and vehicle was under way. Carmack was taking what he had already learned about structuring, reliability, and simplicity in his coding for rockets and applying it to his work in software. Yet he loved the hands-on aspect of rocketry, of taking a bar of solid metal and working on it until he had a gleaming and useful part. He would follow the XPRIZE flights in Mojave, but he was undeterred in his own goal of accessing space. He told his team, "We'll see if we have a one-hundred-kilometer vehicle ready for testing this time next year."

Carmack was also staying true to his hacker roots, and to the hacker creed. He was intent on sharing all that he learned. He would continue to build everything from scratch, from the bottom up. He believed in the beauty that could be attained through working with one's hands, whether on a keyboard or with a soldering iron. Results mattered more than rules, and everything should be free—not free of charge, but free to change. He was determined to play a part in the hacking of aerospace, in taking it away from the government and putting it into the hands of the people.

In the United Kingdom, XPRIZE contender Steve Bennett was in the process of building Nova 2, following his successful launch in 2001 of Nova 1 from Morecambe Bay. The Nova 2 would be bigger and fly higher than Nova 1, standing 57 feet high and reaching a target of 120,000 feet. Bennett saw his new rocket as the key to his future manned space program.

He did ground systems tests on the Nova 2 capsule with a pilot and two passengers sealed into the cabin for two hours at a time. Bennett took the capsule to Florida to have it fitted with parachutes, and then to Arizona for piloted drop tests from 13,000 feet. Everything was working. Around this time, though, he found his work sidelined by the unexpected.

Without warning, a story appeared on the BBC Web site alleging that the self-taught rocket maker was designing a death trap, and that his spaceship was nothing more than "a converted cement mixer." Bennett believed that the author of the BBC piece had a vendetta against him. He spent twenty months and 250,000 pounds on a lawsuit against the BBC. Eventually, the BBC issued an apology to Bennett and paid his legal fees. But that wouldn't bring back the two years lost to litigation. He did find solace in an education program he'd started that was aimed at introducing the Nova 1 to schools across England. He would bring rocket models and rocket scientists into classrooms and talk to the kids about the importance of following their dreams. "It's kind of an American thing," Bennett would tell students. "Americans are brought up to dream big things. It's the land of opportunity. We don't do that enough in this country, and we should. You should."

Even though Bennett had resumed building Nova 2, it was clear to him by late summer of 2004 that Burt Rutan was within arm's reach of the XPRIZE. Rutan was smart, had a great team and a billionaire backer. Bennett had heard that Paul Allen put $25 million into Rutan's space program. No one else came close to having that kind of funding. But everything was still relative, Bennett thought. NASA would have spent that kind of money on blueprints alone.

Bennett had met Rutan in 2003, when the XPRIZE invited teams to Los Angeles to show their models and share some of what they were doing. The visit included a field trip to El Segundo, California, where Elon Musk had started SpaceX in an empty 75,000-square-foot hangar. One of Bennett's favorite moments was on the bus ride to SpaceX, when he overheard Rutan talking on his phone in a low tone about who was attending the event. Bennett smiled when he heard Rutan say, "Bennett's here."

To be sure, the XPRIZE was a competition, but there was also a shared mission and friendships forged. Brian Feeney of the da Vinci Project was there, and had the nickname "Flying Brian," because he was always "just days away from flying." Bennett saw the full-size model that Feeney had built. He talked with Carmack's team of volunteers and thought they had cool hardware. He was also impressed with Pablo de León of Argentina and his plans for suborbital flight. Bennett remembered Dumitru Popescu telling him how they were so short on cash that at one point one of his workers resorted to closing his eyes while welding because he didn't have proper eye protection. Bennett met fellow Brit Graham Dorrington, an aeronautical engineer who was one of the first people to register for the XPRIZE but hadn't gotten far beyond paper studies for a vehicle called the Green Arrow.

All in all, Bennett was proud that he had followed his dream, despite the years of challenges. He could have kept his secure job at the toothpaste plant, avoiding financial troubles, tensions with his family, and media hits. But like an artist in pursuit of a masterpiece, he had no choice but to go on. He was a rocket maker, and that was that.

In September 2004, Peter Diamandis invited Bennett to attend Rutan's first XPRIZE flight. Bennett wished that his own rocket was the one ready to soar, but that didn't change the fact that he had been the first XPRIZE contender to fly hardware, even making history in his country. And the specter of someone else's winning the XPRIZE wasn't going to stop him from completing his own suborbital ship. In the not too distant future, he would shuttle passengers to the start of space, just for the fun of it. For Bennett, and for fellow competitors like Popescu and Carmack, this was just the beginning.

The XPRIZE was already shaping a legacy—even before its first official flight.

30

One for the Money

It was late September 2004, more than ten years after Peter hatched the XPRIZE idea in Montrose, Colorado, and eight years after the private race to space was announced to great fanfare in St. Louis. That memorable night, in the city that Charles Lindbergh put on the aviation map, legendary airplane designer Burt Rutan had stood at the dais and revealed his dream of making a homebuilt spacecraft. Burt was now ready to display that dream—and so was Peter. All eyes were on Mojave, where the XPRIZE competition was finally set to begin.

Peter's army of volunteers descended on the small desert town. The troops were culled from space organizations and university engineering departments and managed by Loretta Hidalgo, who had attended International Space University and also ran Peter's "special ops" team, a smaller group that did everything from pick up dignitaries at the airport to help mainstream journalists understand the technicalities of this space story. Another ISU graduate from years before, Harry Kloor, had reentered Peter's life and was pitching the XPRIZE story to dozens of media and television outlets. It was critical to Peter that the event represent more than a competition and a prize. He wanted to see a global shift in the way people thought about space travel. For help with this, Peter called on his longtime friend Dan Pallotta, a California entrepreneur who pioneered multiday charitable events, including the AIDS Rides. Pallotta knew how to create pop-up tent cities for a massive audience. Pallotta, like Peter, was born in 1961 and reared on the magic of Apollo.

Peter, Bob Weiss, Pallotta, and Stuart Witt, director of the Mojave Air and Space Port, spent long hours going over the message of the XPRIZE that they wanted conveyed to the world and the minute-by-minute schedule of events. They had grids of the airport, VIP and hospitality tents, souvenir booths, food vendors, Jumbotrons, emergency personnel, and contingency plans, including those based on the number of medical emergencies expected given the anticipated crowd size. (Two heart attacks were expected for a crowd of thirty thousand.) Weiss was applying all he'd learned directing Hollywood movies to the scene in the Mojave Desert. Everyone, from the organizers to the onlookers, understood that this reality show could end in victory, defeat, or despair.

Everything was coming together for Peter—and at the same time. After a decade of battling the Federal Aviation Administration—a bureaucratic quagmire—Peter, Byron Lichtenberg, and Ray Cronise had gotten approval for their company, ZERO-G, to offer the first parabolic gravity-free flights to the public. Cola company Diet Rite was sponsoring a twelve-city tour of the ZERO-G plane, with media stops along the way. Peter's life had been divided between introducing the world to the thrills of zero-gravity flights and planning for the XPRIZE.

The race for the XPRIZE was shaping up to be one of the biggest news stories of 2004. Peter, who had gambled before and lost, now had all his chips on Burt Rutan and *SpaceShipOne*. The hole-in-one contract with XL Specialty Insurance would expire in three months. At that point, the $10 million prize money would dry up, and he had no Plan X, Y, or Z. After Scaled's successful June 21 flight, Peter was called to Santa Barbara for a meeting with the insurance executives. The executives, apparently figuring out a tad late that it had been unwise to bet against Burt Rutan, wanted to offer a new deal: they would lower the altitude requirement, from 100 kilometers to 50 kilometers, and proposed that the insurance payout be reduced by half, to $5 million. Peter would have none of it. "You guys are ridiculous," he said on his way out the door. "You are writing a check for $10 million."

Peter arrived in Mojave on the Monday before the Wednesday X1 flight. He was joined by his parents, Tula and Harry, his sister Marcelle, and someone else near and dear—the woman he was falling for, Kristen Hladecek, an artist and a vice president of creative advertising at 20th Century Fox. It was as if Peter had needed to get the XPRIZE to a secure place before he could find a secure relationship. He was joined by his close friends and allies Erik Lindbergh and Gregg Maryniak, Adeo and Elon, Bob Richards and Jim Cameron, Diane Murphy, who was running marketing and publicity for the XPRIZE, and Harry Kloor, his friend from early ISU days, and the Ansaris. With no one else stepping forward as a title sponsor, the XPRIZE had been renamed the Ansari XPRIZE.

Mike Melvill drove from his home in Tehachapi to Mojave. In the months leading up to this big day, he had been doing more media interviews than rigorous training. He thought he was done with *SpaceShipOne* flights. There had been only eight days to prepare after being summoned back into the cockpit. He had spent time in the simulator, an apparatus that the instinctive pilot had never taken seriously. Mike also did strenuous workouts on his bicycle. Luckily, Mike was already in great shape, and this was not exactly his first rodeo. Getting ready now was more mental than physical. He needed to put his test pilot's Teflon suit back on.

As he drove to Mojave, he looked out toward Edwards Air Force Base and wondered, *What will this day bring?* This was the land of coyotes and

Joshua trees, dry lake beds and long runways, where B-24s had thundered in the sky and pilots in P-38s had strafed their practice targets in training before heading off to battle in World War II. The skies were layered with the memories of pilots and planes that had gone before. It was where the turbojet revolution began with the Bell XP-59A and the Lockheed XP-80 Shooting Star, where high-testosterone test pilots competed to see who could fly higher and faster. It was where supersonic flight became almost commonplace and some of the coolest-looking planes shredded the sky. It was also where some machines broke apart, where brave pilots lost their lives in deadly spins. This was the reality of the skies here. Mike wanted Scaled's little rocket program to be a part of that grand history. He was not ready to join the bravely departed.

Just as Mike arrived at Scaled in the early morning hours before his September 29 flight, he saw how dusty Mojave was turned into a tent city. Music pulsated down the flight line. Parties were being held, with dancing and DJs spinning tracks. Mike took it all in and then headed inside. Security crews were already out on the runway, and the sky was dusky pink. The early arrivals delighted in watching private planes come in and land, like birders thrilled to spot exotic feathered species. Among those early arrivals was Peter. The first thing he saw was a bird of a different feather on the runway. It was an airplane splashed with Virgin logos, perfectly positioned in front of all the TV cameras. *Impressive ambush marketing,* Peter thought. He had pitched Virgin chief Richard Branson not once but twice to fund the XPRIZE. Now that they were at the starting line, Branson was all in. Branson had paid Paul Allen $2 million the day before this flight to have the Virgin logo painted on the tail end of *SpaceShipOne.* Then Branson's people had called airport director Stuart Witt late in the night before this first XPRIZE flight with a last-minute request: make sure the Virgin logo was visible from the runway when the sun came up and the television cameras were on. There was only one airplane in Mojave—besides *SpaceShipOne*—with the Virgin logo. That was the *GlobalFlyer*, sponsored by Virgin, designed by Burt, and soon to be flown around the world by Steve Fossett. While Peter was still at his hotel, Witt had called some of the Scaled crew, and together they pushed the *GlobalFlyer* out of the hangar and onto the runway, in prime viewing position for the crowds and media. Witt had anticipated Peter would be startled to see Virgin front and center. Peter was also trying to get the best visibility for his XPRIZE sponsors—something he'd been working on for years.

Out on the runway, Peter looked at the logo-decorated plane with a mix of respect and disbelief. The XPRIZE was spending a small fortune hosting this event—not knowing whether there would even be a second flight—and now it felt like a Virgin commercial. Peter also found it ironic that he had met with Paul Allen's lieutenant, Dave Moore, several times in a bid to get XPRIZE funding, but no one had let Peter in on the little fact that Allen was actually financing *SpaceShipOne*.

But Peter refused to let anything dampen his spirits. He considered Branson brilliant and was cheered by the news that the Virgin chief was investing in commercial space. A few days before the X1 flight, Branson and Burt had held a press conference in London for the British press announcing that Virgin was forging a deal to buy the technology behind *SpaceShipOne* with the goal of sending ticketed passengers to space. Peter wished Branson had come in earlier, before everything was done, but he appreciated that he was in now. Space needed talented and deep-pocketed believers.

Inside Scaled, the preflight briefing was under way and lasted for close to an hour. The hybrid motor was ready. The nitrous oxide had been stirred and was now loaded into the spaceship, with temperatures carefully monitored. Crew chief Steve Losey had gone over the spaceship inside and out. He had made a promise to Sally long before that he would never put Mike in a plane unless it was safe to fly. Losey stayed with the plane through the night, like a groom with a prized racehorse. Even the paint job had been perfected for the big day. Dan Kreigh, an engineer who had structural responsibility for the spaceship—taking Burt's designs, making any fine-tune adjustments, and then instructing the shop on building—was also Scaled's in-house artist. For *SpaceShipOne*, he and Brian Binnie had worked out a patriotic theme. Because the plane was going to space, Kreigh became obsessed with finding a subliminal graphic design idea. He wanted *SpaceShipOne* to look like it had flown through a magical cloud of stars—and brought some back. The *White Knight* would carry the red stripes to complete the image. Kreigh and his wife, Rojana, would touch up the plane in the middle of the night, cutting all the vinyl decals and stripes and painstakingly applying them. Recent nights had brought them to the hangar to apply the Virgin logos. They had also affixed the smaller decals of the various XPRIZE sponsors that had kept the dream alive along the way: the St. Louis Science Center, Champ Car Worldseries, M&Ms, and 7UP.

After the preflight briefing, Mike headed into the locker room to finish getting ready. Sally was there, and soon Burt appeared holding a small vial that looked like a sand dial. He asked Mike to put it in one of his pockets for his flight. Burt said it was a bit of his mother's ashes. Irene Rutan, diminutive but forceful, had always been Burt's biggest supporter, driving him to his model plane competitions when he was a child and tallying how many of his plane designs flew in each year to Oshkosh. Burt wanted his mom to take another journey with him, this time to space. Mike looked up and down the front and sides of his green flight suit. Sally pointed out that the new suits had no pockets, so Mike tucked the vial into his helmet bag. He and Sally then made their way out to the tarmac. Photographers snapped pictures, and well-wishers appeared from out of nowhere to shake Mike's hand and offer salutations. Dick Rutan was there as a commentator for CNN.

It was showtime. Peter took to the stage and looked out at the crowd. His parents were front and center.

"Ladies and gentlemen," he began, "we are at the start of the personal spaceflight revolution, right here, right now. It begins in Mojave, today. What happens here in Mojave is not about technology. It is about a willingness to take risk, to dream, and possibly, to fail." Peter said he believed that *SpaceShipOne* would reach space today, and again within two weeks.

Anousheh Ansari followed Peter onstage. "This is an exciting day for all of us," she said, "a day our dreams will come true. I'm grateful to Burt, and to all of the teams that competed. Without their courage and willingness to achieve the impossible, this dream would never happen."

"I believe we are changing the future of space exploration," she added. "By being here, we are supporting the XPRIZE teams and foundation. We are no longer just dreaming about going to space. We are making it a reality. Each of us has a responsibility." Anousheh knew a thing or two about dreams coming true. Early that morning, she had run into her childhood hero, William Shatner, Captain James T. Kirk. She was speechless when she saw he was wearing an ANSARI XPRIZE baseball cap.

Not far away, out on the runway, Mike and Sally embraced. Sally was shaken by the belief that luck eventually ran out. The two held each other close, and Sally again pinned their lucky horseshoe on Mike's sleeve. After a few more moments, Burt pulled Mike aside for a final pep talk. Like a coach reviewing the game plan with his quarterback, Burt reminded Mike that the rocket had never carried such a heavy load. Mike needed to get the rocket turned vertical as quickly as possible. Mike nodded, and the men hugged.

Mike climbed into the spaceship. He felt the horseshoe pin on his sleeve and closed his eyes for a brief calming moment. Behind his seat was a container holding all sorts of mementos thrown in by Scaled employees. There were dozens of wedding rings, bundles of coins, photos, and personal talismans.

Soon, the *White Knight*, coupled with *SpaceShipOne*, began rolling down the runway. In the *White Knight*, Brian Binnie was again the "bus driver," and Stinemetze was in the passenger seat. A consummate team player, Brian still had mixed emotions about today's flight. He should be in the cockpit of *SpaceShipOne,* he thought; his piloting abilities had been unfairly maligned. Yet he had deep respect for Mike, not only as a pilot but also as a person. Over the past few months, Brian had relentlessly trained in case his name was called, but he couldn't escape from a nagging feeling that his hard work would be for naught—and that Mike would end up piloting both XPRIZE flights.

The crowd went wild when the *White Knight* took to the air. An hour later, as commentators, including Peter and Witt, addressed the crowd from various stages, and onlookers aimed powerful lenses at the sky, *SpaceShipOne* was released from the *White Knight.* X1 was under way, watched by a global Webcast and live television. Even the hotshots at Edwards Air Force Base were tuning in; they were testing a high-resolution missile-tracking camera that would observe the flight from takeoff to landing.

Seconds after the release of *SpaceShipOne,* Mike, in helmet and aviator shades, said, "Armed-fire."

"Good light," he said, thrust back into his seat. The rocket was on its ascent.

In the crowd on the desert floor, a man pointed to *SpaceShipOne*'s contrail and said, "Look at that sucker move! Man!"

Cameras from outside the spaceship beamed images to the Jumbotrons. Witt, onstage, said, "Cross your fingers and say a prayer. Here we go, guys."

In the cockpit, Mike's breathing was heavy.

In Mission Control, Burt sat next to Doug Shane, and Jim Tighe was a row back. Paul Allen and FAA heads Patti Grace Smith and Marion Blakey were also in the room.

"Twenty seconds, Mike," Shane said when the motor was twenty seconds into the burn. "Thirty seconds coming."

"Doing good," Mike offered, sounding upbeat.

"Coming forty seconds," Shane said. "Forty-five seconds."

Mike, wanting to get as close to vertical as he could, pulled hard on the stick, briefly exceeding 90 degrees. He stopped the rapid pitch up at 91.6 degrees by gently pushing the stick forward, which got him back to 90 degrees. Now the wing was generating no lift at all, and the plane began to oscillate slightly in yaw, then it pitched up 8 degrees and simultaneously yawed 15 degrees.

Mike now had a new and serious problem. He was spinning—slowly at first, then faster like a figure skater in a parallel spin. He was going up—fast. But it was roll after intense roll.* Mike told himself that the spinning would eventually stop. If he cut the engine, the bid to win the XPRIZE would be over. He would have to ride it out.

Witt, seeing the rolls in the sky and on the screens, said, "Uh oh, uh oh. He's in a roll." His fellow commentator said, "It does not appear to be a scripted maneuver."

Silence fell over the crowd. Richard Branson looked over at Sally Melvill. Her anguished expression said it all. Branson dreamed of private spaceflight, but knew this was pushing the limits. Paul Allen watched, feeling himself age twenty years in a few stressful moments.

Dick Rutan was talking with Miles O'Brien live on CNN when the rolls began. Dick said to O'Brien, "This is probably not how Mike planned it." Dick looked up at the sky and thought that there was a real chance that Mike would not recover.

From the stage, Peter had the same feeling he'd always had watching manned rockets fly: rapture, awe, and uncertainty. He stood up and said, "Mike is going to be out of the atmosphere soon, and he has a reaction control system he will be able to use to nullify the roll. He'll be okay." Then he said it again, "He'll be okay." Peter's mom, Tula, wasn't so confident. She said a prayer, asking for everything to be okay. A bad outcome

*Detailed analysis by Scaled determined that the rolls resulted from a mild thrust asymmetry, which could not be offset by pilot inputs at a flight condition of low directional stability. This flight condition had not been tested on previous flights. The low directional stability occurs only at high Mach numbers and at very low (zero or negative) angles of attack.

for the pilot would also be a bad outcome for her son. Witt could be heard saying, "Come on, Mike. Come on, Mike." He told the crowd, "Communications with air show center are shut off. These are tense moments, folks. We are waiting to hear Mike is okay." A downlink camera showed the view of Earth from *SpaceShipOne*.

In the cockpit—the crowd couldn't see this, only Mission Control—Mike held the stick and kept his focus on the instruments before him. He was going straight up at more than 2,000 miles an hour. The cockpit was a strobe, spinning dark-light dark-light, with the brightness of the sun and the blackness of space flashing through the cockpit windows. If he looked outside, he got disoriented fast. He had a roll rate of 283 degrees per second. He found that if he concentrated on one thing on the instrument panel, his disorientation abated. Mike stared at the rate of roll indication, a small display of digital numbers showing degrees per second of rotation. The numbers were the size of typewriter keys, and Mike had to look closely to read the display. He held the stick against the left stop, and the left rudder pressed forward as far as it could go, opposing the roll motion to the right.

Reprieve came in the slowing of the strobe, like a figure skater opening her arms after a tucked spin. Now he had longer periods of sun and space, dark and light. Mike had gone out of the atmosphere and was spinning at 160 degrees per second. During his years of flight testing, Mike had deliberately spun planes. He didn't know until now that you could spin while going straight up. Now that *SpaceShipOne* was out of the atmosphere, the aerodynamic controls no longer had any effect. The control stick and rudder pedals became "loose" and were completely ineffective with no air to press against.

But in short order, the spinning began to ease even more, and communication with the air show center was returned. "He held on," Dick Rutan said to O'Brien. "That's courage—and skill."

With the telemetry returning to normal, Shane asked Mike how he was doing. This time, Mike answered tentatively, "Okay."

Jim Tighe said Mike would need a lot of the reaction control system to orient the ship. "Start the feather up," he said.

Seventy-seven seconds after the release of *SpaceShipOne*, the motor shut down at 328,000 feet—the targeted edge of space—and Mike was still shooting up. There was clapping in Mission Control, and cheering on the Mojave floor. Sally clutched her hands together. It wasn't over until Mike was beside her.

Mike put the feather up and activated the reaction control system, which consisted of four pairs of small jetlike nozzles mounted near the nose, with four more pairs on top and under the wingtips. He used up almost all of the compressed air in both the A and B bottles to finally bring the roll rate to near zero.

When he was sure that *SpaceShipOne* had achieved a sufficient altitude, Mike reached under the instrument panel for a small camera and took pictures out the oval-shaped windows. The next three to four minutes, marked by weightlessness, would be the calmest stretch of the trip. As he had done on his first flight to space, Mike admired the blurry blue sapphire below. But this second trip made him marvel, more than ever, at the beauty down on Earth: his wife, his son, everything he held dear.

The feather was up, roll rates down, and the reaction control system was working. *SpaceShipOne* was at 330,000 feet, and finally 337,000 feet. Now it was time to make it back down. He got ready, knowing he was about to fly three times the speed of sound and five times the force of gravity.

On the desert floor, Witt said, "Coming back down. Okay. Here it comes. The gs build. There's a rapid rate of acceleration. He's coming in now at Mach 2.2, higher. Some rumbling and buffeting. G rise. He's ballistic right now."

Moments later, Witt couldn't help but break into a smile, now that *SpaceShipOne* was no longer spinning and Mike appeared back in control. He believed there always came a moment in life when you were asked to pull it all together—all you had learned, trained for, all you had deep inside. When he'd come to Mojave to interview for the job, the offices reminded him of a 1940s Marine Corps operations center with louvered doors and dust flying in. But he saw a place where people could be given permission to take extraordinary risks outside the reach and arm of the federal government.

Looking at Mike flying some 60 miles up, Witt said, "Can you imagine the technology required to bring a spaceship from 328,000 feet back to landing? It's a goose-bumps day. Mike now has tail feathered to reduce the rate of descent."

He continued, "Now they will need to reconfigure it back to a plane. Looking for feather. Feather down. Okay. Now the crowd can pick up sight of *SpaceShipOne*. This space plane, this rocket plane, is now a glider. We now have a glider back inbound for the money run."

After a pause, Witt added, "This is flight test at its purest form. It's risky business. There's the boom—the sonic boom! Folks, Mike Melvill's comin' home."

But Mike wasn't done with his air show. Feeling joyful as hell to still be alive, the sixty-four-year-old wanted to squeeze one last bit of daredevilry out of *SpaceShipOne*. He had already done twenty-nine spins—why not make it an even thirty? Mike pitched the nose of *SpaceShipOne* down and picked up speed. The pilot of the Alpha jet, on his wing, knew exactly what the cowboy pilot was doing. The pilot exclaimed, "Mike's going to do a roll!" Jeff Johnson, sitting in the back of the jet, began to laugh. Mike got to around 190 knots, pulled the nose up, and began the roll.

In Mission Control, Shane could see what Mike was setting up for.

This was *not* on the flight card, and he had the head of the Federal Aviation Administration standing behind him.

At the end of the roll, Mike let out a "YEEHAWWW!"

Shane realized that he needed to make Mike's audacious maneuver appear planned, so as not to arouse the suspicions of the FAA folks. He said coolly, "Roll evaluation complete."

The flight above the start of space was confirmed. X1 was achieved. As soon as the red-white-and-blue space plane was on the runway, Burt was there to greet his wingman. After dashing off to shake hands with spectators pressed against a rope line, Burt returned and jumped in the back of the truck to begin the towing of *SpaceShipOne*. Seated between Richard Branson and Paul Allen, Burt made jokes about what would happen if a billionaire was accidentally bounced off. Mike was back on the top of slippery *SpaceShipOne*—happy to have good balance from his earlier days as a gymnast. He waved to the crowds, and singled out his sweetheart.

When he was back down on the ground, Sally looked at Mike and said, "Now we can grow old together."

Peter and his girlfriend Kristen watched as *SpaceShipOne* was cheered on like the best float in a ticker-tape parade. Kristen didn't have a background in engineering or even an interest in space. When she met Peter, he was living in a small and dimly lit two-bedroom apartment—one bedroom doubled as his office—and driving an old BMW. He told her stories about the XPRIZE and suborbital flight and tried to explain something called hole-in-one insurance. He told her of the CEOs he'd pitched and the endless noes he'd heard. Kristen was drawn to Peter because he refused to give up on a dream that most people said was not going to happen. What she fell for was the purity of his intentions and his almost naïve soul. As the cheers continued for the improbable *SpaceShipOne*, its brave pilot, and ingenious designer, talk on one of the stages turned to the one remaining flight needed to win the $10 million prize. But in Kristen's mind, the star of the day, the one who had won, was Peter.

Burt and his Scaled team did not have much time to celebrate. They worked through the night of the first XPRIZE flight to figure out what had caused the twenty-nine rolls and how to prevent the dangerous situation from happening again. But there was also a new glitch, this one on the ground.

"I'm out," Mike said.

Hours after his X1 triumph, Mike revealed to Burt and Doug—Paul Allen was also present—that he would not be flying X2. He was firm about his decision. He had done enough, and he believed that Brian deserved a chance. "I've been working with Brian," Mike said. "I've trained Brian. He is ready."

Mike noticed that Paul looked incredulous. Brian was the guy who had crashed his spaceship in December, only ten months before. Shane also couldn't believe that their go-to guy would not be available for this critical flight. Burt, to his credit, kept quiet. He had flown in dangerous planes and understood there comes a time when you say enough is enough. Mike felt he had already done it: he had the world's first private astronaut wings. He'd been the first person to fly to space alone since Alan Shepard in 1961. And he'd been the only guy to fly to space stick and rudder. Now he'd done it twice.

The next afternoon, on September 30, the Scaled folks met again to talk about the X2 flight. Burt, fond of staging important flights on anniversaries of big aviation events, had chosen October 4 as the date for the second flight. This was the forty-seventh anniversary of the day the Soviet Union launched Sputnik and started the government-sponsored space race.* Now the goal of the Scaled team was to create a new anniversary, for private spaceflight.

The meeting was winding down at around six P.M. when Steve Losey said he'd like to finish going over the spaceship and then turn it over to the propulsion crew. He needed one more piece of information to complete his logbooks. This required an important bit of data: the weight of the pilot who was going to fly. He asked Burt, "Who is going to fly?"

Brian was at the table, having worked with Jim Tighe and Pete Siebold, reviewing the telemetry and flying the sim. He was unaware of Mike's bombshell the day before. Shane was across the table from Brian, and Burt was at one end, opposite Losey. Burt deferred to Shane.

Brian waited for the answer, feeling his life was in the balance. Would he be given this chance, or would he be remembered for his missed putt? He was fifty-one years old. This opportunity wasn't going to come around again. Yet for all his training and dedication, Brian felt in his gut that the nod would go to Mike: the pilot who has flown gets to fly. This notion was reinforced when Shane wouldn't even look at Brian.

Then Shane said flatly, "Brian's the pilot."

Brian did his best to take a page from the Oracle and show no emotion, to make it appear that he had known all along. But inside, he was sprinting with perfect knees. He had made the perfect landing on the carrier in the dark of night. He had chipped in for a birdie on the eighteenth green of Pebble Beach. He saw his wife's smile, as beguiling now as when they'd met so many years before.

He had three days to make sure he was ready for the most important day of his life.

*The beach-ball-size Sputnik 1, the world's first artificial satellite, was launched by the Soviet Union on October 4, 1957, and jumpstarted the space race with its ninety-eight-minute orbit of Earth. One month later, on November 3, the Soviets launched Sputnik 2 with a dog named Laika inside. The press used terms like "Muttnik" to describe the canine cosmonaut, who died in space.

31

Rocketing to Redemption

In his living room, Brian Binnie watched on television as the irrepressible Burt Rutan talked to CNN about the XPRIZE flight scheduled for the next morning, Monday, October 4. The Scaled team was not just going to hit a "home run," Burt told the CNN interviewer. They were going to hit a "grand slam." Brian couldn't help but think, *Ohmygod, isn't the bar high enough already?*

Brian paced the room, his emotions bouncing between fear and optimism. When the interview ended, Brian turned off the TV. He settled into the couch, where he was spending the night so his in-laws could have the master bedroom upstairs. His wife, Bub, had thought the arrangement would be best for him, allowing him to wake at 2:15 A.M., make his coffee, and be out the door early as needed.

But just as he started to doze off, he was jolted awake with some new worry. The family dog, an oversized golden retriever named Tanner, trying to claim his usual territory on the couch, wasn't helping either. Instead of counting sheep, Brian tried going over the numbers of his flight. He'd been in the simulator almost nonstop since finding out at six P.M. Thursday, three days before, that he was the pilot for X2. There on the couch, he visualized the course. He was inside the cockpit. Dropped from the *White Knight,* he positioned *SpaceShipOne*'s nose at about sixty degrees. To avoid the spins that Mike had endured, Brian would fly at between eighty-one and eighty-seven degrees.

He dozed off again, numbers and images swirling in his head. Then he'd wake up, panicked he'd slept through his alarm, only to find his dog's paw across his face. At some point, he got up and wrote down the goals of the flight:

1. Get to 100km = $10M
2. Don't leave the atmosphere out-of-control = (Richard) Branson's future interest
3. Beat the existing X-15 altitude record = Burt's personal desire
4. Land with grace = My personal salvation

Then he wrote, "1 & 2 are mandatory. 3 is more in the nice to have territory & 4 is all about me."

Even in his half sleep, he was lucid enough to know he was in limbo, caught between reality and dreams, past and future. With the ticking of the clock, he was moving closer to clarity. The gift of this flight was in the possibility it presented. The cloud of sand that had enveloped him after his crash landing had dissipated quickly. The cloud of doubt remained.

His wife had been praying for him more than usual. A devout Catholic, Bub had jumped into action when she learned he was getting a second chance. She started a prayer chain that by now had stretched across the globe. She made sure the prayers being offered were the same, and that they were specific. "God likes specifics," she said to her fellow Catholics.

"Be specific in your prayer." The prayer she sought for Brian was this: "Safe flight up. Safe flight down. Safe landing."

Brian put his faith more in American ideals than in biblical commands, but he welcomed help from any corner. At some point in his fitful night, he even said his own prayer, promising God that he would be "forever grateful" for a successful flight start to finish.

When the alarm beeped at 2:15 A.M., it was almost a relief. Brian was already up. He wanted to be at Scaled by 3:00 A.M. Once outside his house, he inhaled the cool fresh air. The night was clear and silent. He gazed at the stars and spotted the Milky Way. The sky seemed to be beckoning.

On his fifteen-minute drive to Mojave, he turned on the AM radio and landed on a show called *Midnight in the Desert: Late Night Paranormal.* The show's host, Art Bell, was talking about the "other dimension" that exists beyond the reach of most people. As Brian listened to this odd early morning show, his mind revisited an encounter he'd had a few days after his December 17 flight of *SpaceShipOne.* It was one of the strangest things that had ever happened to him. He had been awake in bed, waiting for his 5:45 A.M. alarm to sound, when suddenly the bedroom lit up like a television had blinked on. Only it wasn't the TV. He got out of bed and went to the window, where slivers of bright light peeked through the curtains. His backyard was lit up like daylight, while the rest of the neighborhood was dark. There were bubblelike shapes—the size of beach balls and volleyballs—roaming, airborne, playful looking. After only a minute, this magical performance of translucent shapes faded back into darkness. He stood still for several minutes, not wanting to move or speak. He didn't believe in the paranormal or in extraterrestrial beings, but he knew what he had seen. It was beyond understanding. For whatever reason, the moment gave him hope at the time and buoyed his spirits now. Turning in to the Mojave Air and Space Port, Brian looked again at the sparkly sky. What did the heavens know that he would soon find out?

Before five A.M. in nearby Palmdale, about 1,500 schoolchildren boarded buses from three different pickup sites to be driven to Mojave to watch *SpaceShipOne,* a junket that had been dreamed up by Stuart Witt, Peter, and the XPRIZE planning team. The once-in-a-lifetime field trip was funded and organized—permission slips, insurance, chaperones, and buses—by a local real estate developer, R. Gregg Anderson, who saw an opportunity to place kids in a moment of history that would inspire them for the rest of their lives. Anderson also wanted to see local youth introduced to the aerospace industry and to glimpse how the team at Scaled Composites had achieved a global audience. Witt had said to Peter in one of their early planning meetings, "You see pictures of Kitty Hawk and there are no children." Peter thought the idea would create a "Lindbergh moment" for a new generation. Instead of Le Bourget, they would have Mojave.

Upon arrival at the spaceport, the kids were escorted to the flight line, where they watched, waited, and cheered at the sight of the *White Knight.* The mother ship approached and then did a hairpin turn in the direction of the mountains.

Near the runway, Peter took a moment to appreciate every detail: the crowds, reporters, celebrities, billionaires, kids, astronauts, NASA and FAA administrators, his family, the cloudless blue sky, and the strangely beautiful *White Knight* mated with the spaceship. He didn't know what today's flight would bring, but he was struck by how far they'd come. As his childhood friend Scott Scharfman had reminded him: he had launched a space prize without the prize money; he had wanted to do something privately that only governments had done; he had believed it was possible to make a spaceship that wasn't thrown away after one use; and he believed that if he offered a prize, teams would come. One more flight was all that was needed. The other teams founded to capture the XPRIZE were continuing to build; the dream of space was not going away if the prize was won. The passion would persist. Rockets and hardware were being built in Romania, England, Argentina, Texas, and elsewhere. Richard Branson had secured a deal to develop SpaceShipTwo, using Rutan's technology and feather design. Peter's friend Elon Musk was taking on the aerospace establishment with his private rocket company, and former SEDS chapter head Jeff Bezos was beginning to apply his massive wealth to space. This was the inflection point, Peter believed, in which the downward spiral of manned spaceflight turned around.

Peter watched *SpaceShipOne*. He had three treasured books in the ballast box behind Brian's seat: *The Man Who Sold the Moon; The Spirit of St. Louis,* given to him by Gregg Maryniak; and *Atlas Shrugged,* a gift from Todd Hawley. He and Todd had that favorite line: "All that lunacy is temporary. It can't last. It's demented, so it has to defeat itself. You and I will just have to work a little harder for a while, that's all."

After the preflight briefing, Burt reminded the crew, "This is the money flight." *He keeps adding on the pressure,* Brian thought. *A grand slam! The money flight!* Brian got into his flight suit and headed to the hangar. He stepped on the scales holding a small bag that he planned to bring on board. He was five feet eleven and weighed 165 pounds, having lost weight through all of his running and stress. He had two American flags that he wanted to fly to space—one a heavy-duty cloth flag and the other polyester—and an assortment of things handed to him at the last minute by colleagues. The official ballast boxes were already full and included ten thousand pennies added by Dave Moore to make weight and have as space souvenirs. Brian's weight allocation was 200 pounds, and he was slightly over. He reluctantly opted to bring the lighter polyester flag and, taking a cue from Charles Lindbergh, tore unnecessary pages out of a hefty flight checklist. He was in trouble if he didn't know the flight checklist by now. He wore multiple pairs of socks to protect his feet from the minus-seventy-degree air on the other side of the rocket. Only three layers of carbon fiber separated his toes from the outside world. He wasn't going to pare back on his socks. Finally, when he'd made weight, Brian began the walk to the plane.

On the way, he was intercepted every few feet by a procession of well-wishers. Brian finally gave up trying to stay in his game-day zone, and now understood one reason for sequestering astronauts before a flight. Test pilot Chuck Coleman, who had survived more crashes than anyone could remember, and who would pilot the chase plane today that would guide Brian toward his landing, said, "Meet you at fifteen thousand feet." Robert Scherer, who owned the Starship chase plane, said solemnly to Brian, "The world is with you. The heavens are with you." Jeff Johnson, who knew of Brian's determination and efforts to get back into the cockpit, embraced him. Brian saw Erik Lindbergh and Peter Diamandis, who told him that he was "the day's Charles Lindbergh," the one who was going to make history.

Heading his way next was his mother-in-law, Maria Anderson, looking well rested and holding a cup of McDonald's coffee. The woman had never been shy, and was not easily thwarted. Brian eyed the coffee and his mother-in-law with equal wariness. Before he could say anything, she threw her arms around him in an expression of good-luck and do-right-by-my-daughter. What came next happened in a rush: the tight mother-in-law hug, the hot liquid flowing on his neck and down his back. *Hot,* he thought, *just as advertised.* The contents of the sixteen-ounce cup soaked his white T-shirt under his flight suit. After the shock and awe, as he called it, he somehow found humor in the moment, pointing out that she had generously saved about four ounces of the sweet-smelling vanilla-flavored stuff for herself. Aerodynamicist Jim Tighe assessed the situation in terms of weight added by the coffee and its possible impact on the altitude reached. He informed Brian that he was "wearing about four hundred feet of apogee," close to the margin that Mike had made it to space on his June 21 flight.

Brian continued his march toward the plane. Bub was now at his side, wearing a shirt that she had made with the pattern of the American flag. She kissed him, removed her wedding ring, and tucked it into his pocket, saying, "Think of me being with you up there." She had her crystal blue rosary with her and would feel the beads as Brian flew. Her prayer group was similarly armed and ready.

Soon, Sally Melvill approached. She offered Brian their lucky horseshoe, which he gladly accepted. Then there was Mike, mentor and friend. Brian knew Mike had gone to bat for him to get this flight, but he didn't know that Shane had told Mike this morning, "We'll soon know whether this was a good decision or not." But Mike was confident and told Brian, "You can do this. I know you can do this." That was exactly what he had said when they were out in the Long-EZ. Mike added, "We're going to have a great day." For this flight, Mike was the bus driver, piloting the *White Knight*.

Finally, there appeared Burt—boss, friend, golfing buddy. He looked excited rather than nervous, leaned into the cockpit, and delivered his advice—in golfing terms—for how Brian should fly: "Take out the driver, swing smooth, and go long." Seconds later, the door closed and Brian was alone, back in the cockpit that held his dreams and fears. Instead of experiencing some profound moment of illumination, Brian was overcome by something else—the smell of French vanilla coffee.

At 6:49 A.M. local time, the *White Knight* reached 130 miles per hour and lifted off runway 30. Brian had an hour-long ride underneath the *White Knight* as it flew its approved pattern and climbed to altitude. Brian hadn't known how he'd feel being back in the hot seat, waiting for his moment to fly. But he was ready. And he was calm.

Exactly one hour later, at 7:49 A.M., the *White Knight* was at 47,100 feet. Brian pushed the control stick forward to prepare for the release.

Mike called the release: "Three-two-one—release," and Stinemetze pulled the lever to drop *SpaceShipOne*.

"Released, armed, fire," Brian said.

"Holy crap, that was close!" Stinemetze said of the spaceship's quick turn upward. Brian had charged out of the gate.

A handful of seconds later, Shane in Mission Control said, "Rates look good and low. Doing okay?"

"Doing all right," Brian said. This time, he knew the bull he was riding. He expected to be shaken, tugged, and beat up. He waited for the noises that sounded like the start of World War III. His breathing was steady. He had this.

"Little lateral oscillations," Brian said.

Shane said, "Copy. Thirty seconds. A little nose-up trim. Forty. Trajectory perfect."

Mike added, "Brian looks great."

On the Mojave floor, crowds held cameras to the sky. They had seen the separation, where the *White Knight* veered to the left and the rocket shot straight up, leaving a thick white vertical contrail. The wind was calm, the visibility perfect. Stuart Witt was back providing commentary, as was Gregg Maryniak. Today, the cheers were, "Go Brian!"

It was time for the motor's transition from liquid to gas, when the nitrous oxide begins to run low in the oxidizer tank. Brian focused on exiting the atmosphere without any rolls.*

"Three-fifty, recommend shutdown," Shane said. The altitude predictor showed that Brian would end up at 350,000 feet if the motor was shut down now. This already put him safely past the Karman line.

Brian wanted more. He was going to squeeze every molecule of energy

*The initial thrust of the motor, the "kick," ranged between sixteen thousand and eighteen thousand pounds before tapering off to about eight thousand pounds before entering the liquid to gas transition phase, when the shaking and vibrating commenced at about one minute into the burn.

from the motor. He started a slow-motion response, hoping to see 370,000 feet on the predictor.

At eighty-four seconds, the engine finally shut down. Mike had flown the engine to seventy-seven seconds on X1 and seventy-six seconds on June 21.

Shane formalized it: "Engine shut down."

"Feather up," Brian said. "Feather green." Then, looking out at the black sky, he said, "Wow. I'm upside down." He was in space. And he'd gotten there without a hint of a roll.

During his journey past the Karman line, Brian had felt strangely *guided*, and not by Mission Control. The feeling was as clear as this morning's sky. Now, away from the pull of gravity, he gazed at the pale blue curvature of Earth against a black dome.

"Feel good?" Shane asked.

"I'm feeling great," Brian said. "Wow, it's quiet up here."

"Copy that."

"Better get the camera out."

"Roger that."

Brian took pictures and then released a paper model of *SpaceShipOne* that someone had given him before the flight. The paper spaceship effortlessly took its own gravity-free flight around the cockpit.

Then Brian heard Burt's voice: "X-15 record."

In Mission Control, Burt pumped both fists in the air. Paul Allen patted Burt on the back. Brian was more than 10,000 feet above the highest altitude ever reached by the X-15, which was 354,200 feet in 1963. This was the boss's grand slam.

Burt was studying the numbers. The engine had shut down at 213,000 feet, going Mach 3.09, and *SpaceShipOne* had continued like a ball tossed in the air on its own momentum. But the amount of that momentum was a surprise. The spaceship kept going upward until it reached 367,550 feet.

"Outstanding," Brian said.

Having reached its top altitude, *SpaceShipOne* began its quick descent. Brian could still smell the vanilla-flavored coffee.

"Here come the gs," Shane said.

"Five gs," Brian called.

Burt took notes on a legal pad and alternated between looking at Shane and studying the screen.

"Peak gs are done, coming through seventy-five thousand feet," Shane said to applause in Mission Control.

"Feels a little loosey-goosey now," Brian said of the spaceship.

"Copy," Shane said. "Get the roll trim back to neutral as you defeather."

At around 63,000 feet, Brian retracted the feather. Mission Control watched anxiously as the tail booms went from their upward bend of sixty-three degrees slowly back down to locked position.*

"Feather lock," Brian said to more cheers in Mission Control. Burt's ingenious design, taking inspiration from the dethermalizing model planes of his youth, had performed flawlessly. The moment was not lost on Burt, who wiped tears from the corner of his eyes. After all, Burt knew that his entire space travel program would succeed or fail based on how the feather worked. Many experts had told him that the feather was unworkable, that it was sheer lunacy. But once again, as he had done all his life, Burt found breakthroughs where others saw nonsense.

With only the landing remaining, Burt and Paul Allen headed out onto the tarmac to join Richard Branson, Peter, Erik Lindbergh, and the pilots' families. At one point, Burt, Paul, and Richard could all be seen pointing up at the sky with their left hands.

Nearby, Bub was working her prayer beads, believing all of the prayers were guiding Brian home. Their kids held up signs: "GO DAD!"

The wind socks on the runway were hanging calm. Mojave was green for landing.

---

*Burt reduced the feather angle from sixty-five degrees to sixty-three when they installed a bumper insert to reduce the hammering when the actuator came to a stop.

Brian extended the landing gear for the moment of truth. Would he stumble again at the finish line? In his heart and soul, and after his breathtaking experience in space, Brian knew the answer: He was going to make that perfect stop. He was going to let his confidence take over—and vanquish his doubts once and for all.

Brian focused, not like a quarterback who needs to throw a Hail Mary pass, but like a painter poised to make the perfect last stroke. He heard the call: "Looking good, right down the middle."

SpaceShipOne was gliding in, the motor now spent. There was no wind, no buffeting. Brian didn't see the cheering crowds or satellite trucks or emergency response vehicles. He saw only a centerline. The little spaceship, its nose clouded with stars, was close to touching down.

From the chase plane, Coleman made the calls: "Two hundred mph, one hundred, looking good."

Brian thought about all those practice landings in the Long-EZ, about all the training runs, about all the work in the sim. Brian kept it coming, leveled off, the small shadow of his ship sweeping the runway below. There was no aircraft carrier, no arresting wire. This was his canvas.

Three-two-one-down. Softly, on the centerline. It was perfect.

"Congratulations, Brian!" Shane said. The Oracle was showing some emotion.

In the mother ship, Mike couldn't help himself, offering another "Yee-haw!"—this one for Brian. Then, with his voice choking, Mike added, "Proud of you, man."

Brian responded to his mentor, "Thank you, Mike."

Out on the runway, Burt reached Brian and congratulated him. "How 'bout them apples! You got the X-15! This is so cool."

Sally and Bub jumped into a truck on the flight line—the wife of a test pilot knows to have keys to a response car—and headed straight for Brian.

When their truck got close, Bub jumped out and ran the rest of the way. She climbed into the cockpit and said, "You did it!" over and over. Brian held her close and blinked back tears. He didn't trust himself to speak yet. Elton John's song "Rocket Man" blasted across the spaceport. He'd nailed it. He'd earned his astronaut wings.

As the crowd continued to cheer, the schoolchildren were shepherded to their buses. They still had a day of classes ahead. When a reporter for the local *Antelope Valley Press* stopped and asked a group of middle schoolers whether they wanted to be astronauts, all hands shot up—no hesitation.

There was singing, dancing, and the spraying of champagne. Several of the XPRIZE contenders were there, including Pablo de León, standing next to Loretta Hidalgo and George Whitesides.* "This is the beginning of a new era," de León said, tears in his eyes. "There is a before day and an after day. Things will never be the same after today. This is the end of the government's monopoly over manned launch." The celebration moved to an area in front of Scaled. Peter stood on a makeshift platform with Burt, Paul Allen, and Richard Branson. Erik Lindbergh and the Ansaris were close by. Peter's voice carried to the end of the long runway and back:

"For forty years, we have watched as spaceships have flown," Peter said. "Crowds of people had to be moved five miles away while those few astronauts got on board and ignited those engines. Today, *SpaceShipOne* has landed, and stands not five miles away but five feet away."

Peter continued, "We are at the birth of a new era—the personal spaceflight revolution. It is our pleasure to announce today in Mojave, California, that *SpaceShipOne* has made two flights to one hundred kilometers and has won the Ansari XPRIZE."

When Peter returned to the crowd, he stood with his family and

---

*George Whitesides, Peter's roommate during the Blastoff days, would go on to become Virgin Galactic's CEO.

Kristen. His parents might not understand orbital versus suborbital, and they couldn't begin to know all of the planning, heartache, and passion that went into making the XPRIZE happen. But in their own way, Tula and Harry were the ones who could understand best just what this meant to him, to the boy who had raced around with the energy of a rocket and could not be contained. The boy who sat them down for lectures on space and kept note cards on every episode of *Star Trek*. The teenager who stashed explosives in the house and made experimental rockets that often turned into missiles. The college student who started student space clubs and a space university. The grad student who finished medical school to please them, but always had his own dreams, including one that came true today.

Tula allowed, jokingly, that she should probably stop asking Peter when he was going to practice medicine. Harry told Peter that he had brought great pride to the family name. To Peter, this day was a beginning. As he listened to Burt, he kept thinking one thing: *We lit the fuse of a new Space Age.*

Burt addressed the crowd and said, "If you look at the twelve months after Yuri Gagarin was flown to space by the Russians in 1961, that first year, there were five manned spaceflights. Now, this year—*forty-three years later*—how many spaceflights are there? There are five. We did three of them with our little program, and the Russians did two. Our little team was able to show American exceptionalism."*

Brian, standing next to Burt, was up next. He spoke forcefully: "I wake up every morning and thank God I live in a country where all of this is possible. Where you have the Yankee ingenuity to roll up your sleeves,

---

*Burt believed that the *SpaceShipOne* program would be historic and wanted every Scaled employee to be able to tell his or her grandkids that they played some part in helping design, build, or test it. There were eight people involved in the initial design and planning; twenty-five people involved in building *White Knight* and designing and testing the rocket; and about sixty people working on both *White Knight* and *SpaceShipOne*, the avionics and simulator. During the year leading up to the flights to space, there were thirty people directly involved in the space program. In all, Burt estimates that about eighty people had a role in the success of the world's first private manned space program.

get a band of people who believe in something, and go for it and make it happen." Just hours earlier, Brian's fate was uncertain. Now he was the 434th human to go to space.

Later that day, after most of the crowd had pulled out of Mojave, Peter, Paul Allen, Burt, Mike, and Brian and the Scaled team gathered in a conference room. They had a call coming in, from President George W. Bush. The president, aboard Air Force One, noted that his plane was not nearly as cool as *SpaceShipOne*, and not nearly "as exciting as Mike's and Brian's flights."

Mike and Brian were side by side—Mike the world's 433rd astronaut—across from Burt and Peter. The rest of the crew huddled close. After some pleasantries about the program, President Bush said, "The sky of Mojave is very big. And you've got very big dreams there." He added, "Thank you for dreaming the big dream."

That night, Stinemetze, Losey, and the others lingered over beers. There was a quiet and calm they hadn't felt in years.

"People talk about the magic of the early days of the Apollo program," Stinemetze said. "That's what it feels like happened here. All the right people just showed up to play a part. Every person was key to the success. We had magic."

Not far away, as *SpaceShipOne* was being put away for the night, Burt had told the team, "You put your hearts and your talents into this. This is not an end. It's just a *very good* beginning."

In a month, Burt, Paul Allen, and the team at Scaled would head to St. Louis for the awarding of the $10 million check. Then *SpaceShipOne* would take to the skies one more time, under the mother ship's wings, toward its final destination.

32

Hallowed Company

It was an hour before closing time in the Milestones of Flight Hall in the Smithsonian National Air and Space Museum. Peter was there by himself, having stolen some time while in Washington, D.C.

He had come here over the years seeking inspiration, often following frustrating meetings with officials at NASA or the Federal Aviation Administration to discuss International Microspace, Angel Technologies, or ZERO-G. He would arrive at the Air and Space Museum numb from battles with bureaucracy, in need of a reminder of the rewards that came with risk.

While the Infinite Corridor at his alma mater remained Peter's place of limitless possibilities, this hall of flight was a validation of dreams achieved.

As soon as he walked into the sunlit hall, he lost himself in the stories behind the machines. There was much he knew, much more he wanted to learn. So many untold stories of late sleepless nights, fights over funding, arguments over design. Standing under the orange Bell X-1, he thought of what it must have taken to build this bullet-shaped craft and then find the right Air Force pilot to fly it faster than the speed of sound when it had never been done before. There was the rocket-powered North American X-15, built to explore man's role in space. It wasn't known at the time that pilot and plane could leave the atmosphere and safely return.

Peter made his way over to the Mercury spacecraft, imagining the courage of John Glenn in becoming the first American to orbit Earth. Looking at the Apollo 11 command module *Columbia* always made Peter euphoric. The craft had changed his life, mesmerizing him when he was eight years old with the grainy black-and-white images of man's first steps on the Moon.

Peter looked up and saw the world's fastest jet-propelled aircraft, the titanium-alloy SR-71 Blackbird, built in Lockheed's clandestine "Skunk Works" division. He wandered toward Burt's *Voyager* aircraft, which answered the question of whether a pilot could make it around the world nonstop without refueling. Each milestone probed deeper into the unknown, each built on what had been learned before.

Peter watched how people paused for a moment, if at all, in front of these heroic achievements. He wondered whether they thought of the designer, engineers, funders, pilots, materials, setbacks, heartaches, and breakthroughs. With everything he looked at, he imagined the decade before its debut. He thought of the improbable odds of these vehicles coming into existence at all. He wondered who had to be pitched, who had to be convinced, how the funding came in, how many doors were closed, and who was there to keep opening the next one.

In a more personal way, the hall held memories representing gifts and losses. This was where he and Bob Richards came with Todd Hawley in the spring of 1995 after formalizing their International Space University charter. He kept a picture of the three of them—*Peterbobtodd*—standing in front of the *Spirit of St. Louis* and the Bell X-1, a space in between the planes. Three months later, Todd had died. His best friend—the man he called his brother—was memorialized here. Everyone had stood in line to touch the Moon rock, as if seeking a final connection with Todd. It was also in this hall where he'd met Reeve Lindbergh, who would introduce him to the "flying Lindbergh." Erik's life had been transformed by the XPRIZE, and Erik in turn had helped rescue Peter's dream when things looked bleak.

Peter made his way back to the front entrance, continuing to stop along the way and pose his silent questions: How many people lost their lives for this? How many marriages were ruined? What suffering was endured to build these crazy vehicles? Now for the first time, he had all of the answers to one icon of flight: *SpaceShipOne*. The little rocket's final destination was here. It was installed a few months earlier, in October 2005, on the one-year anniversary of the winning XPRIZE flight. This was one story Peter knew well—crazy, traumatic, glorious, exhausting, thrilling, and almost-didn't-happen.

Where there had been the unoccupied space between the *Spirit of St. Louis* and Chuck Yeager's X-1, there was now the world's first privately made, financed, and flown spaceship. The *Spirit of St. Louis* was wing to wing with the wonder from Mojave. The two vehicles came to life seventy-seven years apart, but shared contrarian ideas, confident creators, and a race for a prize. And just as Charles Lindbergh had asked, "Why shouldn't I fly to Paris?" Burt Rutan had said, "Why shouldn't I fly to space?"

As the museum was preparing to shut down for the night, Peter took one last look at *SpaceShipOne*, the rocket with the star-spangled nose that now shared rarefied air with the greatest achievements in aeronautical history. It was Peter's 16,157th day on Earth—and he had proof that the impossible was possible.

Epilogue: Where Are They Now?

Peter H. Diamandis continues to run the XPRIZE Foundation as founder and executive chairman. The XPRIZE has expanded into a global organization creating incentive prizes to solve some of the world's biggest problems in fields ranging from energy, environment, space, and oceans to education, health, and global development. To date, the foundation has awarded more than $34 million in purses, is currently offering $82 million in active prize purses, and has more than $100 million in prizes under development. The Foundation's largest prize, the $30 million Google Lunar XPRIZE, will be awarded to the first team to land a robot on the Moon, travel 500 meters on the surface, and transmit video back to Earth. The Google Lunar XPRIZE "is Blastoff reincarnated," Peter says of the ill-fated dot-com company he ran in Pasadena. Today both SEDS and International Space University (ISU) continue to thrive. Peter has also taken what he learned from his days running ISU and cofounded Singularity University (SU), an educational think tank and incubator headquartered in Silicon Valley for entrepreneurs and executives to launch or refocus companies. SU focuses on the application of exponential technologies and operates globally. Most recently Peter has cofounded two new companies that continue his duality between space and medicine. The first, Planetary Resources, Inc., is building deep-space drones for the prospecting of asteroid resources (as well monitoring terrestrial agricultural and energy resources). The second, Human Longevity, Inc. (HLI), is a genomics-, stem-cell-, and machine-learning-driven company focused on "extending the healthy human lifespan." HLI is Peter's mechanism for achieving longevity—his Harvard Medical School redux—so he can one day be assured his trip to space. In addition, the ZERO-G Corporation has flown more than fifteen thousand people, ages nine to ninety-three, into weightlessness. In 2007, Peter and Byron Lichtenberg flew Professor Stephen Hawking, the world's expert on gravity, into zero gravity. ZERO-G is now the sole provider of parabolic flight services to NASA. Peter lives in Santa Monica, is the best-selling coauthor of *Abundance* and *Bold*, and travels the globe talking to Fortune 500 CEOs and advising entrepreneurs. He and his wife, Kristen, have twin boys, Jet and Dax.

—Erik Lindbergh's lifelong quest to escape from the gravity of life inspired him to start a new venture called Escape from Gravity. His new company is about helping people claim strength and grace in the face of aging. Lindbergh explained, "I'm delighted to have gotten a second chance at life, survived to the ripe old age of fifty-one, and I'm still rockin' it—this is the best time of my life. Escape from Gravity is about sharing stories of people rockin' their spirit in life and using that passion for social good. *My* Escape from Gravity is through aerospace, art, and adventure. How do *you* escape?"

—After retiring from Scaled Composites in April 2011, Burt Rutan and his wife, Tonya, moved to Coeur d'Alene, Idaho. Burt came out of his short-lived retirement to build his forty-seventh new type of aircraft—in his garage. The amphibious plane, called the SkiGull—with retractable skis for landing—was designed to be able to fly from California to Hawaii without refueling and be capable of landing on water, snow, grass, or hard surfaces. Burt and Tonya, who received her private pilot's license in a seaplane, plan to use the SkiGull to explore the world in their own Walter Mitty-like adventure. Burt insists the SkiGull, which is saltwater resistant, is the last plane he will ever design and build himself. Six of his planes, including *SpaceShipOne*, hang in the Smithsonian National Air and Space Museum.

—Mike Melvill retired from Scaled Composites in October 2007. He still flies his Long-EZ and Pitts biplane, logging at least 120 flight hours a year. He is also privileged to fly a friend's collection of World War I fighter aircraft. Sally retired at the same time as Mike, and has since found her calling—volunteering in a kindergarten classroom two days a week. Mike recently said, "We are happy we left when we did. Burt was the best boss either of us ever had. He was the most generous and the most exciting guy to work for. Never a dull moment, always something neat or cool going on. When we started working for Burt in September 1978, the whole company consisted of Burt and Sally and me!"

—After the success of *SpaceShipOne*, Scaled Composites began work on SpaceShipTwo for Richard Branson and his spaceship company. Brian Binnie spent the decade after Scaled won the XPRIZE working through the many issues related to scaling *SpaceShipOne* into the considerably larger SpaceShipTwo, which will have two pilots and six passengers. Brian left Scaled in 2014 to work on a completely different suborbital spaceship called the *Lynx*, being built by XCOR Aerospace. As the senior test pilot and engineer, he has worked to ready the vehicle to take off from the runway under its own rocket power, fly to space, and return (like many of its predecessors) as a glider. The *Lynx* is about the same size as *SpaceShipOne* and has room for carrying one space participant who will sit side by side in the cockpit with the pilot/astronaut. Brian considers his days at Scaled as the most creative and rewarding years of his professional life. He said, "I've never been in an organization that was so flooded with incredible talent and genuinely nice people. It is an environment that reflects the genius and humanity of Burt Rutan." Brian and his wife, Bub, eventually hope to retire somewhere along the western coastline, where the more rustic weather and temperatures remind him of his Scottish roots.

—Paul Allen, who spent around $26 million on *SpaceShipOne*, said that seeing *SpaceShipOne* hang in the Smithsonian was a day he'll never forget. "I haven't had any days prouder than that one," Allen said. "The whole experience was something I'd dreamed about. It was a very singular, peak experience to be a part of that amazing team." Allen remains involved in space but was happy to hand the manned spaceflight aspect over to Richard Branson. "I was not so enamored of being in the commercial spaceflight business," Allen said, noting that sooner or later, errors—or "deviations from a plan"—are inevitable. Allen's company Vulcan Aerospace contracts with Scaled to build Stratolaunch, the largest aircraft (by wingspan at about 117 meters) ever made, designed to launch payloads and vehicles to orbit. Burt Rutan, responsible for the initial concept of Stratolaunch, remains on the board but has little interface with the Scaled team working on the craft.

—The $10 million XPRIZE purse was shared 50–50 between Burt Rutan's company and Paul Allen. Burt took the money and handed bonuses out to every employee who had some role in making *SpaceShipOne*, from the guys who cleaned the shop to the engineers and pilots. The bonuses were the equivalent of the employee's salary for a year. Scaled Composites is now fully owned by Northrop Grumman Corp. A number of people have left Scaled to work at Virgin Galactic, including Doug Shane and Steve Losey, the former crew chief of *SpaceShipOne*. Virgin Galactic is just down the flight line from Scaled Composites. Stratolaunch is being built in an enormous hangar in a far corner of the Mojave Air and Space Port.

—Richard Branson's Virgin Galactic was established to build on the work of *SpaceShipOne* and open access to space for more people and payloads, starting with commercial suborbital flights. The company faced a setback and tragedy during a test flight in October 2014. SpaceShipTwo (named VSS *Enterprise* after the *Star Trek* craft) violently broke apart when the feather was unlocked at the wrong time. The test pilots were Pete Siebold and Mike Alsbury. The cause, according to the National Transportation Safety Board, was an error by Alsbury, who inexplicably unlocked the feather *before* reaching Mach 1 instead of after, as Burt had intended and the pilots had been trained. The breaking apart of the craft killed Alsbury instantly. Siebold, who found himself hurtling through the air still strapped into his pilot seat after the plane broke apart, lost and then regained consciousness in time to unclip from his seat and pull his parachute. He landed with only a shoulder injury. Alsbury was Mike Melvill's protégé; he trained him, and he was planning to leave him his Long-EZ. In February 2016, Richard Branson held an event in Mojave to unveil the second SpaceShipTwo, named VSS *Unity*, based on the same Rutan feathering design—though now with a feather lock safety mechanism. Theoretical physicist Stephen Hawking said at the unveiling, "I would be very proud to fly on this spaceship." Branson noted, "I thought we would be where we are today a lot quicker than we have been, and I was hoping we'd get there with a lot less pain. But once again we are not far off. If it hadn't been for the XPRIZE, we wouldn't be here today. Yes, it's taking people to space, but I believe that so many fascinating things will emerge from what we are doing—and what Jeff Bezos is doing and Elon Musk is doing. People want to do extraordinary things and push the limits."

—In September 2006, Anousheh Ansari became the world's first female space tourist, the first Iranian woman in space, and the fourth space tourist to fly to the International Space Station. Her dream of flying to space became a reality thanks to her association with Peter Diamandis. The company he cofounded with Eric Anderson, Space Adventures, brokered the deal.

—John Carmack and his team at Armadillo Aerospace went on to build piloted rocket-powered airplanes for the Rocket Racing League and launch computer-controlled rockets close to 100 kilometers. They even had a year when they showed an operating profit. After the team went from all-volunteer to full-time salaried employees in 2006, Carmack began to notice what he calls "creeping professionalism." He explained, "Once it became people's jobs, they had other hobbies—go-karts and model airplanes and other distractions. When it was two days a week, everyone was much more focused on really getting things done. We started doing work with NASA and there were blueprints and diagrams and technical reviews. It slowed the team down." Carmack, who is now the CTO of Oculus Rift, says, "There are still aerospace ideas that interest me, so there is a decent chance that I will return to try again after virtual reality is all sorted out. I don't regret any of the work. We didn't achieve all the goals we wanted, but we took a good shot at it."

—When the XPRIZE was won and entrepreneurs including Branson, Musk, and Bezos were starting private space companies, Argentinian Pablo de León looked around and realized there would be a need for spacesuits for the private sector. Building on his previous work, he became a recognized spacesuit expert, founding a company to develop commercial spacesuits. In 2004, he began working at the Department of Space Studies at the University of North Dakota, first as a researcher in human spaceflight and, since 2013, as a professor. De León secured several NASA grants to develop new-generation spacesuits for the Moon and Mars and secured more than $2 million of external funding for the university. He currently has a NASA grant to develop an inflatable Mars base prototype.

—Dumitru Popescu left Romania in 2014 to establish the headquarters of ARCA Space Corporation in New Mexico. The company now makes large drones, including electric-powered ones that can fly higher than commercial aircraft, high-altitude balloons, and suborbital vehicles.

Popescu recently unveiled the ArcaBoard, an all-composite mattresslike board that "surfs" a foot above the ground, giving the rider the feeling of flying. Popescu said his participation in the XPRIZE changed his life. "I knew it would make the history books from the moment I first read about it. It never crossed my mind to give up. We created a company, a brand. We had inertia on our side."

—Steve Bennett of Starchaser Industries in England said, "For me, the XPRIZE legitimized what I was trying to do. Before the XPRIZE, people humored me. After the XPRIZE, the same people began knocking on my door, asking when I will be able to send them into space." In the years following the XPRIZE, Bennett's company focused on the development of its liquid oxygen/kerosene rocket engines. In 2007 Starchaser won a competition to land a European Space Agency (ESA) contract entitled "Study of European Privately Funded Vehicles for Commercial Human Space Flight." Between 2008 and 2009 Starchaser won UK Development Agency funding to carry out the research and development on an eco-friendly rocket engine suitable for space tourism vehicles. As part of this project, the Starchaser team successfully designed, built, and test-fired a series of hybrid rocket engines. Starchaser's educational outreach program continues to grow, providing rocket- and space-related workshops, shows, and presentations for around two hundred schools every year. Bennett is working his way toward launching the company's Nova 2 man-rated rocket in 2017. "2017 will be a special year for us," Bennett predicted. "It will mark the twenty-fifth anniversary of the founding of Starchaser and the sixtieth anniversary of Sputnik and the dawn of the space age. The launch of Nova 2 will set the stage for flying our first Starchaser astronauts into outer space."

—Other XPRIZE contenders, including Jim Akkerman in Texas, continue to work on manned spaceflight programs and applaud the pervasiveness of private enterprise in the field of space. Jeff Bezos's company Blue Origin has had consistent success with its New Shepard suborbital rocket, which has a propulsive vertical landing system. The New Shepard is achieving reusable and relatively low-cost spaceflight. Bezos is starting with suborbital on his way to orbit. Elon Musk's SpaceX, founded in an old hangar in El Segundo, California, and given little chance of succeeding, has repeatedly made history. Starting with a team of thirty, SpaceX now has more than four thousand employees and some of the most advanced rockets and engines in the world. After three highly publicized rocket failures early on—a fourth would have put Musk into bankruptcy—SpaceX succeeded and went on to become the first private company to send a rocket to orbit, the first private company to deliver cargo to the International Space Station, and the first private company to land an orbital booster back on the launchpad. SpaceX has a $1.6 billion contract with NASA to fly cargo resupply missions to the ISS and to carry crew. NASA has contracted with SpaceX and Boeing to fly astronauts to the International Space Station. Both companies say they are on track for manned launches in 2017.

Afterword: Space, Here I Come!

by Stephen Hawking

I have no fear of adventure. I have taken daredevil opportunities when they presented themselves. Years ago I barreled down the steepest hills of San Francisco in my motorized wheelchair. I travel widely and have been to Antarctica and Easter Island and down in a submarine.

On April 26, 2007, three months after my sixty-fifth birthday, I did something special: I experienced zero gravity. It temporarily stripped me of my disability and gave me a feeling of true freedom. After forty years in a wheelchair, I was floating. I had four wonderful minutes of weightlessness, thanks to Peter Diamandis and the team at the Zero Gravity Corporation. I rode in a modified Boeing 727 jet, which traveled over the ocean off Florida and did a series of maneuvers that took me into this state of welcome weightlessness.

It has always been my dream to travel into space, and Peter Diamandis told me, "For now, I can take you into weightlessness." The experience was amazing. I could have gone on and on.

Now I have a chance to travel to the start of space aboard Richard

Branson's Virgin Galactic SpaceShipTwo vehicle, VSS *Unity.* SpaceShipTwo would not exist without the XPRIZE or without Burt Rutan, who shared a vision that space should be open to all, not just astronauts and the lucky few. Richard Branson is close to opening spaceflight for ordinary citizens, and if I am lucky, I will be among the early passengers.

I immediately said yes to Richard when he offered me a seat on SpaceShipTwo. I have lived with ALS, amyotrophic lateral sclerosis, for fifty years. When I was diagnosed at age twenty-one, I was given two years to live. I was starting my PhD at Cambridge and embarking on the scientific challenge of determining whether the universe had always existed and would always exist or had begun with a big explosion. As my body grew weaker, my mind grew stronger. I lost the use of my hands and could no longer write equations, but I developed ways of traveling through the universe in my mind and visualizing how it all worked.

Keeping an active mind has been vital to my survival. Living two thirds of my life with the threat of death hanging over me has taught me to make the most of every minute. As a child, I spent a lot of time looking at the sky and stars and wondering where eternity came to an end. As an adult, I have asked questions, including Why are we here? Where did we come from? Did God create the universe? What is the meaning of life? Why does the universe exist? Some questions I have answered; others I am still asking.

Like Peter Diamandis and those who fill the pages of this book, I believe that we need a new generation of explorers to venture out into our solar system and beyond. These first private astronauts will be pioneers, and I hope to be among them. We are entering a new space age, one in which we will help to change the world for good.

I believe in the possibility of commercial space travel—for exploration and for the preservation of humanity. I believe that life on Earth is at an ever-increasing risk of being wiped out by a disaster, such as a sudden nuclear war, a genetically engineered virus, or other dangers. I think the human race has no future if it doesn't go to space. We need to inspire the next generation to become engaged in space and in science in general, to ask questions: What will we find when we go to space? Is there alien life, or are we alone? What will a sunset on Mars look like?

My wheels are here on Earth, but I will keep dreaming. It is my belief, and it is the message of this book, that there is no boundary of human endeavor. Raise your sights. Be courageous and kind. Remember to look up at the stars and not at your feet. Space, here I come!

Stephen Hawking

Theoretical physicist, cosmologist,
and bestselling author of six science books
and five children's books coauthored
with his daughter, Lucy

Author's Note

I met Peter Diamandis in the spring of 2014, when I interviewed him for a story in the *San Francisco Chronicle.* I remember asking Peter at the time a seemingly simple question, "How did the XPRIZE begin?" Peter laughed and asked how much time I had. I was immediately drawn into Peter's story—of the space geek kid who can't let go of an out-of-this-world dream—and the even bigger tale of how the private race to space was launched. I love underdog stories, and this embodies a great David versus Goliath struggle.

I had just finished doing publicity for my last book, on Larry Ellison and his quest for the America's Cup—and his unlikely partnership with a radiator repairman who was the commodore of a blue-collar boating club. I realized Peter's story, and the story of the men and women who went after the Ansari XPRIZE, would make for a compelling book. Peter said he had been waiting to find the right writer to tell this story of how history was made. I began researching this subject in October 2014, on the tenth anniversary of the winning of the XPRIZE. But I didn't commit full time to it until early spring of 2015.

I traveled to Florida to meet with Peter's mom, dad, and sister, and to sift through boxes of photos, albums, and newspaper clippings. I made several reporting trips to the Mojave Desert, including a memorable visit with my son, Roman, then nine, who marveled at the intricacies of a rocket engine. He said he had never seen anything that looked so complicated. I spent countless hours with the engineers at Scaled Composites and had long and wonderful talks with Brian Binnie and his wife, Bub. Brian's story of getting knocked down and of his determination to get back into the proverbial ring wowed me. I pored over flight logs of *SpaceShipOne*, watched hours of video, listened to audio recordings, and read notes taken during the test flights and XPRIZE flights. I interviewed spectators and competitors as well as aviation and space historians.

I spent a great deal of time with Mike and Sally Melvill and got to fly with Mike in his Long-EZ. The experience was enlivened when Mike casually relayed—as we were high above the mountains of Tehachapi—that he was going to do some rolls and other high-adrenaline maneuvers to give me a feel for the plane. It was an amazing experience to fly with the world's first commercial astronaut, who also is a world-class guy: humble, brave, self-taught, and kind.

I had the privilege of long talks with Dick Rutan and Burt Rutan. These brothers were born to build, test, and fly planes. They're pioneers and mavericks. Burt, especially, is still this kid whose eyes light up with the joy of a new idea and tear up with memories of milestone moments. He is revered in the world of aviation, but my hope is that his story and his genius will also become known to those who are not aviators. Burt is truly one of America's great innovators. It all started for him as a kid making model planes.

There are many others I interviewed for this book, from Elon Musk, Richard Branson, and Paul Allen to NASA's Dan Goldin and the FAA's Marion Blakey. I did well over one hundred interviews and many of these people put up with my returning again and again with seemingly infinite questions. I traveled to Dallas to meet with Oculus Rift CTO John Carmack, who was both considerate and exceedingly smart. I met with Russell Blink and the team from Armadillo Aerospace, driving to remote corners of the Lone Star State to find these persistent rocket makers and to see the remnants of their XPRIZE vehicle. I visited Seattle to meet Erik Lindbergh. What a story he has! I found Erik to be smart and soulful, and was struck by how his fate was inextricably tied to the XPRIZE. I even went to St. Louis to visit the Racquet Club where Erik's grandfather Charles Lindbergh met with his supporters and where, decades later, the XPRIZE found its backers.

And, of course, there is Peter Diamandis, a true force of nature and probably the most tenacious person I've ever met. Beyond sharing his time over the last year and a half, Peter shared his contacts, photos, and video and audio archives. I listened to hours of his recorded thoughts and was even a fly on the wall in certain meetings that he recorded. I was entrusted with scrapbooks from his earliest SEDS and ISU days. I interviewed his childhood friends and a handful of his influential teachers and professors. Peter exhumed long-sealed boxes containing his personal, handwritten journals, spanning from 1979, his senior year in high school, to 2006, two years after the XPRIZE was won. These journals were as private and revelatory as any diary. They were raw in their expression of dreams, desires, frustrations, failures, and successes as the boy grew into the man. Collectively, they revealed someone with a sincere and unrelenting belief in the beauty and bounty of space. And they helped bring to life his herculean effort to open space to private industry—a dream that is a reality today.

Acknowledgments

First, I want to thank David Lewis, my friend and editor who has guided me through all of my books and got me through this one—which brought with it a complex cast of characters spread across the globe, a lengthy passage of time, and sophisticated science, technology, and aerospace history, all written under a tight deadline. Next, I want to thank my technical adviser, Paul Pedersen, who is as smart as he is inquisitive, and helped me understand a range of intellectual challenges, was never daunted by my odd or arcane questions, and happily forged ahead with formulating many of the book's footnotes.

I'm grateful and humbled to have the foreword of my book written by the irrepressible Virgin founder Richard Branson and the afterword by Professor Stephen Hawking. I am awed by the life stories of these two men. Fitting with the theme of this book, they believe that rules are meant to be broken and limits transcended.

Special thanks go to my longtime agent, Joe Veltre of the Gersh Agency, who believed in me before my first book was sold. Thank you; what a fun ride we're on! A special note of gratitude goes to Scott Moyers at Penguin Press, for embracing this story from the moment he read my proposal. It was Scott who came up with the book title, *How to Make a Spaceship*. We had gone back and forth with dozens of titles over many weeks, and even crowdsourced a half dozen different titles. Nothing felt just right. *How to Make a Spaceship* fit for a number of reasons: As I see it, people *make* things while governments *build* things. There is a great renaissance under way with the maker movement and do-it-yourself culture. This book is about rolling up your sleeves and making things yourself. It is about making your own spaceship-like dreams come true. I'm grateful for the opportunity to work with such a savvy and inspired editor as well as with the talented team at Penguin, notably: Christopher Richards, Yamil Anglada, Chris Holmes, Matt Boyd, and Sabila Kahn.

I have been lucky to have worked with the XPRIZE team and owe a thank-you to Marcus Shingles, Esther Count, Eric Desatnik, Greg O'Brien, Cody Rapp, Maxx Bricklin, Joe Polish, and Diane Murphy. Thank you to others who spent a great deal of time talking with me, including Gary Hudson, Gregg Maryniak, Dezso Molnar, and Byron Lichtenberg.

Last but not least, thank you to my family—my amazing mom, Connie Guthrie; my smart and creative brother, David Guthrie, and his kids, Wayne, Lauren, and Garrett; and my special and forever friend Martin Muller. I now have a tradition of ending my books with a note to my son, Roman, who watched this process of reporting and writing up close and personal, who put up with my late nights, long days, and vacations spent working. Roman, this is a story you and your peers should love: full of rockets and rebels, mind-blowing innovations, a major dare followed by a huge prize, and high-adrenaline moments that came together in California's high desert. Roman, take inspiration from these pages: follow your dreams, ignore the naysayers, and make really cool stuff. And listen to Stephen Hawking, who says that the best way to transcend limits is "with our minds and our machines."

Index

INDEX

INDEX

INDEX

INDEX

INDEX

INDEX

INDEX

ABOUT THE AUTHOR

Julian Guthrie is an award-winning journalist who spent twenty years at the *San Francisco Chronicle* and has been published by *The Wall Street Journal, The Huffington Post,* and others. Her most recent book is *The Billionaire and the Mechanic,* a bestselling 2014 account of Oracle CEO Larry Ellison's pursuit of the America's Cup.